Everyday Mathematics®

The University of Chicago School Mathematics Project

Student Math Journal
Volume 2

Chicago, IL • Columbus, OH • New York, NY

The University of Chicago School Mathematics Project (UCSMP)

Max Bell, Director, UCSMP Elementary Materials Component; Director, *Everyday Mathematics* First Edition; James McBride, Director, *Everyday Mathematics* Second Edition; Andy Isaacs, Director, *Everyday Mathematics* Third Edition; Amy Dillard, Associate Director, *Everyday Mathematics* Third Edition; Rachel Malpass McCall, Associate Director, *Everyday Mathematics* Common Core State Standards Edition

Authors

Max Bell, John Bretzlauf, Sarah R. Burns‡, Amy Dillard, Robert Hartfield, Andy Isaacs, James McBride, Ann McCarty*, Kathleen Pitvorec, Peter Saecker, Robert Balfanz†, William Carroll†

**Third Edition only* *†First Edition only* *‡Common Core State Standards Edition only*

Technical Art
Diana Barrie

Third Edition Teacher in Residence
Denise Porter

UCSMP Editorial
Kathryn M. Rich
Laurie K. Thrasher

Contributors

Mollie Rudnick, Ann Brown, Sarah Busse, Terry DeJong, Craig Dezell, John Dini, James Flanders, Donna Goffron, Steve Heckley, Karen Hedberg, Deborah Arron Leslie, Sharon McHugh, Janet M. Meyers, Donna Owen, William D. Pattison, Marilyn Pavlak, Jane Picken, Kelly Porto, John Sabol, Rose Ann Simpson, Debbi Suhajda, Laura Sunseri, Jayme Tighe, Andrea Tyrance, Kim Van Haitsma, Mary Wilson, Nancy Wilson, Jackie Winston, Carl Zmola, Theresa Zmola

Photo Credits

Cover (l)Stuart Westmoreland/CORBIS, (r)Kelly Kalhoefer/Botanica/Getty Images, (bkgd)Digital Vision/Getty Images; **Back Cover Spine** Kelly Kalhoefer/Botanica/Getty Images; **iii** The McGraw-Hill Companies; **iv** Dana Tezarr/Photonica/Getty Images; **v** (t)C Squared Studios/Getty Images, (b)Tom Grill/CORBIS; **vi** (t)Tom Grill/CORBIS, (b)Tim O'Hara Photography; **vii** Martin Barraud/Stone/Getty Images; **viii** Adam Crowley/Photodisc/Getty Images; **233 267 322** The McGraw-Hill Companies.

This material is based upon work supported by the National Science Foundation under Grant No. ESI-9252984. Any opinions, findings, conclusions, or recommendations expressed in this material are those of the authors and do not necessarily reflect the views of the National Science Foundation.

everyday**math**.com

STEM McGraw-Hill is committed to providing instructional materials in Science, Technology, Engineering, and Mathematics (STEM) that give all students a solid foundation, one that prepares them for college and careers in the 21st century.

Send all inquiries to:
McGraw-Hill Education
STEM Learning Solutions Center
P.O. Box 812960
Chicago, IL 60681

ISBN: 978-0-07-657644-9
MHID: 0-07-657644-2

Printed in the United States of America.

1 2 3 4 5 6 7 8 9 QDB 17 16 15 14 13 12 11

The McGraw·Hill Companies

Contents

UNIT 6 Number Systems and Algebra Concepts

UNIT 7 Probability and Discrete Mathematics

UNIT 8 Rates and Ratios

UNIT 9 More about Variables, Formulas, and Graphs

UNIT 10 Geometry Topics

Projects

References

Activity Sheets

Date Time

LESSON 6•1

Applying Properties of Multiplication

Math Message

1. Write a general pattern in words for the group of three special cases.

$19 * 1 = 19$

$\frac{2}{7} * 1 = \frac{2}{7}$

$0.084 * 1 = 0.084$

General pattern: ______

2. Write a general pattern in words for the group of three special cases.

$\frac{58}{58} = 1$

$\frac{\frac{3}{8}}{\frac{3}{8}} = 1$

$\frac{7.02}{7.02} = 1$

General pattern: ______

3. Multiply. Write your answers in simplest form.

a. $4 * \frac{19}{19} =$ ______ b. $\frac{2}{3} * \frac{6}{6} =$ ______ c. $0.5 * \frac{2}{2} =$ ______

Multiply. Write your answers in simplest form. When you and your partner have finished solving the problems, compare your answers.

4. $\frac{5}{6} * \frac{3}{10} =$ ______
5. $6 * \frac{2}{3} =$ ______
6. $7 * \frac{3}{7} =$ ______
7. $2\frac{3}{4} * \frac{4}{1} =$ ______
8. $2\frac{3}{5} * 1\frac{2}{3} =$ ______
9. $\frac{7}{3} * \frac{1}{3} =$ ______
10. $\frac{1}{4} * \frac{2}{5} =$ ______
11. $3\frac{3}{8} * \frac{3}{4} =$ ______
12. $1\frac{5}{6} * 4\frac{2}{3} =$ ______
13. $\frac{7}{10} * 2\frac{3}{5} =$ ______
14. $\frac{4}{1} * \frac{1}{4} =$ ______
15. $\frac{1}{100} * \frac{100}{1} =$ ______
16. $\frac{7}{8} * \frac{8}{7} =$ ______
17. $1\frac{5}{6} * \frac{6}{11} =$ ______
18. Write three special cases for the general pattern $x * \frac{1}{x} = 1$.

______ ______ ______

Date Time

LESSON 6•1

Reciprocals

Reciprocal Property

If a and b are any numbers except 0, then $\frac{a}{b} * \frac{b}{a} = 1$.

$\frac{a}{b}$ and $\frac{b}{a}$ are called reciprocals of each other.

$a * \frac{1}{a} = 1$, so a and $\frac{1}{a}$ are reciprocals of each other.

Find the reciprocal of each number. Multiply to check your answers.

1. 6 ________

2. 17 ________

3. $\frac{3}{4}$ ________

4. $\frac{1}{3}$ ________

5. $\frac{3}{8}$ ________

6. $\frac{13}{16}$ ________

7. $8\frac{1}{2}$ ________

8. $3\frac{5}{6}$ ________

9. $4\frac{2}{3}$ ________

10. $6\frac{1}{4}$ ________

11. 0.1 ________

12. 0.4 ________

13. 0.75 ________

14. 2.5 ________

15. 0.375 ________

16. 5.6 ________

Solve mentally.

17. $3\frac{1}{2} * 4 * \frac{1}{4}$ ____________

18. $\frac{1}{6} * \frac{2}{5} * 6$ ____________

19. $\frac{5}{7} * \frac{7}{5} * 4\frac{5}{7}$ ____________

20. $2 * 8\frac{1}{2} * \frac{1}{2}$ ____________

21. $3\frac{1}{3} * \frac{3}{10} * \frac{7}{10} * 1\frac{3}{7}$ ____________

22. $3.875 * 2.5 * 0.4$ ____________

Date Time

LESSON 6•1

Math Boxes

1. Rename each mixed number as a fraction.

a. $3\frac{7}{8}$ = ________

b. ________ = $5\frac{8}{9}$

c. ________ = $8\frac{5}{6}$

d. ________ = $6\frac{9}{7}$

e. $14\frac{2}{3}$ = ________

2. Multiply.

a. $3\frac{1}{2} * 4$ = ________

b. $\frac{1}{2} * 2\frac{1}{3}$ = ________

c. $\frac{1}{8} * \frac{2}{9} * 8$ = ________

d. ________ = $\frac{1}{5} * \frac{1}{2} * 10$

3. Circle the number sentence that describes the numbers in the table.

A. $y = x + 10$

B. $(2 * x) + 5 = y$

C. $y - 2 = (5 - x)$

D. $y - 8 = x$

x	y
3	11
5	15
0	5
10	25

4. Write each number in standard notation.

a. $(5 * 10^1) + (3 * 10^0) + (4 * 10^{-2})$

____

b. $(9 * \frac{1}{10}) + (7 * \frac{1}{100}) + (6 * \frac{1}{1,000})$

____

5. Write a percent for each fraction.

a. $\frac{4}{5}$ = ________

b. $\frac{1}{8}$ = ________

c. $\frac{7}{8}$ = ________

d. $1\frac{3}{4}$ = ________

e. $\frac{3}{100}$ = ________

6. a. Use your Geometry Template to draw a spinner with colored sectors so the chances of landing on these colors are as follows:

red: $\frac{3}{10}$

blue: 0.33

green: 20%

b. On this spinner, what is the chance of *not* landing on red, blue, or green? ________

Date Time

LESSON 6·2 Dividing Fractions and Mixed Numbers

Math Message

SRB 91 93

1. How many 3-centimeter pieces of string can you cut from a piece that is 12 centimeters long? ______

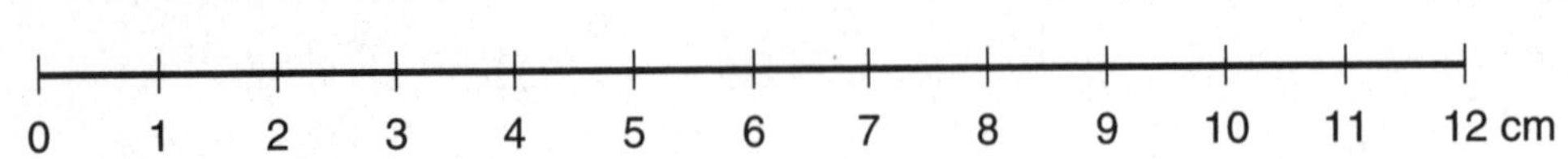

2. How many $\frac{1}{2}$-inch pieces of string can you cut from a piece that is 4 inches long? ______

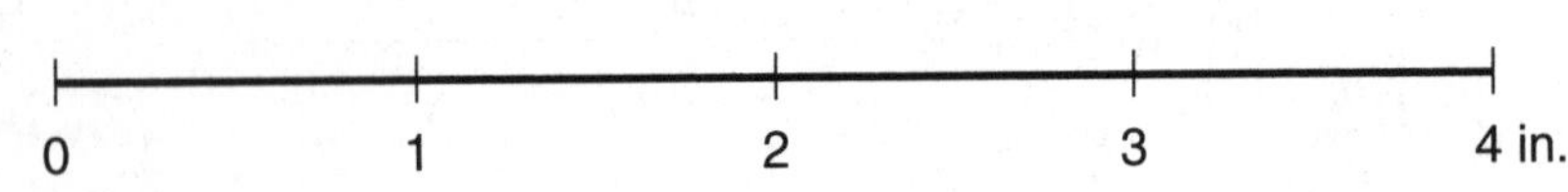

3. How many $\frac{3}{4}$-inch pieces of string can you cut from a piece that is 3 inches long? ______

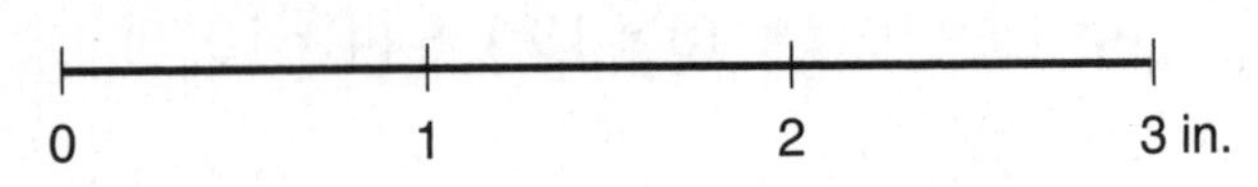

4. How many $\frac{3}{4}$-inch pieces of string can you cut from a piece that is $4\frac{1}{2}$ inches long? ______

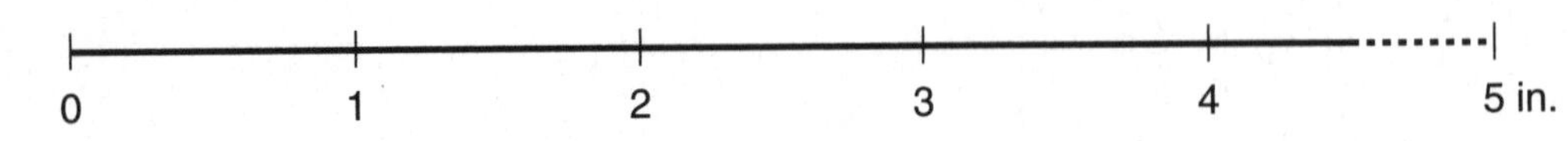

Division of Fractions Algorithm

$$\frac{a}{b} \div \frac{c}{d} = \frac{a}{b} * \frac{d}{c}$$

Divide. Show your work. Write your answers in simplest form.

5. $\frac{3}{8} \div \frac{5}{6} =$ ______

6. $\frac{4}{7} \div \frac{2}{3} =$ ______

7. $\frac{3}{10} \div \frac{3}{5} =$ ______

8. $\frac{11}{12} \div \frac{8}{5} =$ ______

Date Time

LESSON 6•2

Dividing Fractions and Mixed Numbers *cont.*

Divide. Show your work. Write your answers in simplest form.

9. $\frac{7}{8} \div \frac{4}{9} =$ ______

10. $\frac{7}{12} \div \frac{1}{3} =$ ______

11. $\frac{5}{9} \div \frac{1}{10} =$ ______

12. $\frac{3}{4} \div \frac{7}{8} =$ ______

13. $\frac{5}{3} \div \frac{3}{5} =$ ______

14. $\frac{9}{10} \div \frac{2}{3} =$ ______

15. $1\frac{5}{8} \div \frac{4}{6} =$ ______

16. $\frac{3}{8} \div \frac{8}{2} =$ ______

17. $\frac{5}{4} \div \frac{16}{8} =$ ______

18. $1\frac{2}{3} \div 2\frac{1}{4} =$ ______

19. $\frac{8}{9} \div \frac{8}{9} =$ ______

20. $3\frac{7}{8} \div 1\frac{3}{4} =$ ______

21. Explain how you found your answer to Problem 19.

______

______

______

22. How is dividing 5 by $\frac{1}{5}$ different from multiplying 5 by $\frac{1}{5}$?

______

______

______

Date Time

LESSON 6•2

Math Boxes

1. Write the reciprocal.

   a. $\frac{3}{8}$ ______

   b. $\frac{5}{9}$ ______

   c. $1\frac{3}{4}$ ______

   d. 0.68 ______

2. Divide. Simplify if possible.

   a. $8 \div \frac{4}{5} =$ ______

   b. $5\frac{1}{5} \div \frac{2}{5} =$ ______

   c. ______ $= \frac{2}{9} \div \frac{1}{3}$

   d. ______ $= \frac{9}{14} \div \frac{3}{7}$

3. Lines *l* and *m* are parallel. Without using a protractor, find the degree measure of each numbered angle. Write each measure on the drawing.

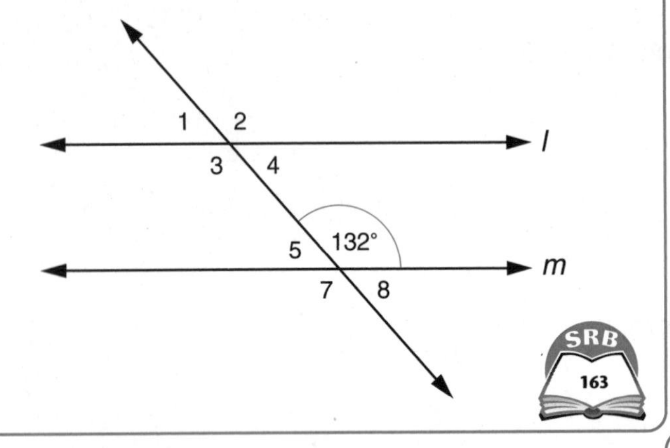

4. There are 30.48 centimeters in 1 foot.

   Complete each statement.

   a. ______ mm = 1 ft

   b. ______ cm = 1 yd

   c. 304.8 cm = ______ ft

   d. ______ cm = 1 in.

SRB 371

5. Express each decimal as a percent.

   a. 0.82 = ______

   b. ______ = 0.4375

   c. 0.077 = ______

   d. ______ = 0.009

6. If you randomly pick a date in April, how many equally likely outcomes are there?

   ______

   Explain your answer.

   ______

   ______

Date Time

LESSON 6•3

Negative Numbers on a Calculator

Math Message

Read the section "Negative Numbers" on page 271 in your *Student Reference Book.* Study the key sequence for the calculator you are using.

1. Enter each number into your calculator. Record the calculator display.

Enter	-4.85	$-(\frac{2}{3})$	-0.006	$(-4)^2$	-8^5
Display	______	______	______	______	______

You can use the negative sign (−) or OPP to represent the phrase "the opposite of." For example, "the opposite of 12" is written as −12 or OPP(12). In the same way, "the opposite of −5" is written as −(−5) or OPP(−5).

2. Enter the first number into your calculator. Record the calculator display. Clear the calculator before entering the next number.

Enter	OPP(75)	OPP(−89)	OPP$(-3)^2$	OPP(15 − 21)
Display	______	______	______	______

Use a calculator to add or subtract. Remember, the term OPP means "the opposite of."

3. −26 − 17 = ______

−26 + OPP(17) = ______

4. −34 − 68 = ______

−34 + (−68) = ______

5. 56 − 24 = ______

56 + OPP(24) = ______

6. 18 − 84 = ______

18 + (−84) = ______

7. 43 − (−97) = ______

43 + OPP(−97) = ______

43 + 97 = ______

8. 31 − (−13) = ______

31 + (−(−13)) = ______

31 + 13 = ______

9. −130 − (−62) = ______

−130 + OPP(−62) = ______

−130 + 62 = ______

10. −2 − (−22) = ______

−2 + (−(−22)) = ______

−2 + 22 = ______

Date Time

LESSON 6•3

Subtracting Positive and Negative Numbers

One way to subtract one number from another number is to change the subtraction problem into an addition problem.

Subtraction Rule

To subtract a number b from a number a, add the opposite of b to a.
Thus, for any numbers a and b, $a - b = a + \text{OPP}(b)$, or $a - b = a + (-b)$.

Examples:

$6 - 9 = 6 + \text{OPP}(9) = 6 + (-9) = -3$

$-15 - (-23) = -15 + \text{OPP}(-23) = -15 + 23 = 8$

Rewrite each subtraction problem as an addition problem. Then solve the problem.

1. $22 - (15) =$ $22 + \text{OPP}(15) = 7$
2. $-35 - 20 =$ ______
3. $-3 - (-4.5) =$ ______
4. $-27 - (-27) =$ ______

Subtract.

5. $-23 - (-5) =$ ______
6. $9 - (-54) =$ ______
7. $-(\frac{4}{5}) - 1\frac{1}{5} =$ ______
8. $\$1.25 - (-\$6.75) =$ ______
9. $-76 - (-56) =$ ______
10. $-27 - 100 =$ ______
11. Explain how you solved Problem 9. ______

Fill in the missing numbers.

12. ______ $+ 5 = -10$

    $-10 - 5 =$ ______
13. ______ $+ (-5) = -10$

    $-10 - (-5) =$ ______
14. $-9 +$ ______ $= 0$

    $0 - (-9) =$ ______
15. $16 +$ ______ $= -7$

    $-7 - 16 =$ ______
16. $-25 +$ ______ $= 15$

    $15 - (-25) =$ ______
17. ______ $+ 13 = -8$

    $-8 - 13 =$ ______

Date Time

LESSON 6•3

Math Boxes

1. Rename each fraction as a mixed number.

a. $\frac{320}{25}$ = ____________

b. ____________ = $\frac{43}{7}$

c. ____________ = $\frac{101}{5}$

d. ____________ = $\frac{75}{8}$

e. $\frac{147}{4}$ = ____________

2. Multiply.

a. $\frac{1}{3} * 12\frac{2}{9} * 3 =$ ____________

b. $\frac{4}{5} * \frac{1}{2} * 4 =$ ____________

c. $6\frac{2}{3} * \frac{1}{5} * \frac{3}{20} =$ ____________

d. ____________ $= \frac{2}{5} * \frac{3}{8} * \frac{15}{6}$

3. Circle the number sentence that describes the numbers in the table.

A. $p = m * 2$

B. $(3 - m) = p + 8$

C. $p = (3 * m) - 8$

D. $m - 8 = p$

m	*p*
8	16
0	−8
4	4
10	22

4. Which of the following is $(3 * 10^3) + (7 * 10^0) + (5 * 10^{-2})$ written in standard notation? Choose the best answer.

- 30.2
- 30.50
- 307.005
- 3,007.05

5. Write a percent for each fraction.

a. $\frac{10}{50}$ = ____________

b. $\frac{2}{3}$ = ____________

c. $\frac{24}{25}$ = ____________

d. $\frac{11}{8}$ = ____________

e. $\frac{2}{1,000}$ = ____________

6. a. Use your Geometry Template to draw a spinner with colored sectors so the chances of landing on these colors are as follows:

red: 1 out of 4

blue: $\frac{3}{8}$

b. On this spinner, what is the chance of *not* landing on red or blue?

____

Date Time

LESSON 6•4

Multiplication Patterns

In each of Problems 1–4, complete the patterns in Part a. Check your answers with a calculator. Then circle the word in parentheses that correctly completes the statement in Part b.

1. a. $6 * 4 = 24$

$6 * 3 = 18$

$6 * 2 =$ ________

$6 * 1 =$ ________

$6 * 0 =$ ________

b. Positive * Positive Rule:

When a positive number is multiplied by a positive number, the product is a

(positive or negative) number.

2. a. $5 * 2 = 10$

$5 * 1 = 5$

$5 * 0 = 0$

$5 * (-1) =$ ________

$5 * (-2) =$ ________

b. Positive * Negative Rule:

When a positive number is multiplied by a negative number, the product is a

(positive or negative) number.

3. a. $2 * 3 = 6$

$1 * 3 = 3$

$0 * 3 = 0$

$-1 * 3 =$ ________

$-2 * 3 =$ ________

b. Negative * Positive Rule:

When a negative number is multiplied by a positive number, the product is a

(positive or negative) number.

4. a. $-4 * 1 = -4$

$-4 * 0 = 0$

$-4 * (-1) = 4$

$-4 * (-2) =$ ________

$-4 * (-3) =$ ________

b. Negative * Negative Rule:

When a negative number is multiplied by a negative number, the product is a

(positive or negative) number.

5. a. Solve.

$-1 * 6 =$ ________

$-1 * (-7.7) =$ ________

$-1 * -(-\frac{1}{2}) =$ ________

$-1 * m =$ ________

b. Multiplication Property of −1:

For any number *a*, $-1 * a = a * -1 = \text{OPP}(a)$, or $-a$. The number *a* can be a negative number, so OPP(*a*) or $-a$ can be a positive number. For example, if $a = -5$, then $-a = \text{OPP}(-5) = 5$.

Date Time

LESSON 6•4

Fact Families for Multiplication and Division

A fact family is a group of four basic, related multiplication and division facts.

Example: The multiplication and division fact family for 6, 3, and 18 is made up of the following facts:

$6 * 3 = 18$ $18 / 6 = 3$
$3 * 6 = 18$ $18 / 3 = 6$

As you already know, when a positive number is divided by a positive number, the quotient is a positive number. Problems 1 and 2 will help you discover the rules for division with negative numbers. Complete the fact families. Check your answers with a calculator. Then complete each rule.

1. a. $5 * (-3) =$ ______
   $-3 * 5 =$ ______
   $-15 / (-3) =$ ______
   $-15 / 5 =$ ______

   b. $6 * (-8) =$ ______
   $-8 * 6 =$ ______
   $-48 / (-8) =$ ______
   $-48 / 6 =$ ______

   c. $5 * (-5) =$ ______
   ______
   ______
   ______

   d. **Negative / Negative Rule:** When a negative number is divided by a negative number, the quotient is a ______ (positive or negative) number.

   e. **Negative / Positive Rule:** When a negative number is divided by a positive number, the quotient is a ______ (positive or negative) number.

2. a. $-4 * (-3) =$ ______
   $-3 * (-4) =$ ______
   $12 / (-3) =$ ______
   $12 / (-4) =$ ______

   b. $-7 * (-5) =$ ______
   ______
   ______
   ______

   c. $-2 * (-10) =$ ______
   ______
   ______
   ______

   d. **Positive / Negative Rule:** When a positive number is divided by a negative number, the quotient is a ______ (positive or negative) number.

3. Solve. Check your answers with a calculator.

   a. ______ $* (-4) = 24$ (*Think:* What number multiplied by -4 is equal to 24?)

   b. ______ $* 9 = -81$

   c. $-6 *$ ______ $= 48$

   d. ______ $* (-3) = -27$

   e. $-81 / 9 =$ ______

   f. $48 / (-6) =$ ______

   g. $-27 / (-3) =$ ______

Date Time

LESSON 6•4

*, / of Positive and Negative Numbers

A Multiplication Property	A Division Property
For all numbers *a* and *b*, if the values of *a* and *b* are both positive or both negative, then the product $a * b$ is a positive number. If one of the values is positive and the other is negative, then the product $a * b$ is a negative number.	For all numbers *a* and *b*, if the values of *a* and *b* are both positive or both negative, then the quotient a / b is a positive number. If one of the values is positive and the other is negative, then the quotient a / b is a negative number.

Solve. Use a calculator to check your answers.

1. $-7 * 8 =$ _______
2. $73 * (-45) =$ _______
3. _______ $\div (-10) = 70$
4. $\frac{1}{2} * (-\frac{3}{4}) =$ _______
5. $0.5 * (-15) =$ _______
6. _______ $* 3.3 = -3.3$
7. $-3 * 4 * (-7) =$ _______
8. _______ $* (-8) * (-3) = -48$
9. $-54 / 9 =$ _______
10. $36 / (-12) =$ _______
11. $-\frac{3}{5} \div (-\frac{4}{5}) =$ _______
12. $45 / (-5) / (-3) =$ _______
13. _______ $\div 15 = -6$
14. $72 / (-8) =$ _______
15. $-99 /$ _______ $= -11$
16. $\frac{1}{2} \div (-\frac{3}{4}) =$ _______
17. $-3 * (-4 + 6) =$ _______
18. $32 \div (-5 - 3) =$ _______
19. $(-9 * 4) + 6 =$ _______
20. $(-75 / 5) + (-20) =$ _______
21. $(-6 * 3) + (-6 * 5) =$ _______
22. $(4 * (-7)) - (4 * (-3)) =$ _______

Evaluate each expression for $y = -4$.

23. $3 - (-y) =$ _______
24. $-y/(-6) =$ _______
25. $y - (-7 + 3) =$ _______
26. $y - (y + 2) =$ _______
27. $(-8 * y) - 6 =$ _______
28. $(-8 * 6) - (-8 * y) =$ _______

Date Time

LESSON 6•4 Math Boxes

1. Write the reciprocal.

a. 5 ______

b. $\frac{2}{3}$ ______

c. $2\frac{4}{7}$ ______

d. 9.64 ______

2. Divide. Simplify if possible.

a. $\frac{3}{5} \div 1\frac{1}{4} =$ ______

b. $\frac{5}{6} \div 2\frac{1}{3} =$ ______

c. ______ $= 6\frac{4}{5} \div 2\frac{1}{2}$

d. ______ $= 20\frac{2}{5} \div 10\frac{1}{5}$

3. Lines *a* and *b* are parallel. Without using a protractor, find the degree measure of each numbered angle. Write each measure on the drawing.

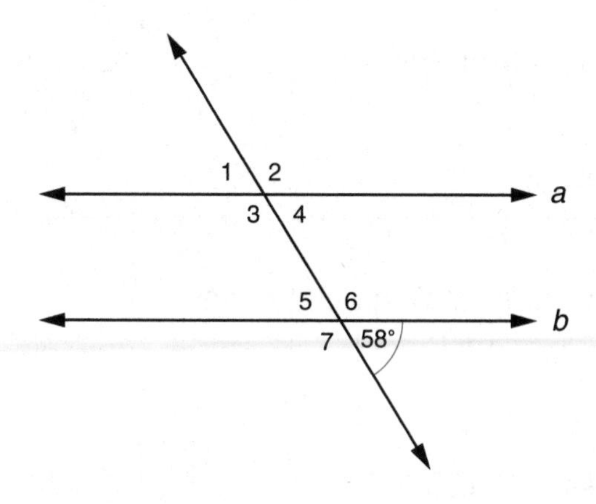

SRB 163

4. If 1 kilogram (kg) is about $2\frac{1}{5}$ pounds (lb), then 5 kg is about ______ lb.

Circle the best answer.

A. $\frac{11}{25}$

B. $7\frac{1}{5}$

C. 10

D. 11

5. Express each decimal as a percent.

a. 1.04 = ______

b. ______ = 0.0825

c. 0.0035 = ______

d. ______ = 4.0

6. Suppose you toss a penny and a nickel together. How many equally likely outcomes are there?

Complete the table.

Penny	Nickel
H	H

______ equally likely outcomes

Date Time

LESSON 6•4a Absolute Value

Solve.

1. $|16| =$ ______
2. $|-0.003| =$ ______
3. ______ $= |-64|$
4. ______ $= |\frac{1}{2}|$
5. $|0.72| =$ ______
6. ______ $= |-4\frac{1}{3}|$
7. $|-3.09| =$ ______
8. $|0| =$ ______
9. ______ $= |21.741|$
10. ______ $= |8{,}906|$

11. What do you notice about the absolute value of positive numbers and zero?

______

______

12. What do you notice about the absolute value of negative numbers?

______

______

Use algebraic notation to write the general pattern for each group of 3 special cases.

13. $|4.78| = 4.78$
    $|0| = 0$
    $|19{,}722| = 19{,}722$

    General pattern:

    If $x \geq 0$, ______.

14. $|-270| = 270$
    $|-\frac{15}{16}| = \frac{15}{16}$
    $|-12.125| = 12.125$

    General pattern:

    If $x < 0$, ______.

Answer the following questions. Use an inequality involving absolute value to justify your answer.

15. Jayson charged $172 on his credit card for materials to start his lawn-mowing business. Rita charged $98 on her credit card for materials to start her tutoring business.

    Who has the greater debt? ______

16. The number of home sales in Illinois dropped 24% in one year while the number of home sales in Indiana dropped 20%.

    Which state had the larger decrease in home sales? ______

Date Time

LESSON 6•4a

Taxi Rides

Jerissa and her aunt have planned a trip to the city. A map of the city is shown below. Each grid square represents one block. They will be traveling around the city by taxi. The taxi meter calculates the distance the taxi has traveled from the starting point (s) to the ending point (e) using the formula $d = |s - e|$.

Jerissa has planned to visit different locations in the order shown below. Find the distance that they will travel on each part of their trip. Use a number sentence to justify each answer.

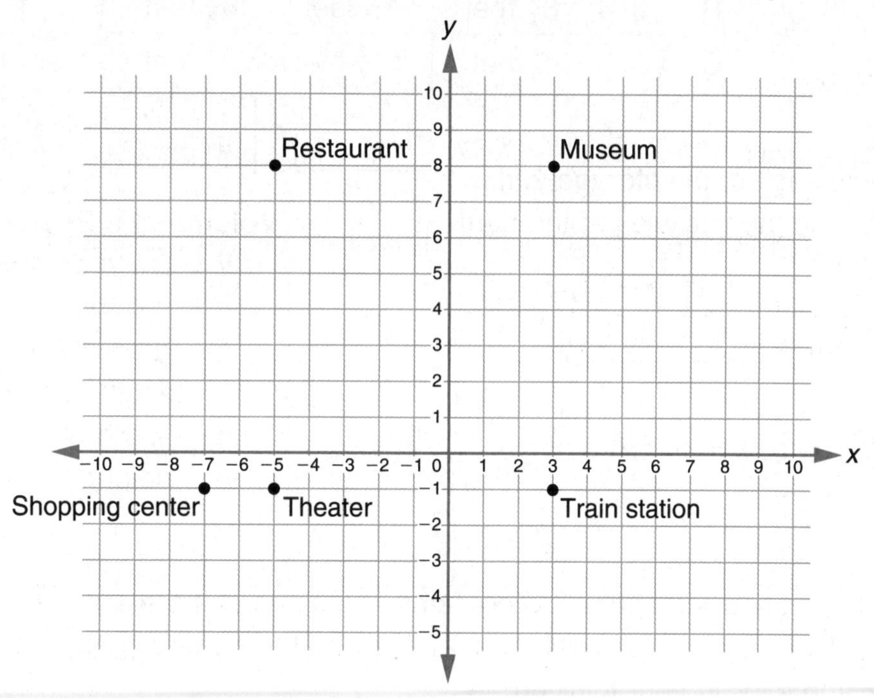

1. Train station to the museum: ______________________

   Number sentence: ______________________

2. Museum to the restaurant: ______________________

   Number sentence: ______________________

3. Restaurant to the theater: ______________________

   Number sentence: ______________________

4. Theater to the shopping center: ______________________

   Number sentence: ______________________

5. Shopping center to the train station: ______________________

   Number sentence: ______________________

Date Time

LESSON 6•4a

Volume Problems

Leroy's Moving and Storage has three different storage units for rent. The table below shows the dimensions of each unit.

Unit Type	Length	Width	Height
A	$12\frac{1}{2}$ feet	$10\frac{1}{2}$ feet	$8\frac{1}{2}$ feet
B	$8\frac{1}{2}$ feet	15 feet	$8\frac{1}{2}$ feet
C	9 feet	$13\frac{3}{4}$ feet	$8\frac{1}{2}$ feet

1. Find the volume of each of the storage units. Show your work. Use the formulas at the right to help you.

   a. Volume of Unit A = ______________

   b. Volume of Unit B = ______________

   c. Volume of Unit C = ______________

Volume of a Rectangular Prism

$V = B * h$ or $V = l * w * h$

V = volume

B = area of base

l = length of base

w = width of base

h = height

2. Jacqueline needs to rent a storage unit. The items she plans to store are currently in a trailer with dimensions of $14\frac{1}{2}$ feet by $8\frac{1}{3}$ feet by 9 feet. Which storage unit should Jacqueline rent? Show your work and explain your thinking.

____

____

____

____

3. Jarrett has packed all his items into boxes. The boxes are $2\frac{1}{2}$ feet in length, $2\frac{1}{2}$ feet in width, and 2 feet in height. Jarrett has rented Unit B. How many boxes will Jarrett be able to fit in Unit B? Explain your thinking.

____

____

____

____

Date Time

LESSON 6•4a

Math Boxes

1. Simplify.

a. $\frac{2}{3}$ of 123 ________

b. $\frac{7}{6}$ of 72 ________

c. $\frac{3}{4}$ of $2\frac{1}{2}$ ________

d. $\frac{7}{8}$ of $\frac{4}{7}$ ________

SRB 87–89

2. Divide. Simplify if possible.

a. $\frac{4}{5} \div \frac{1}{2} =$ ________

b. $\frac{8}{9} \div \frac{2}{5} =$ ________

c. $\frac{3}{4} \div \frac{4}{5} =$ ________

d. $\frac{9}{18} \div \frac{3}{18} =$ ________

3. Give a ballpark estimate for each quotient.

a. 487.8 ÷ 3 ________

b. 619.725 / 5 ________

c. 1,824 ÷ 60 ________

d. $\frac{5{,}478}{92}$ ________

4. Complete each sentence using an algebraic expression.

a. Raven is 3 years older than her brother. If Raven's brother is *b* years old, then

Raven is ________ years old.

b. If every box of crayons contains 8 crayons, then *c* boxes of crayons

contain ________ crayons.

5. Write each fraction as a decimal.

a. $\frac{32}{100} =$ ________

b. $\frac{1}{4} =$ ________

c. $\frac{47}{50} =$ ________

d. $\frac{8}{20} =$ ________

6. Suppose you toss a fair coin 10 times. What is the probability of getting HEADS on:

a. the first toss? ________

b. the third toss? ________

c. the tenth toss? ________

SRB 148 149

Date Time

LESSON 6•5

Scavenger Hunt

Use *Student Reference Book,* pages 2–24 and 94–106 to find answers to as many of these questions as you can. Try to get as high a score as possible.

1. How many rational numbers are there? (10 points) ______________________

2. Give an example of each of the following. (5 points each)

   a. A counting number __________

   b. A negative rational number __________

   c. A positive rational number __________

   d. A real number __________

   e. An integer __________

   f. An irrational number __________

3. Name two examples of uses of negative rational numbers. (5 points each)

   ____

   ____

4. Explain why numbers such as 4, $\frac{3}{5}$, and 3.5 are rational numbers. (10 points)

   ____

   ____

   ____

5. Explain why numbers such as π and $\sqrt{2}$ are irrational numbers. (10 points)

   ____

6. $n + n = n$ What is n? __________ (15 points)

7. $k = \text{OPP}(k)$ What is k? __________ (15 points)

8. $j * j = j$ Which two numbers could j be? __________ (15 points each)

9. $a + (-a) =$ __________ (15 points)

10. $b * \frac{1}{b} =$ __________ (15 points)

Date Time

LESSON 6•5

Scavenger Hunt *continued*

11. Match each sentence in Column 1 with the property in Column 2 that it illustrates. (5 points each)

Column 1

A. $a + (b + c) = (a + b) + c$

B. $a + b = b + a$

C. $a * (b + c) = (a * b) + (a * c)$

D. $a * (b - c) = (a * b) - (a * c)$

E. $a * b = b * a$

F. $a * (b * c) = (a * b) * c$

Column 2

_______ Distributive Property of Multiplication over Subtraction

_______ Commutative Property of Addition

_______ Distributive Property of Multiplication over Addition

_______ Associative Property of Multiplication

_______ Commutative Property of Multiplication

_______ Associative Property of Addition

12. $-a > 0$. How can that be? (15 points)

____

____

13. Complete. (2 points each, except the last problem, which is worth 25 points)

OPP(1) = _______

OPP(OPP(1)) = _______

OPP(OPP(OPP(1))) = _______

OPP(OPP(OPP(OPP(1)))) = _______

OPP(OPP(OPP(OPP(OPP(OPP … (OPP(OPP(1)))))))) = _______

100 OPPs

Explain how you found the answer to the last problem. _______________________

____

____

LESSON 6•5

Scavenger Hunt *continued*

14. Is 5^{-2} a positive or negative number? Explain. (15 points)

15. Two numbers are their own reciprocals. What are they? ______ (15 points each)

16. What number has no reciprocal? ______ (15 points)

Number Stories

1. Diana wants to make a 15 ft by 20 ft section of her yard into a garden. She will plant flowers in $\frac{2}{3}$ of the garden and vegetables in the rest of the garden. How many square feet of vegetable garden will she have?

______

Explain how you got your answer. ______

2. Leo is in charge of buying hot dogs for his school's family math night. Out of 300 parents and children, he expects about $\frac{3}{5}$ of them to attend. Hot dogs are sold 8 in a package, and Leo figures he will need to buy 22 packages so that each person can have 1 hot dog.

a. How do you think he calculated to get 22 packages?

b. Will Leo have enough hot dogs? ______

Date Time

LESSON 6•5 Math Boxes

1. Simplify.

a. $\frac{3}{4}$ of 80 _______

b. $\frac{9}{8}$ of 2 _______

c. $\frac{2}{3}$ of $3\frac{1}{2}$ _______

d. $\frac{3}{8}$ of $\frac{4}{9}$ _______

SRB 87–89

2. Divide. Simplify if possible.

a. $\frac{8}{9} \div \frac{3}{4} =$ _______

b. $\frac{7}{8} \div \frac{1}{3} =$ _______

c. $\frac{6}{9} \div \frac{1}{2} =$ _______

d. $\frac{8}{24} \div \frac{4}{24} =$ _______

3. Give a ballpark estimate for each quotient.

a. 643.27 ÷ 5 _______

b. $\frac{728.09}{7}$ _______

c. 432.67 ÷ 82 _______

d. 2,091.05 / 53 _______

SRB 261

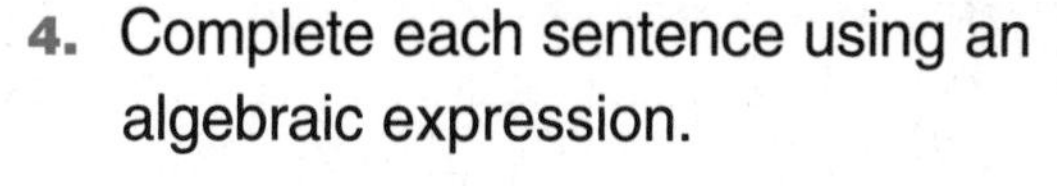

4. Complete each sentence using an algebraic expression.

a. If Mark earns *t* dollars for each hour he tutors, then he earns _______ dollars when he tutors for $3\frac{1}{2}$ hours.

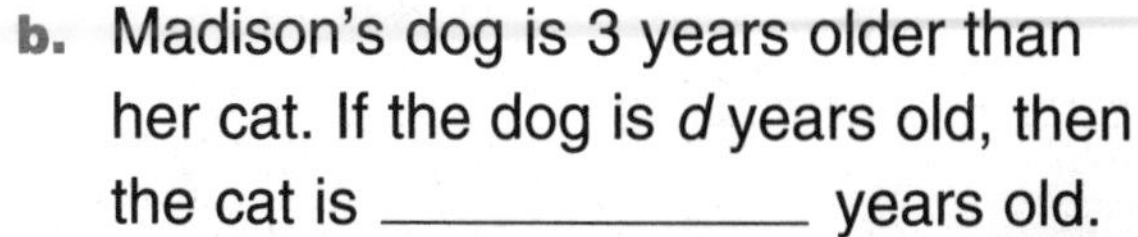

b. Madison's dog is 3 years older than her cat. If the dog is *d* years old, then the cat is _______ years old.

5. Write each decimal as a fraction in simplest form.

a. 0.06 = _______

b. 0.52 = _______

c. 0.09 = _______

d. 0.64 = _______

6. Roll two 6-sided dice, one red, one green. Give the probability of rolling the following totals.

a. 11 _______ **b.** 7 _______

c. 0 _______ **d.** 3 or 4 _______

e. an even number _______

SRB 148 149

Date Time

LESSON 6·6

Order of Operations

Evaluate each expression. Show your work. Then compare your results to those of your partner.

1. $4 * 6 + 3 =$ ________

2. $33 - 16 / 4 =$ ________

3. $4 * 7 - (3 + 5) =$ ________

4. $24 / 6 * 4 =$ ________

5. $8 * 4 + 49 \div 7 =$ ________

6. $9 * 6 \div 3 + 28 =$ ________

7. $7 - 5 + 13 - 23 - 17 =$ ________

8. $100 - 50 \div 2 + 4 * 5 =$ ________

9. $7 / 7 * 4 + 3^2 =$ ________

10. $12 * 2^2 - 3^3 =$ ________

Date Time

LESSON 6•5

Math Boxes

1. Simplify.

a. $\frac{3}{4}$ of 80 ______

b. $\frac{9}{8}$ of 2 ______

c. $\frac{2}{3}$ of $3\frac{1}{2}$ ______

d. $\frac{3}{8}$ of $\frac{4}{9}$ ______

2. Divide. Simplify if possible.

a. $\frac{8}{9} \div \frac{3}{4} =$ ______

b. $\frac{7}{8} \div \frac{1}{3} =$ ______

c. $\frac{6}{9} \div \frac{1}{2} =$ ______

d. $\frac{8}{24} \div \frac{4}{24} =$ ______

3. Give a ballpark estimate for each quotient.

a. 643.27 ÷ 5 ______

b. $\frac{728.09}{7}$ ______

c. 432.67 ÷ 82 ______

d. 2,091.05 / 53 ______

4. Complete each sentence using an algebraic expression.

a. If Mark earns *t* dollars for each hour he tutors, then he earns ______ dollars when he tutors for $3\frac{1}{2}$ hours.

b. Madison's dog is 3 years older than her cat. If the dog is *d* years old, then the cat is ______ years old.

5. Write each decimal as a fraction in simplest form.

a. 0.06 = ______

b. 0.52 = ______

c. 0.09 = ______

d. 0.64 = ______

6. Roll two 6-sided dice, one red, one green. Give the probability of rolling the following totals.

a. 11 ______ b. 7 ______

c. 0 ______ d. 3 or 4 ______

e. an even number ______

Date Time

LESSON 6•6

Order of Operations

Evaluate each expression. Show your work. Then compare your results to those of your partner.

1. $4 * 6 + 3 =$ ________

2. $33 - 16 / 4 =$ ________

3. $4 * 7 - (3 + 5) =$ ________

4. $24 / 6 * 4 =$ ________

5. $8 * 4 + 49 \div 7 =$ ________

6. $9 * 6 \div 3 + 28 =$ ________

7. $7 - 5 + 13 - 23 - 17 =$ ________

8. $100 - 50 \div 2 + 4 * 5 =$ ________

9. $7 / 7 * 4 + 3^2 =$ ________

10. $12 * 2^2 - 3^3 =$ ________

Date Time

LESSON 6•6

Order of Operations *continued*

11. $10^{-1} + 16 - 0.5 * 12 =$ ________

12. $((\frac{1}{2} \div \frac{1}{4}) + 3) * 6 - 3^3 =$ ________

13. $-(-8) - (-4) * 6 - (-12) / 4 =$ ________

14. $-4 + (-18) / 6 + (-3 * -3 - 5) =$ ________

Try This

15. $-5(-6 - (-3)) / 7.5 =$ ________

16. $-(\frac{3}{4} \div \frac{1}{2}) + \frac{1}{2} - (\frac{1}{2} * (-\frac{1}{2})) =$ ________

17. Evaluate the following expressions for $x = -2$.

a. $x * -x + 14 / 2 =$ ________

b. $-x * (6 + x) - 3^3 / 9 =$ ________

Date ____________________ Time ____________________

LESSON 6·6

Math Boxes

1. Solve. Simplify your answers.

a. ________ $= 8 \div 10\frac{2}{3}$

b. $4\frac{1}{2} \div 1\frac{5}{7} =$ ________

c. ________ $= 7\frac{3}{10} \div 5$

2. Multiply or divide.

a. $-10(-14.35) =$ ________

b. $4 * 3 * (-5) =$ ________

c. ________ $= \frac{280}{-4}$

3. Triangles *JKL* and *MNO* are congruent.

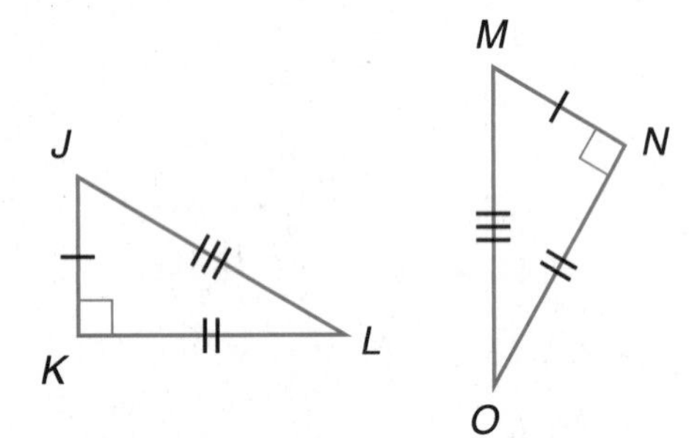

Which side corresponds with $\overline{JL}$?

____

SRB 178

4. Label the axes of this mystery graph and describe a situation it might represent.

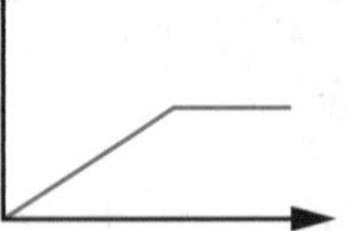

x-axis ________________________

y-axis ________________________

Situation ________________________

____

____

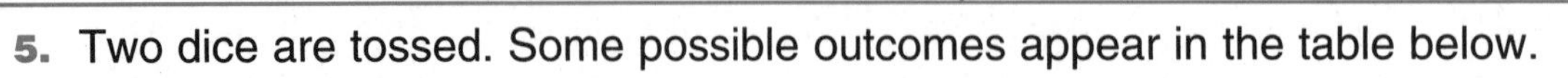

5. Two dice are tossed. Some possible outcomes appear in the table below.

Complete the table.

(1,1)	(1,2)	(1,3)	(1,4)	(1,5)	(1,6)
			(3,4)		
(4,1)					
				(6,5)	

a. How many equally likely outcomes are there? ________

b. What is the probability of tossing a multiple of 2 on both dice? ________

c. What is the probability of tossing a composite number on the first die and a prime number on the second die? ________

Date Time

LESSON 6·7 Math Boxes

1. Simplify.

a. $\frac{4}{7}$ of 84 ________

b. $\frac{1}{20}$ of 35 ________

c. $\frac{8}{3}$ of $9\frac{3}{4}$ ________

d. $\frac{1}{3}$ of $\frac{51}{68}$ ________

SRB 87–89

2. Divide. Simplify if possible.

a. $\frac{7}{12} \div \frac{2}{5} =$ ________

b. $2\frac{5}{8} \div 3 =$ ________

c. $3 \div 6\frac{1}{4} =$ ________

d. $4\frac{1}{2} \div 2\frac{7}{10} =$ ________

3. Give a ballpark estimate for each quotient.

a. 137.8 ÷ 15 ________

b. 248.19 / 12 ________

c. 4,507.08 ÷ 89.76 ________

d. 0.6 / 14.7 ________

4. Complete each sentence using an algebraic expression.

a. If each bag of potatoes weighs at least p pounds, then 6 bags weigh at least

________ pounds.

b. Jack is 6 inches taller than Michael. If Jack is h inches tall, then Michael is

________ inches tall.

5. Which fraction is equivalent to 2.015? Choose the best answer.

- ◯ $\frac{2,015}{10,000}$
- ◯ $\frac{4,030}{20,000}$
- ◯ $\frac{403}{200}$
- ◯ $\frac{2,015}{100}$

6. You draw one card at random from a regular deck of 52 playing cards (no jokers). What is the chance of drawing:

a. a 4? ________

b. a card with a prime number? ________

c. a face card (jack, queen, or king)? ________

d. an even-numbered black card? ________

Date Time

LESSON 6·7

Number Sentences

Translate the word sentences below into number sentences. Study the first one.

1. Three times five is equal to fifteen. $3 * 5 = 15$

2. Nine increased by seven is less than twenty-nine. ______

3. Thirteen is not equal to nine more than twenty. ______

4. The product of eight and six is less than or equal to the sum of twenty and thirty.

______

5. Thirty-seven increased by twelve is greater than fifty decreased by ten.

______

6. Nineteen is less than or equal to nineteen. ______

Tell whether each number sentence is true or false.

7. $3 * 21 = 63$ ______

8. $(3 * 4) + 7 = 19$ ______

9. $42 - 12 / 6 > 5$ ______

10. $8 \geq 7 + 1$ ______

11. $24 / 4 + 2 = 8$ ______

12. $9 / (8 - 5) \leq 3$ ______

13. $21 > (7 * 3) + 5$ ______

14. $8 * 7 \leq 72$ ______

15. $63 / 7 \neq 8$ ______

16. $35 + 5 * 8 = 320$ ______

Insert parentheses so each number sentence is true.

17. $5 * 8 + 4 - 2 = 42$

18. $7 * 9 - 6 = 21$

19. $10 + 2 * 6 < 24$

20. $9 - 7 / 7 = 8$

21. $33 - 24 / 3 \geq 25$

22. $36 / 7 + 2 * 3 = 12$

23. $3 * 4 + 3 > 5 * 3 + 3$

24. $48 / 8 + 4 \neq 100 / 10$

LESSON 6•7 Number Sentences *continued*

25. Write three true and three false number sentences. Trade journals with your partner and determine which sentences are true and which are false.

Number Sentence	True or false?
______________________	________
______________________	________
______________________	________
______________________	________
______________________	________
______________________	________

Try This

26. The word HOPE is printed in shaded block letters inside a 15 ft by 5 ft rectangular billboard. What is the area of the unshaded portion of the billboard?

____

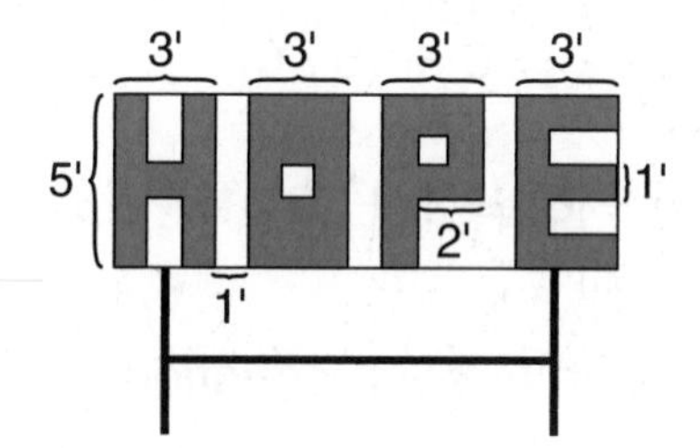

27. Square corners, 6 centimeters on a side, are removed from a 36 cm by 42 cm piece of cardboard. The cardboard is then folded to form an open box. What is the surface area of the inside of the box?

____

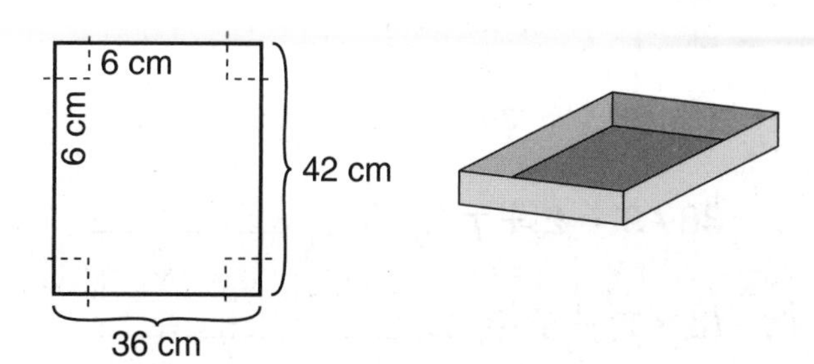

28. Pennies tossed onto the gameboard at the right have an equal chance of landing anywhere on the board. If 60% of the pennies land inside the smaller square, what is the length of a side *s* of the smaller square to the nearest inch?

____

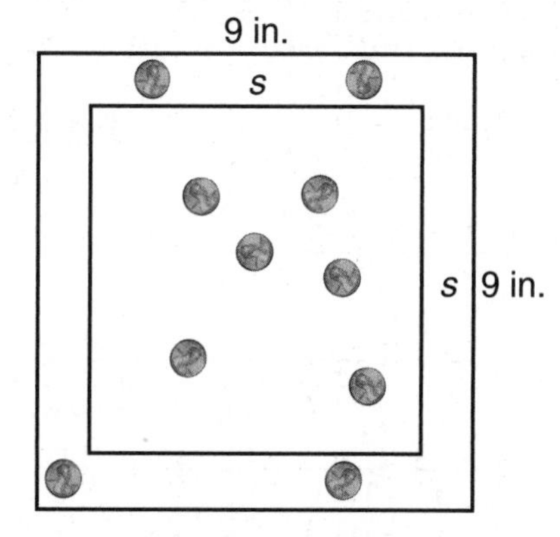

Try the Penny Toss!

Date Time

LESSON 6•8

Solving Equations

Find the solution to each equation. Write a number sentence with the solution in place of the variable. Check that the number sentence is true.

	Equation	Solution	Number Sentence
1.	$12 + x = 32$	______	______
2.	$y + 89 = 93$	______	______
3.	$b - 32 = 15$	______	______
4.	$m * 8 = 35 - 19$	______	______
5.	$p + (4 * 9) = 55$	______	______
6.	$42 = 7 * (a - 4)$	______	______
7.	$(9 + w) / 2 = 6 + (6 / 6)$	______	______
8.	$4 + (3n - 6) = 1 + (3 * 6)$	______	______

Find the solution to each equation.

9. $4 * 6 = 35 - t$ ______

10. $9 * (11 - c) = 81$ ______

11. $17 - 11 = k / 8$ ______

12. $(m + 14) / 4 = 6$ ______

13. $36 / 9 = 2 + p$ ______

14. $23 - a = 15$ ______

15. $(3 * p) + 5 = 26$ ______

16. $2 - d = 3 * 4$ ______

17. Make up four equations whose solutions are whole numbers. Ask your partner to solve each one.

	Equation	Solution
a.	______	______
b.	______	______
c.	______	______
d.	______	______

Date Time

LESSON 6•8

Math Boxes

1. Solve. Simplify your answers.

   a. ______ $= -5\frac{1}{2} \div \frac{3}{4}$

   b. $6 \div (-1\frac{5}{16}) =$ ______

   c. ______ $= -3\frac{1}{8} \div (-4\frac{1}{8})$

2. Multiply or divide.

   a. $(-3)^3 =$ ______

   b. $0.4(-0.5) =$ ______

   c. ______ $= \frac{-352}{-11}$

3. Triangles *DAB* and *BCD* are congruent.

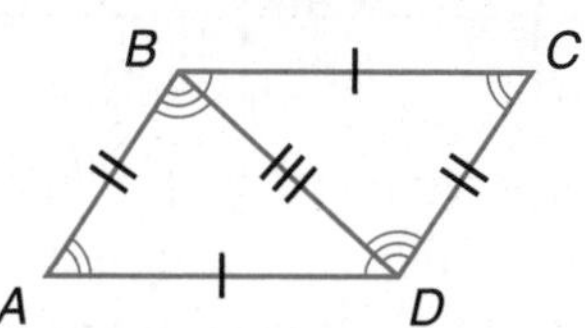

Which is a pair of corresponding angles? Circle the best answer.

A. $\angle ABD$ and $\angle ABC$

B. $\angle BAD$ and $\angle DCB$

C. $\angle BCD$ and $\angle CDB$

D. $\angle ADC$ and $\angle ADB$

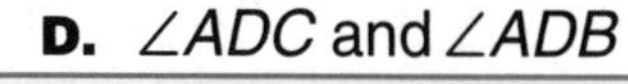

4. Label the axes of this mystery graph and describe a situation it might represent.

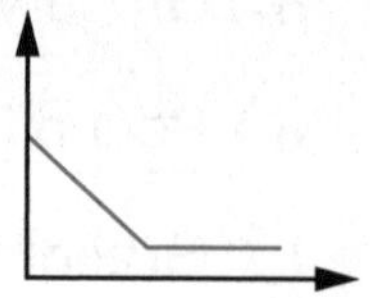

x-axis ______

y-axis ______

Situation ______

______

5. You spin the spinner shown at the right.

   a. How many equally likely outcomes are there? ______

   b. What is the probability that the spinner will land on a factor of 10? ______

   c. What is the probability that the spinner will land on a multiple of 3 or a multiple of 2? ______

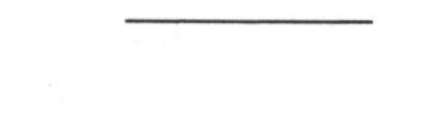

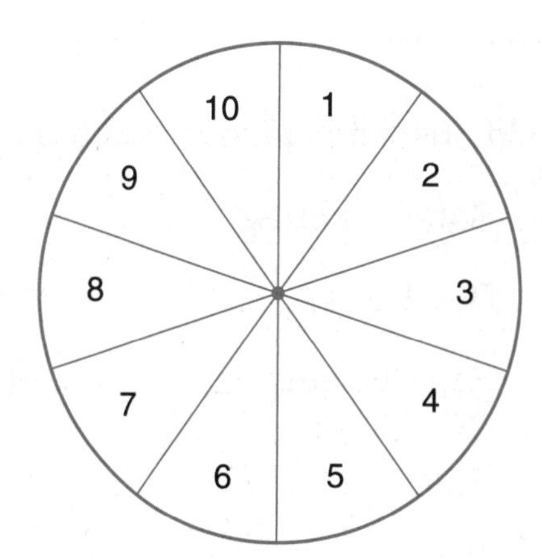

Date Time

LESSON 6•8

Absolute Value on a Map

Two cousins, Sarah and Bev, live in the same town. A map with the locations of some important places in the town is shown below. Each square represents one block.

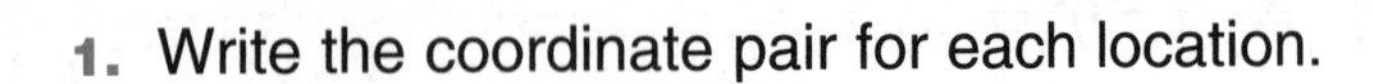

1. Write the coordinate pair for each location.

a. Sarah's house ________

b. Bev's house ________

c. Park ________

d. Grocery store ________

e. Post office ________

f. Grandma's house ________

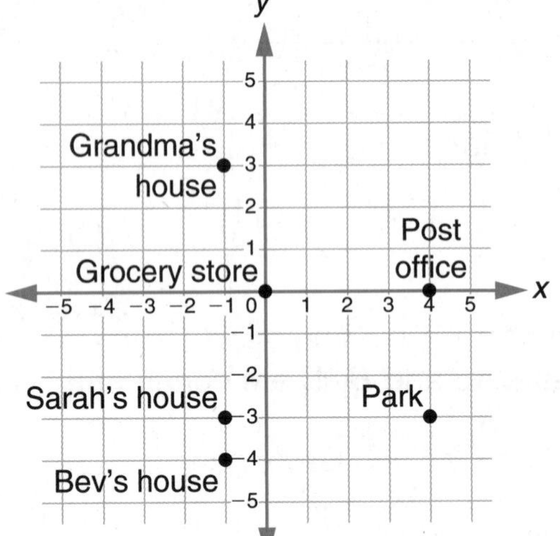

2. The two major roads in Sarah's town are State Street and Bay Road. State Street is represented by the *x*-axis and Bay Road is represented by the *y*-axis.

How could you use absolute value to find the distance of the locations shown on the map from each of the major roads? (*Hint:* Sarah's house is 1 block from Bay Road, because $|-1| = 1$.)

____

____

____

____

3. Use absolute value to find the distance of each location from Bay Road and State Street.

a. Sarah's house:

Bay Road ________ blocks

State Street ________ blocks

b. Bev's house:

Bay Road ________ blocks

State Street ________ blocks

c. Park:

Bay Road ________ blocks

State Street ________ blocks

d. Grocery store:

Bay Road ________ blocks

State Street ________ blocks

e. Post office:

Bay Road ________ blocks

State Street ________ blocks

f. Grandma's house:

Bay Road ________ blocks

State Street ________ blocks

Date Time

LESSON 6•8

Absolute Value on a Map *continued*

4. Bev told Sarah, "My house is closer to State Street than Grandma's house, because $-4 < 3$." Do you agree with Bev? Use absolute value to help justify your answer.

____

____

____

____

5. Use absolute value to find the distance between the following locations. Write a number sentence to show how you found your answer.

   a. Sarah's house and Bev's house: ________ blocks

      Number sentence: ________________________

   b. Sarah's house and the park: ________ blocks

      Number sentence: ________________________

   c. The park and the post office: ________ blocks

      Number sentence: ________________________

   d. The grocery store and the post office: ________ blocks

      Number sentence: ________________________

Try This

6. The streets in Sarah and Bev's town are shown by the gridlines. To walk from one location to the next, they must walk along the streets. Use absolute value to find the distance they would travel if they walked the following routes. Write a number sentence to show how you got your answer.

   a. From Bev's house to the park: ________ blocks

      Number sentence: ________________________________

   b. From the park to Grandma's house: ________ blocks

      Number sentence: ________________________________

   c. From Grandma's house to the grocery store: ________ blocks

      Number sentence: ________________________________

   d. From the grocery store to Sarah's house: ________ blocks

      Number sentence: ________________________________

LESSON 6•9

Pan-Balance Problems

A pan balance can be used to compare the weights of objects or to weigh objects. If the objects in one pan weigh as much as those in the other pan, the pans will balance.

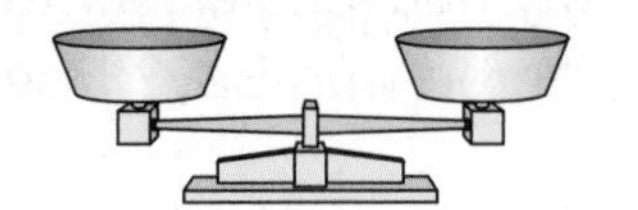

The diagram at the right shows a balanced pan balance.

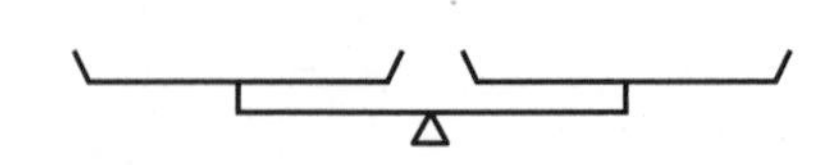

Example: In each of the diagrams below, the pans are balanced. Your job is to figure out how many marbles weigh as much as an orange. When moving the oranges and marbles, follow these simple rules:

- Whatever you do, the pans must always remain balanced.
- You must do the same thing to both pans.

The pan balance shows that 3 oranges weigh as much as 1 orange and 12 marbles.

If you remove 1 orange from each pan, the pans remain balanced.

If you then remove half of the objects from each pan, the pans will still be balanced.

Success! One orange weighs as much as 6 marbles.

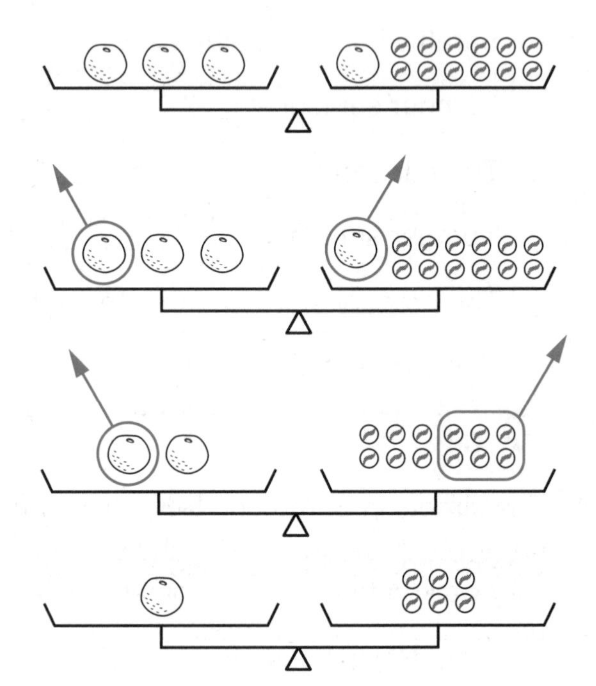

Solve the pan-balance problems with a partner. Be ready to share your strategies with the class.

1. One pencil weighs as much as ________ paper clips.

2. One *p* (pencil) weighs as much as ________ *c*s (paper clips).

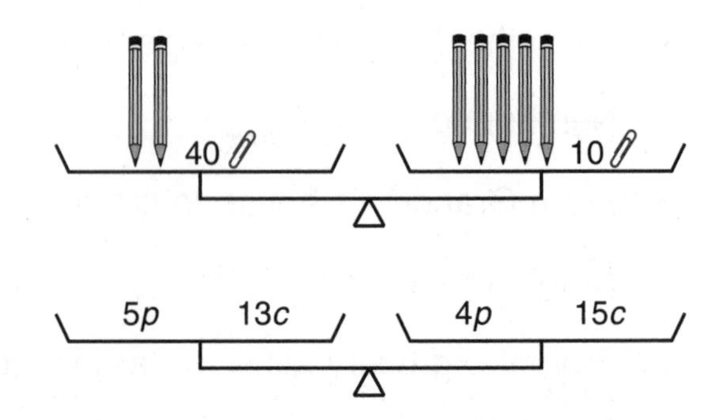

Date Time

LESSON 6•9

Pan-Balance Problems *continued*

Solve these pan-balance problems. In each figure, the two pans are balanced.

3. One banana weighs as much as ________ marbles.

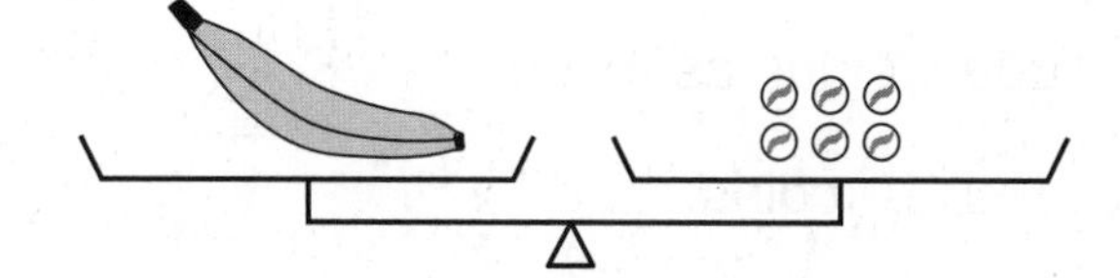

4. One cube weighs as much as ________ paper clips.

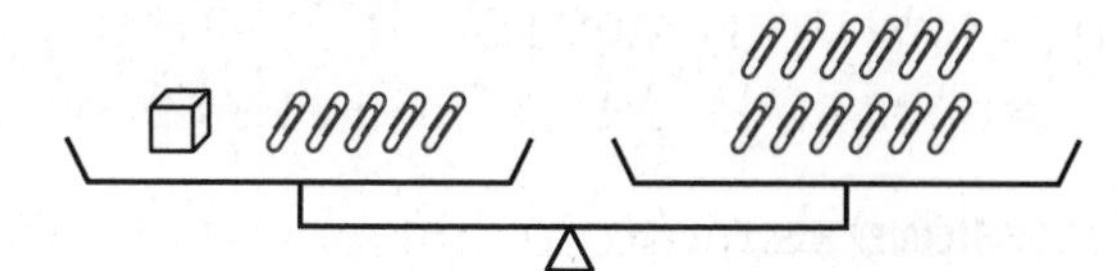

5. One cube weighs as much as ________ marbles.

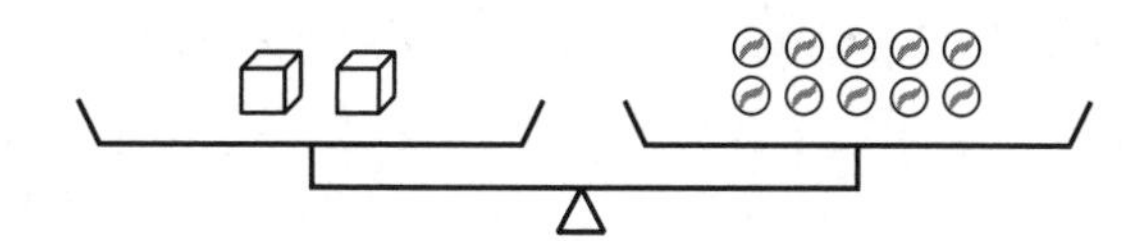

6. One triangle weighs as much as ________ squares.

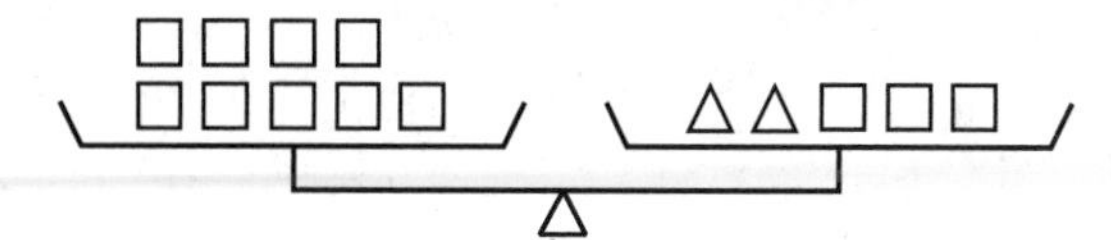

7. One orange weighs as much as ________ paper clips.

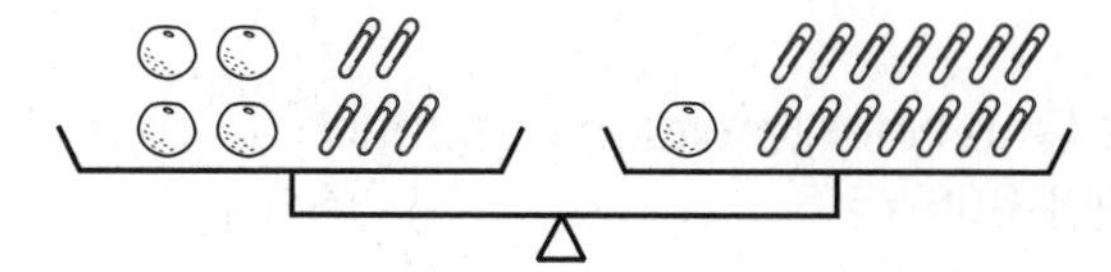

8. One and one-half cantaloupes weigh as much as ________ apples.

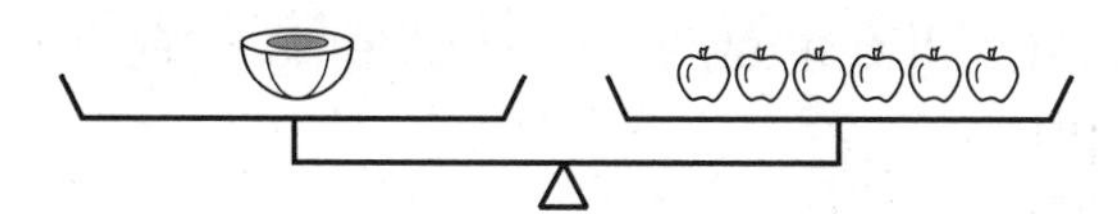

Date Time

LESSON 6•9

Pan-Balance Problems *continued*

Reminder: 4□ or 4 ∗ □ are other ways to write □ + □ + □ + □.

9. One cube weighs as much as ________ coins.

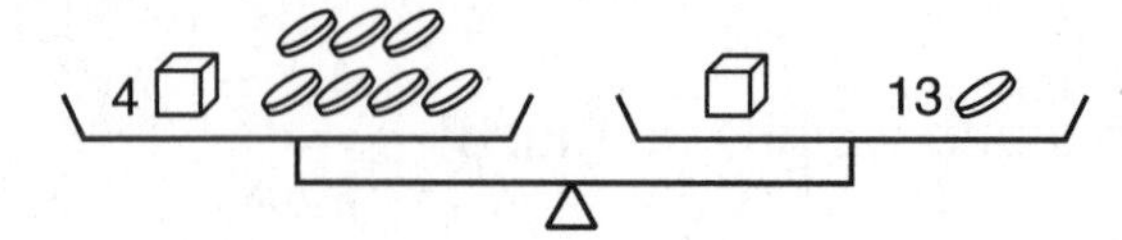

10. One p weighs as much as ________ balls.

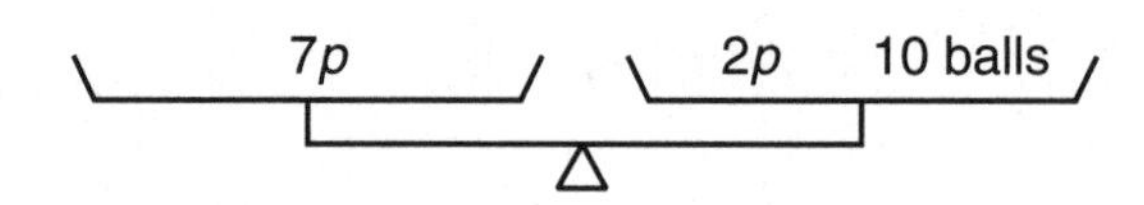

11. One b weighs as much as ________ ks.

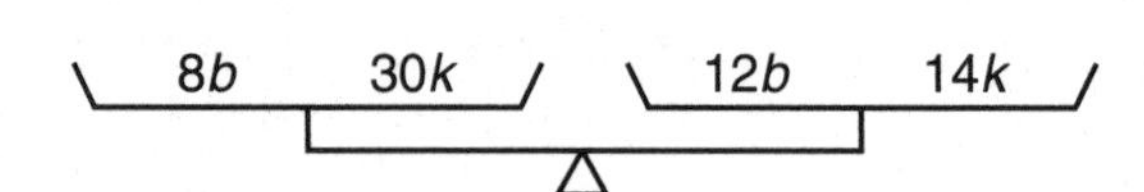

Check your answers.

- The sum of the answers to Problems 3 and 6 is equal to the square root of 81.
- The product of the answers to Problems 8 and 9 is 36.
- The sum of the answers to Problems 4, 5, and 11 is the solution to the equation $4n = 2^6$.
- The product of the answers to Problems 7, 9, and 11 is 24.

Date Time

LESSON 6•9

Math Boxes

1. Use the order of operations to evaluate each expression.

   a. $\frac{1}{2} * (-26) - 3^2 =$ ________

   b. $-3 - 14 \div 7 =$ ________

   c. ________ $= (-8) * (-8) - (-8)$

   d. $-5 - (-30) \div 3 =$ ________

2. Tell whether each statement is true or false.

   a. If a $< b$, then $a - 3 < b - 3$. ________

   b. If $m < n$, then $m + 8 > n * 8$. ________

   c. If $x > y$, then $x * 0 > y * 0$. ________

3. You can use a formula to calculate about how long it will take a falling object to reach the bottom of a well.

   Formula: $t = \frac{1}{4} * \sqrt{d}$

   (This formula does not account for air resistance.)

   - d is the distance in feet the object falls.
   - t is the time in seconds it takes the object to reach the bottom.

   About how long would it take a bowling ball to hit the bottom of a well 100 ft deep?

   ____________ seconds

4. Lines a and b are parallel.

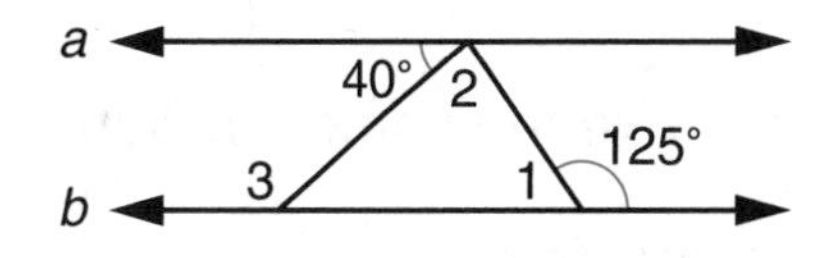

   m∠1 = ________

   m∠2 = ________

   m∠3 = ________

5. Solve mentally.

   a. 10% of 82 = ________

   b. 5% of 44 = ________

   c. 15% of 90 = ________

6. Suppose you toss a coin 10 times and get 10 HEADS. What is the probability of getting HEADS on the 11th toss?

   Probability = ________

Date Time

LESSON 6·10 Pan-Balance Equations

1. Start with the original pan-balance equation. Do the first operation on both sides of the pan balance and write the result on the second pan balance. Do the second operation on both sides of the second pan balance and write the result on the third pan balance. Complete the fourth pan balance in the same way.

Original pan-balance equation $x = 10$

Operation

(in words)	(abbreviation)
Multiply by 3.	M 3
Add 18.	A 18
Add 2x.	A 2x

Equations that have the same solution are called **equivalent equations.**

2. Check that the pan-balance equations above are equivalent equations by making sure that 10 is the solution to each equation.

3. Now do the opposite of what you did in Problem 1. Record the operation used to obtain the results on each pan balance.

Original pan-balance equation $5x + 18 = 2x + 48$

Operation

(in words)	(abbreviation)	
______	______	$3x + 18 = 48$
______	______	$3x = 30$
______	______	$x = 10$

LESSON 6·10 Pan-Balance Equations *continued*

4. Record the results of the operation on each pan, as in Problem 1.

SRB 250–252

Original pan-balance equation

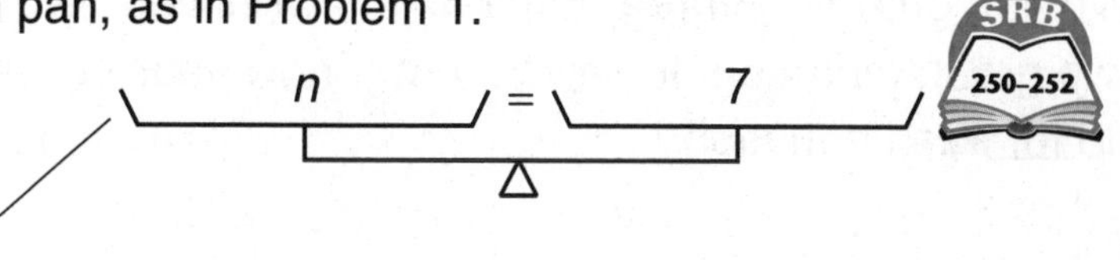

Operation

(in words)	(abbreviation)
Subtract 2.	S 2
Multiply by 4.	M 4
Add 2*n*.	A 2*n*

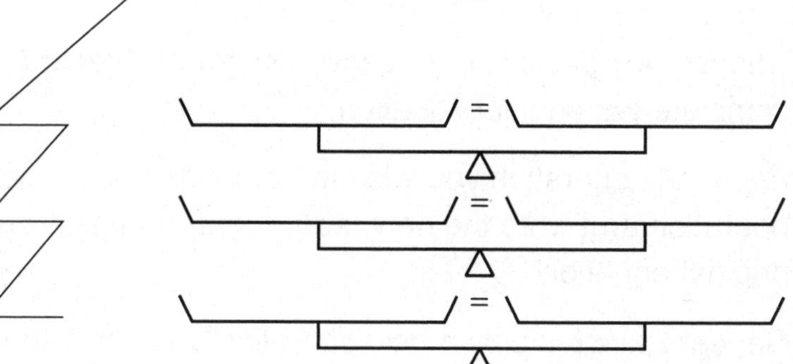

5. Check that 7 is the solution to each pan-balance equation in Problem 4.

In Problems 6 and 7, record the operation that was used to obtain the results on each pan balance, as you did in Problem 3.

6. Original pan-balance equation

Operation

(in words)	(abbreviation)
______	______
______	______
______	______

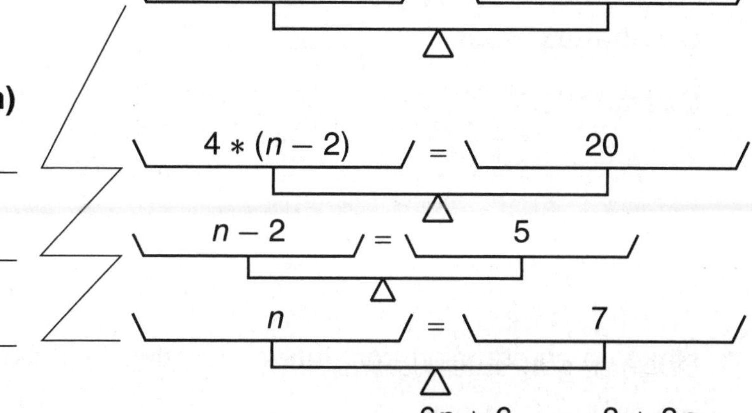

7. Original pan-balance equation

Operation

(in words) (abbreviation)

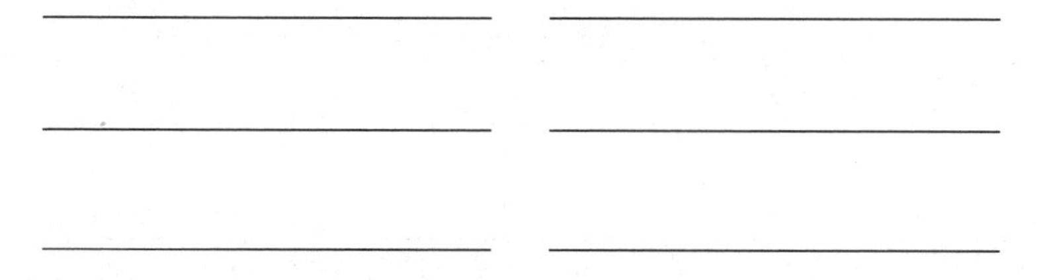

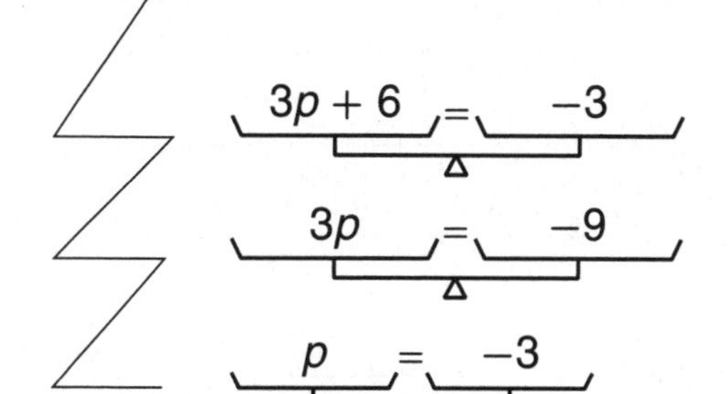

8. Check that 7 is the solution to each pan-balance equation in Problem 6 and that −3 is the solution to each pan-balance equation in Problem 7.

LESSON 6•10

Inventing and Solving Equations

Work in groups of three. Each of you will invent two equations and then ask the other two group members to solve them. Show your solutions on page 237. Here is what to do for each equation.

Step 1 Choose any positive or negative integer and record it on the first line to complete the original equation.

Step 2 Apply any operation you wish to both sides of the equation. Record the operation and write the new (equivalent) equation on the lines below the original equation.

Step 3 Repeat Step 2. Apply a new operation and show the new equation that results.

Step 4 Check your work by substituting the original value of x in each equation you have written. You should get a true number sentence every time.

Step 5 Give the other members of your group the final equation to solve.

1. Make up an equation from two equivalent equations.

Original equation x = ________
Selected integer

Operation

________ ________ = ________

________ ________ = ________

2. Make up an equation from three equivalent equations.

Original equation x = ________
Selected integer

Operation

________ ________ = ________

________ ________ = ________

________ ________ = ________

Date Time

LESSON 6·10 Inventing and Solving Equations *continued*

Use this page to solve your group members' equations.

First, record the equation. Then solve it. For each step, record the operation you use and the equation that results. Check your solution by substituting it for the variable in the original equation. Finally, compare the steps you used to solve the group member's equation to the steps he or she used in inventing the equation.

1. Member's equation ______ = ______

   Operation

   ______ ______ = ______

   ______ ______ = ______

   ______ ______ = ______

2. Member's equation ______ = ______

   Operation

   ______ ______ = ______

   ______ ______ = ______

   ______ ______ = ______

3. Member's equation ______ = ______

   Operation

   ______ ______ = ______

   ______ ______ = ______

   ______ ______ = ______

4. Member's equation ______ = ______

   Operation

   ______ ______ = ______

   ______ ______ = ______

   ______ ______ = ______

LESSON 6•10

Math Boxes

1. Find the solution to each equation.

   a. $-9 + d = 15$ $d =$ ________

   b. $52 = -48 + m$ $m =$ ________

   c. $4 * (12 - x) = -4$ $x =$ ________

   d. $p + (-9) = -5$ $p =$ ________

2. Solve the pan-balance problem.

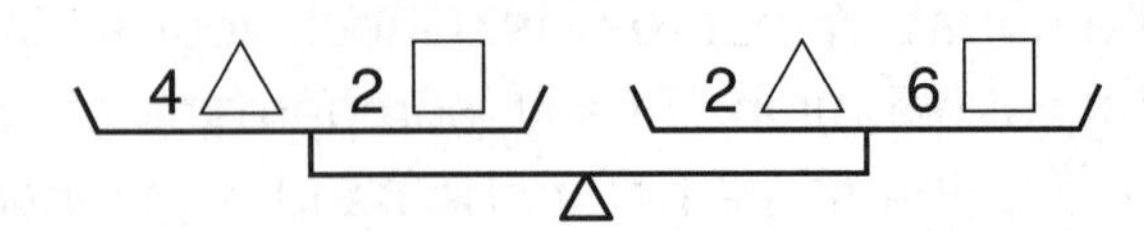

One △ weighs as much as ________ □s.

3. Complete the table. Then graph the data and connect the points.

Heather earns $0.35 for each paper flower she makes for the school fun fair.

Rule:
Earnings =
$0.35 * number of flowers

Flowers (*f*)	Earnings ($) (0.35 * *f*)
1	
2	
	1.05
5	
	2.10

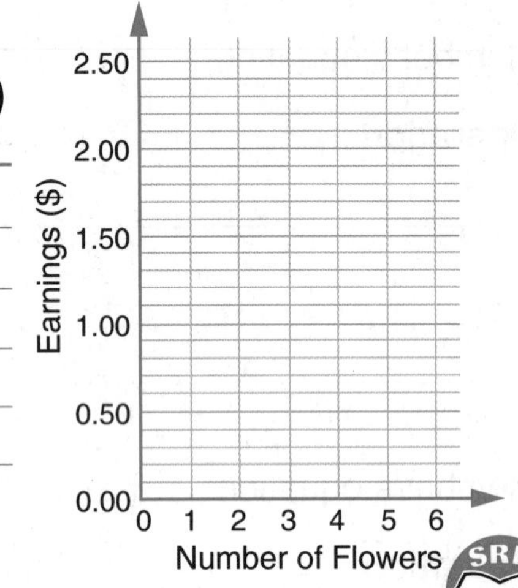

4. Suppose you roll a regular 6-sided die 90 times. About how many times would you expect to roll a 3?

About ________ times

5. How many arrangements of the letters *B*, *O*, and *X* are possible if you use each letter only once in each arrangement? List the arrangements.

________ possible arrangements

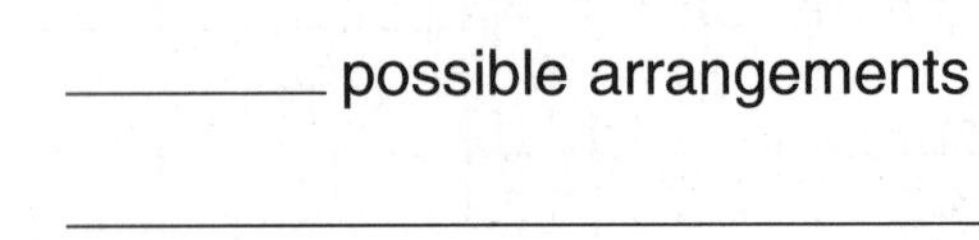

LESSON 6•11

Math Boxes

1. Use the order of operations to evaluate each expression.

   a. $(-3 * 7) + (-2 * 4) =$ ________

   b. $(-4)^3 \div 4 - (-2) =$ ________

   c. ________ $= \frac{1}{3}(9 * -3) + 18$

   d. $\frac{14-40}{-2} =$ ________

2. Which of the following statements is *not* true? Circle the best answer.

   A The order in which two numbers are multiplied does not change the product.

   B The product of any number and -1 is a negative number.

   C The product of any number and -1 is the opposite of the number.

3. Suppose N is a 2-digit whole number that ends in 5, such as 15. It is easy to find the square of N by using the formula

$$N^2 = (t * (t + 1) * 100) + 25$$

where t is the tens digit of the number you are squaring. Use this formula to find the following. (*Hint:* In Problem a, $t = 3$.)

   a. $35^2 =$ ____________

   b. $75^2 =$ ____________

   c. $95^2 =$ ____________

4. Lines l and m are parallel.
Lines p and q are parallel.

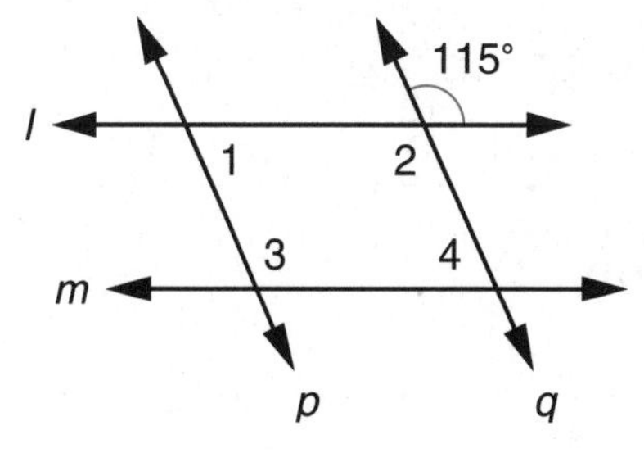

$m\angle 1 + m\angle 4 =$ ________

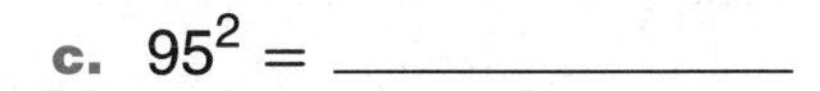

5. Solve mentally.

   a. 59% of 100 = ______

   b. 25% of 30 = ______

   c. 75% of 88 = ______

6. A bag contains 4 black marbles, 2 white marbles, and 1 gray marble. If a marble is drawn at random, what is the probability of:

   a. drawing a black marble? ______

   b. drawing a white marble? ______

   c. drawing a gray or a white marble?

   ____ + ____ = ____

SRB 153

Date Time

LESSON 6•11 Solving Equations

Solve the following equations.

SRB 250–252

Example:

$3x + 5 = 14$

Original equation $3x + 5 = 14$

Operation

S 5 $3x = 9$

D 3 $x = 3$

Check $(3 * 3) + 5 = 14$; true

1. $11y - 4 = 9y$

Original equation ______

Operation

______ ______

______ ______

______ ______

Check ______

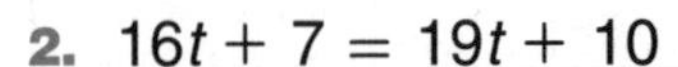

2. $16t + 7 = 19t + 10$

Original equation ______

Operation

______ ______

______ ______

______ ______

Check ______

3. $12n - 5 = 9n - 2$

Original equation ______

Operation

______ ______

______ ______

______ ______

Check ______

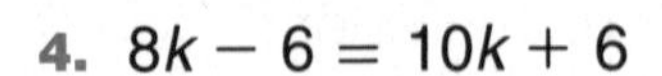

4. $8k - 6 = 10k + 6$

Original equation ______

Operation

______ ______

______ ______

______ ______

Check ______

5. $3b + 7.1 = 2.5b + 11.5$

Original equation ______

Operation

______ ______

______ ______

______ ______

Check ______

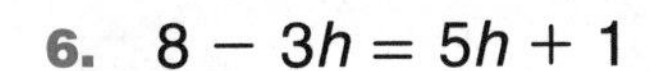

LESSON 6·11 Solving Equations *continued*

SRB 250–252

6. $8 - 3h = 5h + 1$

Original equation

____

Operation

________ ______________

________ ______________

________ ______________

Check ______________________

7. $-2p - 6 = 12 - 4p$

Original equation

____

Operation

________ ______________

________ ______________

________ ______________

Check ______________________

8. $\frac{1}{4}r + 9 = 10 - \frac{3}{4}r$

Original equation

____

Operation

________ ______________

________ ______________

Check ______________________

9. $\frac{2}{3}u - 7 = 9 - \frac{2}{3}u$

Original equation

____

Operation

________ ______________

________ ______________

________ ______________

Check ______________________

Try This

Two equations are equivalent if they have the same solution. Circle each pair of equivalent equations. Write the solution to the equations if they are equivalent.

10. $z = 5$

$3z - 8 = 2z - 3$

Solution ______________

11. $d + 5 = 8$

$6 - 2d = 9 - 3d$

Solution ______________

12. $v + 1 = 2v + 2$

$3v - 8 = 2v - 3$

Solution ______________

13. $t = 4$

$(5t + 3) - 2(t + 3) = 29 - 5t$

Solution ______________

Date Time

LESSON 6•12

Dividing Decimals by Decimals: Part 1

When you multiply the numerator and denominator of a fraction by the same nonzero number, you rename the fraction without changing its value.
For example: $\frac{3}{5} * \frac{10}{10} = \frac{30}{50}$, so $\frac{3}{5} = \frac{30}{50}$

In general, you can think of a fraction as a division problem. The fraction $\frac{3}{5}$ equals $3 \div 5$, or $5\overline{)3}$. The numerator of the fraction is the dividend; the denominator is the divisor. As you do with a fraction, you can multiply the dividend and divisor by the same number without changing the value of the quotient.

Study the patterns in the table below.

Fraction	Division Problem	Quotient
$\frac{6.08}{0.08}$	$0.08\overline{)6.08}$	76
$\frac{6.08}{0.08} * \frac{10}{10}$	$0.8\overline{)60.8}$	76
$\frac{6.08}{0.08} * \frac{100}{100}$	$8\overline{)608}$	76

What do you notice about the quotient when you multiply the dividend and the divisor by the same number?

____

Rename each division problem so the divisor is a whole number. Then solve the equivalent problem using partial-quotients division or another method.

Example:

Equivalent Problem

$0.005\overline{)0.015}$ = $5\overline{)15}$ Quotient = 3

Equivalent Problem **Equivalent Problem**

1. $0.004\overline{)2.05}$ = ________________ **2.** $0.3\overline{)7.08}$ = ________________

Quotient = ________ Quotient = ________

Date Time

LESSON 6•12

Dividing Decimals by Decimals *continued*

Rename each division problem so the divisor is a whole number. Then solve the equivalent problem using partial-quotients division or another method.

Equivalent Problem

3. $0.14\overline{)294}$ = ______________________

Quotient = ____________

Equivalent Problem

4. $0.013\overline{)6.24}$ = ______________________

Quotient = ____________

Equivalent Problem

5. $0.46\overline{)33.58}$ = ______________________

Quotient = ____________

Equivalent Problem

6. $1.67\overline{)13.36}$ = ______________________

Quotient = ____________

Inequalities

1. Name two solutions of each inequality.

a. $15 > r$ ______________

b. $8 < m$ ______________

c. $t \geq 56$ ______________

d. $15 - 11 \leq p$ ______________

e. $\frac{21}{7} > y$ ______________

f. $w > -3$ ______________

g. $6.5 > 3 * d$ ______________

h. $g < 0.5$ ______________

2. Name two numbers that are *not* solutions of each inequality.

a. $(7 + 3) * q > 40$ ______________

b. $\frac{1}{2} + \frac{1}{4} < t$ ______________

c. $y \leq 2.6 + 4.3$ ______________

d. $6 / g > 12$ ______________

3. Describe the solution set of each inequality.

Example: $t + 5 < 8$

Solution set: All numbers less than 3

a. $8 - y > 3$ Solution set ______________________________

b. $4b \geq 8$ Solution set ______________________________

4. Graph the solution set of each inequality.

a. $x < 5$

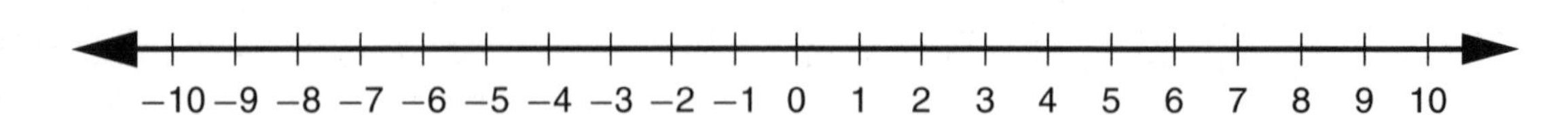

b. $6 > b$

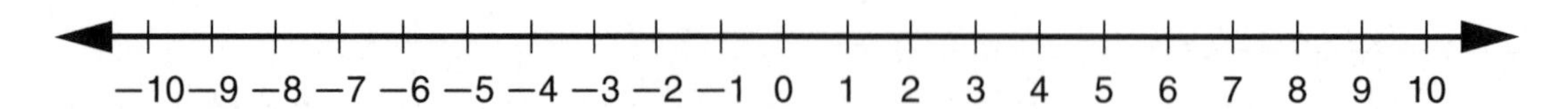

c. $1\frac{1}{2} \geq h$

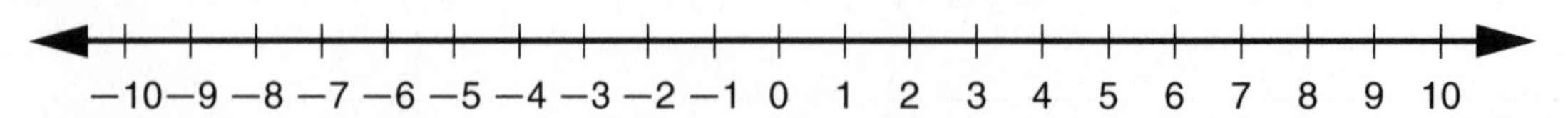

Date Time

LESSON 6•12 Inequalities in Real-World Contexts

Each situation below can be described by an inequality. Follow the steps to identify an appropriate inequality and draw a graph that fits the situation.

1. A city ordinance says that fences can have a maximum height of 6 feet.

   a. List several allowable fence heights. ______

   b. Describe in words the set of all allowable fence heights.

   ______

   c. Use an inequality or two inequalities to describe the set of allowable heights.

   ______

   d. Graph the set of all allowable heights on the number line below.

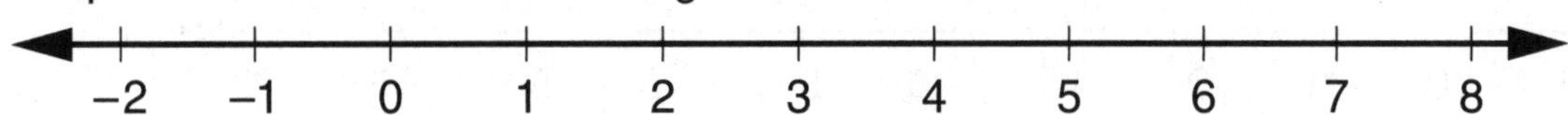

   e. Do the values represented on the graph make sense in the situation? Explain your answer.

   ______

   ______

   ______

2. A CD player can hold up to 8 CDs.

   a. List several possible numbers of CDs in the player. ______

   b. Describe in words the set of possible numbers of CDs in the player.

   ______

   c. Use an inequality or two inequalities to describe the possible numbers of CDs.

   ______

   d. Graph the set of all possible numbers of CDs on the number line below.

   0 1 2 3 4 5 6 7 8 9 10

   e. Do the values represented on the graph make sense in the situation? Explain your answer.

   ______

   ______

LESSON 6•12

Inequalities in Real-World Contexts *continued*

3. The manager of an apartment building turns on the heat when the outside temperature drops below 50°F.

a. List several possible outside temperatures when the heat is on.

____

b. Describe in words the set of outside temperatures when the heat is on.

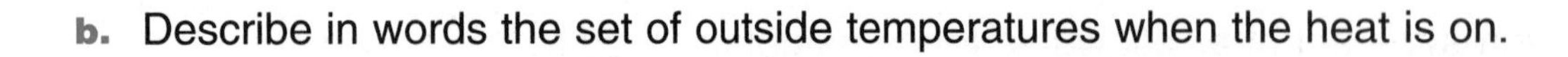

____

c. Use an inequality or two inequalities to describe the possible outside temperatures when the heat is on. ______________________________

d. Graph the set of all possible temperatures on the number line below.

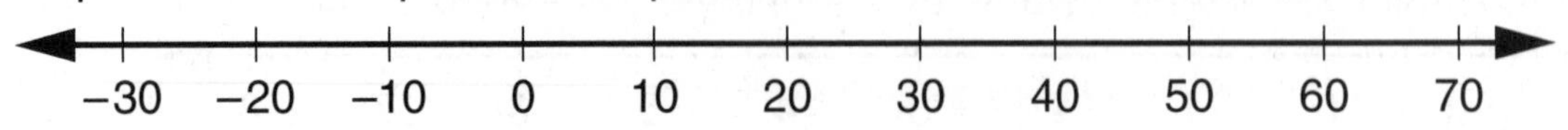

e. Do the values represented on the graph make sense in the situation? Explain your answer.

____

____

4. a. Describe your own real-world situation that can be modeled by an inequality.

____

b. List several possible values for your situation.

____

c. Describe the set of possible values in words and with an inequality or two.

____

d. Graph the set of all the possible values on the number line below.

e. Do the values represented on the graph make sense in the situation? Explain your answer.

____

____

Date Time

LESSON 6•12

Math Boxes

1. Find the solution to each equation.

a. $y + 6 - 28 = 40$ $y =$ ______

b. $45 + 3n = -45$ $n =$ ______

c. $7p + 19 = -2p + 55$ $p =$ ______

d. $7 - 4w = 10w$ $w =$ ______

2. Solve the pan-balance problem.

One □ weighs as much as ______ marbles.

3. Complete the table. Then graph the data and connect the points.

Rebecca walks at an average speed of $3\frac{1}{2}$ miles per hour.

Rule:
Distance =
$3\frac{1}{2}$ mph * number of hours

Time (hr) (h)	Distance (mi) ($3\frac{1}{2} * h$)
1	
2	
	$17\frac{1}{2}$
7	
	35

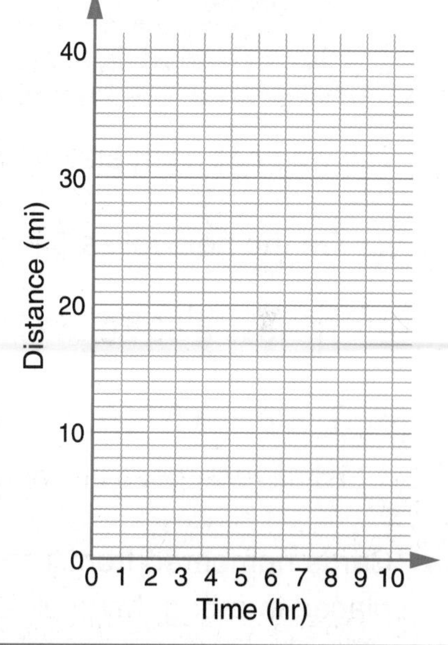

4. The spinner at the right is divided into 5 equal parts.

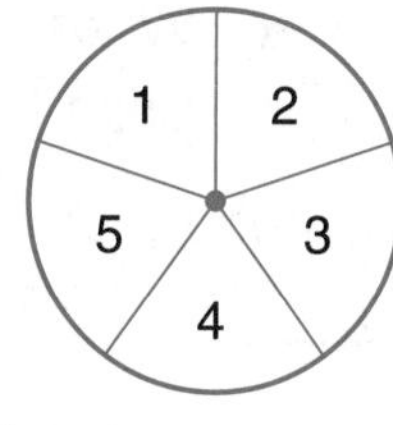

Suppose you spin this spinner 75 times. About how many times would you expect the spinner to land on a prime number?

______ times

SRB 88 149 150

5. Serena keeps 4 stuffed animals lined up on a shelf over her bed. How many different arrangements of the stuffed animals are possible? (*Hint:* Label the animals *A, B, C,* and *D.* Then make a list.)

______ arrangements

______

______

______

SRB 156

Date Time

LESSON 6·13

Math Boxes

1. Solve. Write your answer in simplest form.

   a. $\frac{4}{52} * \frac{13}{28} =$ ____________

   b. ____________ $= \frac{4}{25} * \frac{75}{100}$

   c. $\frac{5}{12} + \frac{1}{4} =$ ____________

   d. $\frac{3}{5} + \frac{3}{8} =$ ____________

2. Rewrite each fraction as a percent.

   a. $\frac{35}{50} =$ ____________

   b. $\frac{18}{24} =$ ____________

   c. $\frac{7}{8} =$ ____________

   d. $\frac{15}{75} =$ ____________

3. The table shows the results of rolling a 6-sided die 50 times.

Number Showing	1	2	3	4	5	6
Number of Times	10	5	11	12	4	8

Tell whether each statement below is true or false.

   a. On the next roll, a 5 is more likely to come up than a 1. ____________

   b. There is a 50-50 chance of rolling a prime number. ____________

   c. There is a 50-50 chance of rolling a composite number. ____________

4. Cards numbered from 1 to 50 are placed in a bag. Myra picks one card without looking.

   a. What is the probability that Myra will pick a 2-digit number card?

   ____

   b. What is the probability that she will pick a number that is not a multiple of 10?

   ____

5. The manufacturer of Vita Munch cereal puts a prize in 120 boxes out of every 600.

   a. What is the probability of getting a prize if you buy a box of Vita Munch?

   ____

   b. Suppose a store has 1,800 boxes of Vita Munch in stock. About how many boxes might you expect to contain prizes?

   About ____________ boxes

Date Time

Probability Concepts

Math Message

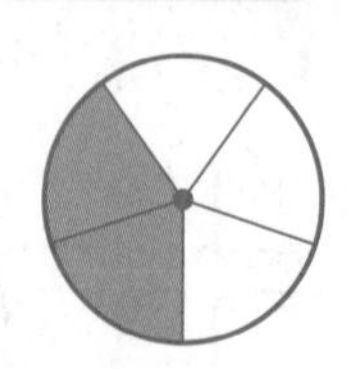

The spinner at the right has 5 equal sections. Two sections are blue. If you spin it many times, the spinner is likely to land on blue about $\frac{2}{5}$ of the time. Therefore, the probability of landing on blue is $\frac{2}{5}$, or 40%.

Using the spinners shown below, write the letter(s) of the spinner next to the statement that describes it. A spinner may be matched with more than one statement.

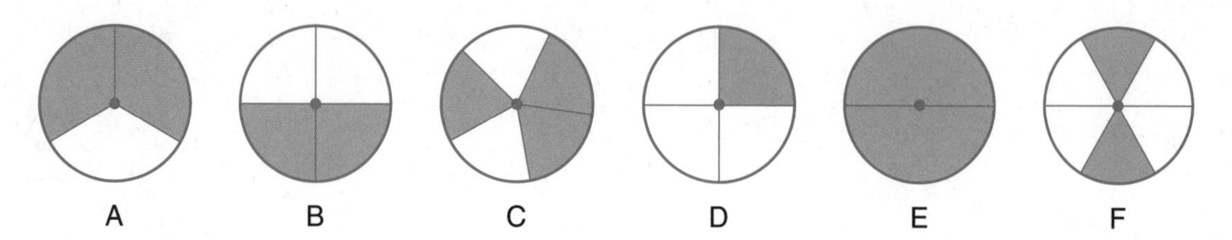

Example:

This spinner will land on blue about 2 out of 3 times. ___A___

1. There is about a $\frac{1}{4}$ chance that the spinner will land on blue. ___________
2. This spinner will land on blue 100% of the time. ___________
3. There is about a 50-50 chance that this spinner will land on white. ___________
4. This spinner will never land on white. ___________
5. The probability that this spinner will land on blue is $\frac{3}{5}$. ___________
6. This spinner will land on white about twice as often as on blue. ___________
7. This spinner will land on white a little less than half the time. ___________
8. The probability that this spinner will land on white is 75%. ___________
9. Suppose you spin Spinner A 4 times and it lands on white every time. What is the probability that the spinner will land on white on the fifth spin? ___________
10. If you spin Spinner A 90 times, how many times would you expect the spinner to land on blue? ___________

Date Time

LESSON 7·1 Domino Probabilities

A set of double-6 dominoes is shown below.

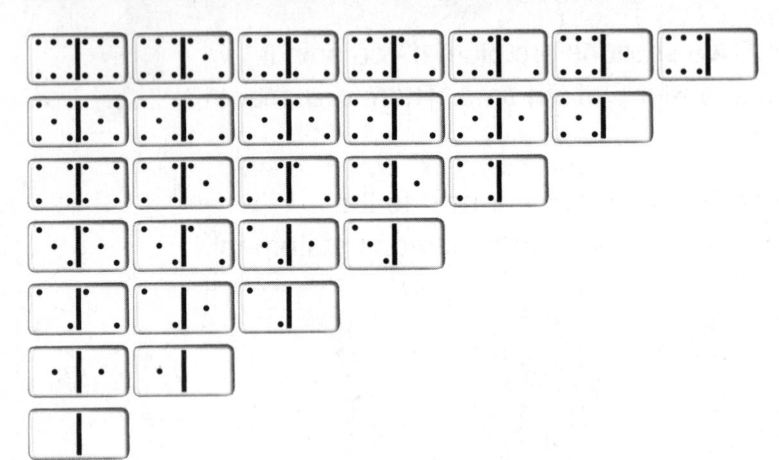

Suppose all the dominoes are turned facedown and mixed thoroughly.
You select one domino and turn it faceup.

1. How many possible outcomes are there? ________ possible outcomes

2. Are the outcomes equally likely? ________

When the possible outcomes are equally likely, the following formula is used to find the probabilities:

A favorable outcome is the outcome that makes an event happen.

$$\text{Probability of an event} = \frac{\text{number of favorable outcomes}}{\text{number of possible outcomes}}$$

3. What is the probability of selecting each domino? ________

Find the probability of selecting each domino described below.

4. A double ________

5. Exactly one blank side ________

6. No blank sides ________

7. The sum of the dots is 7. ________

8. The sum of the dots is greater than 7. ________

9. Exactly one side is a 3. ________

10. Both sides are odd numbers. ________

Date Time

LESSON 7•1

Math Boxes

1. Fill in the missing equivalents. Write fractions in simplest form.

Fraction	Decimal	Percent
$\frac{9}{10}$		
	0.98	
		60%
$\frac{7}{25}$		
		12.5%

2. Solve the following equations. Check each solution by substituting it for the variable in the original equation.

a. $8x - 1 = 11$

Solution: $x =$ ________

b. $\frac{2}{5}y + 6 = 10$

Solution: $y =$ ________

c. $-20m + 20 = -20$

Solution: $m =$ ________

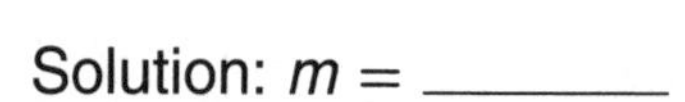

3. Indicate whether each inequality is true or false.

a. $\frac{2}{3} * 9 > 8$ ________

b. $-4 \leq -3 - 1$ ________

c. $48 - (6 * 4) > 20$ ________

d. $8 - 10 \neq 13 - 15$ ________

4. Graph the solution set for $k > -2$ on the number line below.

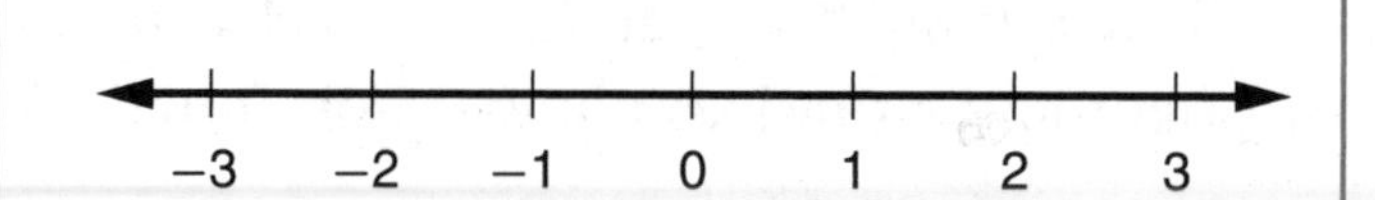

5. Complete.

a. ________ m = 368 mm

b. ________ cm = 0.245 m

c. 32 mm = ________ m

d. 45.2 cm = ________ mm

e. 0.25 mm = ________ cm

6. Solve.

Solution

a. $\frac{w}{8} = 16$ ________

b. $\frac{60}{p} = 5$ ________

c. $\frac{3}{7} = \frac{t}{28}$ ________

d. $\frac{d}{18} = \frac{4}{6}$ ________

Date Time

LESSON 7·2

Generating Random Numbers

Math Message

Suppose you have a deck of number cards, one card for each of the numbers 1 through 5. When you shuffle the cards and pick a card without looking, the possible outcomes are 1, 2, 3, 4, and 5. These outcomes are equally likely. Numbers found in this way are **random numbers.**

Suppose you continue finding random numbers using these steps.

- Shuffle
- Pick a card
- Replace the card
- Repeat

1. If you did this many times, about what percent of the time would you expect to pick the number 5? About ______ percent

An Experiment

2. Work with a partner in a group of 4 students. Use a deck of 5 number cards, one card for each of the numbers 1 through 5.

3. One of you shuffles the deck of 5 number cards and fans them out facedown. Your partner then picks one without looking. The pick **generates a random number** from 1 to 5. The number is an **outcome.**

4. The person picking the card tallies the outcome in the table below while the person with the deck replaces the card and shuffles the deck. Generate exactly 25 random numbers.

Outcome	Tally	Number of Times Picked
1		
2		
3		
4		
5		
	Total Random Numbers	**25**

LESSON 7•2

Generating Random Numbers *continued*

5. Record the results in the table below. In the My Partnership column, write the number of times each of the numbers 1 through 5 appeared.

6. In the Other Partnership column, record the results of the other partnership in your group.

7. For each outcome, add the two results and write the sum in the Both Partnerships column.

8. Convert each result in the Both Partnerships column to a percent and record it in the % of Total column. For example, 10 out of 50 would be 20%.

Outcome	My Partnership	Other Partnership	Both Partnerships	% of Total
1			______ out of 50	
2			______ out of 50	
3			______ out of 50	
4			______ out of 50	
5			______ out of 50	
Total	**25**	**25**	**50 out of 50**	**100%**

Date Time

LESSON 7·2

More Probability

1. The table below shows the results of a survey of 200 health club members. The members were asked to name the area of the club in which they did the most exercise.

Exercise Area	Number of People
Swimming pool	80
Aerobics studio	28
Racquetball court	70
Weight room	22

What is the probability, written as a percent, that a member spends most of his or her exercise time in

a. the aerobics studio? ______________

b. the swimming pool? ______________

c. an area that is *not* the racquetball court? ______________

d. the weight room or the racquetball court? ______________

2. Suppose the value of x is chosen at random from the following set of numbers: {1, 2, 3, 4, 5}.

a. What is the probability that the area of the rectangle at the right is greater than 65 in.^{2}? Express the probability as a percent.

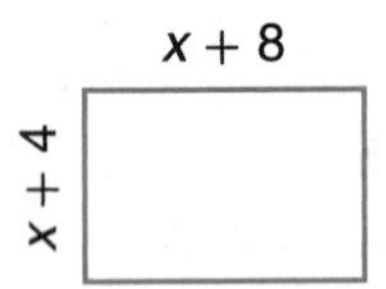

____

b. What is the probability that the perimeter of the rectangle is less than 25 in.? Express the probability as a percent.

____

Date Time

LESSON 7·2

Math Boxes

1. Suppose one of the squares is chosen at random. What is the probability that the square is

a. white? ________

b. checkered? ________

c. checkered or white? ________

2. Complete.

a. $\frac{1}{3}$ of 27 = ______

b. ______ = $\frac{5}{6}$ of 30

c. ______ = $\frac{4}{7}$ of 42

d. $\frac{3}{8}$ of 56 = ______

3. Complete the table for the formula below. Then plot the points to make a graph.

Formula: $2s - 5 = t$

s	*t*
1	
2	
5	
	11
	15

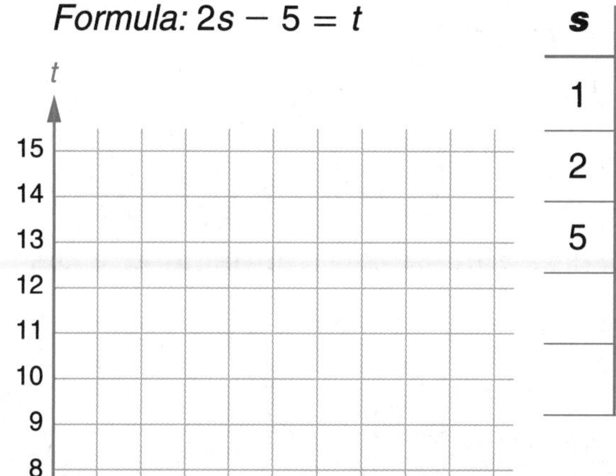

SRB
254

4. Name two solutions for each inequality.

Solutions

a. $1.1 - 0.38 < w$ ________

b. $10\frac{1}{3} - 6\frac{2}{3} \geq g$ ________

5. Write each ratio in simplest form.

a. 8 to 10 ________

b. 35 out of 100 ________

c. 60 wins to 90 losses ________

d. $144 in 12 hours ________

LESSON 7·3

Using Random Numbers

Suppose two evenly matched teams play a fair game that cannot end in a tie—one team must win. Because the teams have an equal chance of winning, you could get about the same results by tossing a coin. If the coin lands on HEADS, Team 1 wins; if it lands on TAILS, Team 2 wins. In this way, tossing a coin **simulates** the outcome of the game. In a **simulation,** an object or event is represented by something else.

Suppose Team 1 and Team 2 play a best-of-5 tournament. The first team to win 3 games wins the tournament. Use a coin to simulate the tournament as follows:

Games 1–3 If the coin lands on HEADS, Team 1 wins. If the coin lands on TAILS, Team 2 wins.

Games 4 and 5 Play only if necessary. Repeat the instructions for Game 1.

Sample results:

If the coin tosses are HEADS, HEADS, HEADS, Team 1 wins the tournament.

If the coin tosses are HEADS, HEADS, TAILS, HEADS, Team 1 wins.

If the coin tosses are TAILS, HEADS, HEADS, TAILS, TAILS, Team 2 wins.

1. Fill in the table as described on the next page.

Number of Games Needed to Win the Tournament	Winner	Tally of Tournaments Won	Total Tournaments Won
3	Team 1		
	Team 2		
4	Team 1		
	Team 2		
5	Team 1		
	Team 2		
		Total	**25**

LESSON 7•3

Using Random Numbers *continued*

2. Use coin tosses to play a best-of-5 tournament. Make a tally mark in the Tally of Tournaments Won column of the table on page 254. The tally mark shows which team won the tournament and in how many games.

3. Play exactly 24 more tournaments. Make a tally mark to record the result for each tournament. Then convert the tally marks into numbers in the Total Tournaments Won column.

4. Use the table on page 254 to estimate the chance that a tournament takes

   a. exactly 3 games. ________%

   b. exactly 4 games. ________%

   c. exactly 5 games. ________%

   d. fewer than 5 games. ________%

Discuss the following situations with a partner. Record your ideas.

5. Suppose there is a list of jobs that need to be done for your class (such as distributing supplies, collecting books, and taking messages to the office). How might you use random numbers to assign the jobs, without using any pattern or showing favoritism?

____

____

____

____

6. You want to play a game. The directions are: *Roll 2 dice and add the numbers. Move your marker ahead that many spaces.* You do not have any dice. How can you use number cards to play the game?

____

____

____

____

Date Time

LESSON 7•3

Math Boxes

1. Fill in the missing equivalents. Write fractions in simplest form.

Fraction	Decimal	Percent
$\frac{7}{8}$		
	$0.\overline{6}$	
$\frac{7}{10}$		
		62.5%
	0.005	

2. Solve the following equations. Check each solution by substituting it for the variable in the original equation.

a. $k = 8k + 28$

Solution: $k =$ ______

b. $20n - 28 = 10n + 2$

Solution: $n =$ ______

c. $-p - 1 = p - 21$

Solution: $p =$ ______

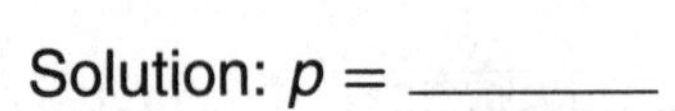

SRB 251 252

3. Which inequality is false when $a = 3$, $b = -2$, and $c = 6$?
Fill in the circle next to the best answer.

Ⓐ $b^2 - 4ac < 0$

Ⓑ $0 < \frac{c}{a} + b$

Ⓒ $a(b + c) \geq ab + ac$

Ⓓ $a - b \leq c$

4. Graph the solution set for $b \neq 0$ on the number line below.

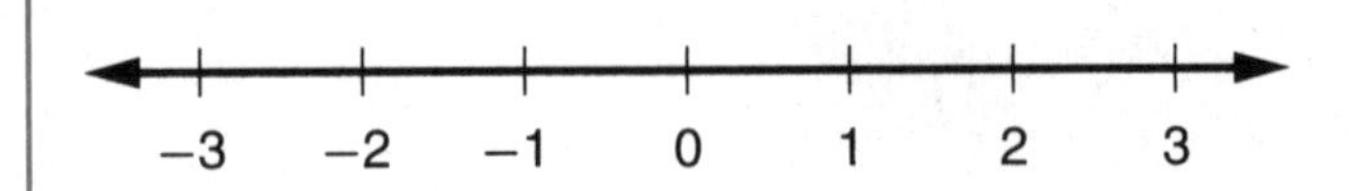

5. Complete.

a. 15 m = ______ cm

b. 25 cm = ______ m

c. 143 mm = ______ cm

d. 2.06 cm = ______ mm

e. ______ mm = 1.43 cm

6. Solve.

Solution

a. $\frac{n}{6} = 4$ ______

b. $\frac{42}{b} = 6$ ______

c. $\frac{5}{8} = \frac{g}{32}$ ______

d. $\frac{k}{21} = \frac{2}{14}$ ______

Date Time

Mazes and Tree Diagrams

Math Message

The diagram at the right shows a maze. A person walking through the maze does not know in advance how many paths there are or how they divide.

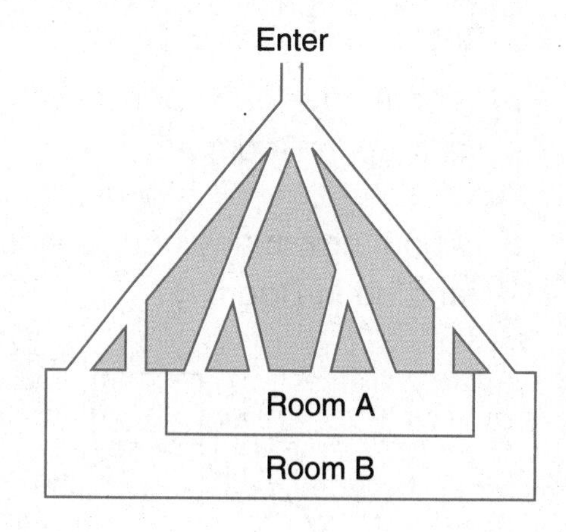

Pretend that you are walking through the maze. Each time the path divides, you select your next path at random. Each path at an intersection has the same chance of being selected. You may not retrace your steps.

Depending on which paths you follow, you will end up in Room A or Room B.

1. In which room are you more likely to end up—Room A or Room B? ____________

2. Suppose 80 people took turns walking through the maze.

   a. About how many people would you expect to end up in Room A? ____________

   b. About how many people would you expect to end up in Room B? ____________

Your teacher will show you how to complete the following tree diagram. You can read about tree diagrams on pages 154 and 155 in the *Student Reference Book.*

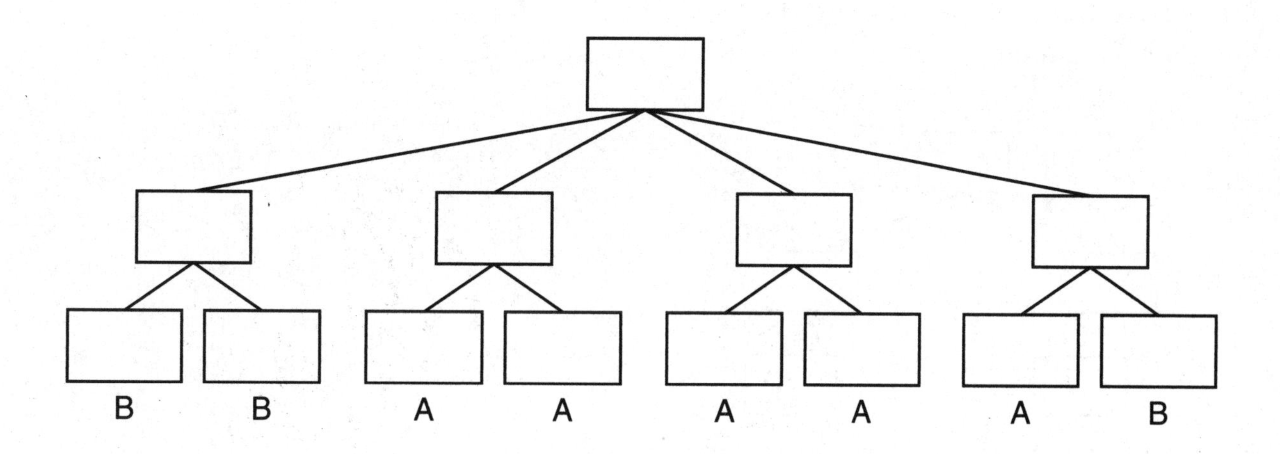

LESSON 7•4

Maze Problems

1. Use the tree diagram below to help you solve the following problem. Suppose 60 people walk through the maze below.

   a. About how many people would you expect to end up in Room A? ____________

   b. About how many people would you expect to end up in Room B? ____________

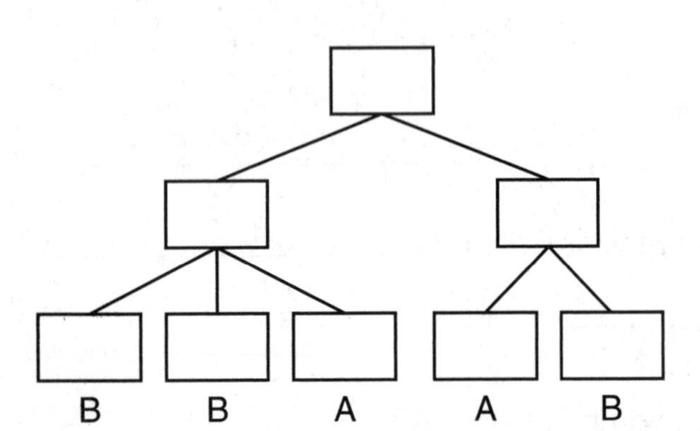

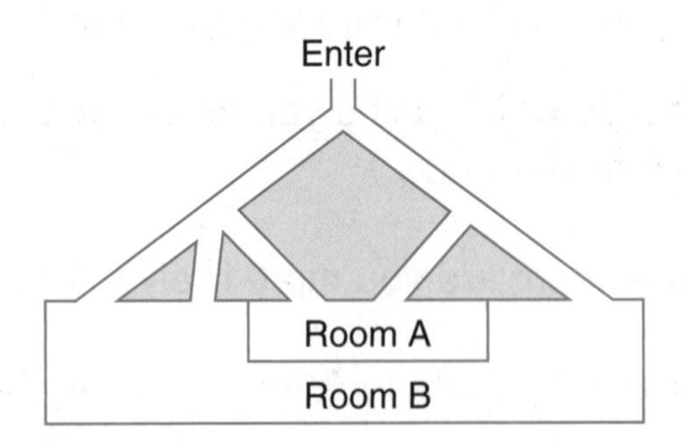

2. Make a tree diagram to help you solve the following problem. Suppose 120 people walk through the maze below.

   a. About how many people would you expect to end up in Room A? ____________

   b. About how many people would you expect to end up in Room B? ____________

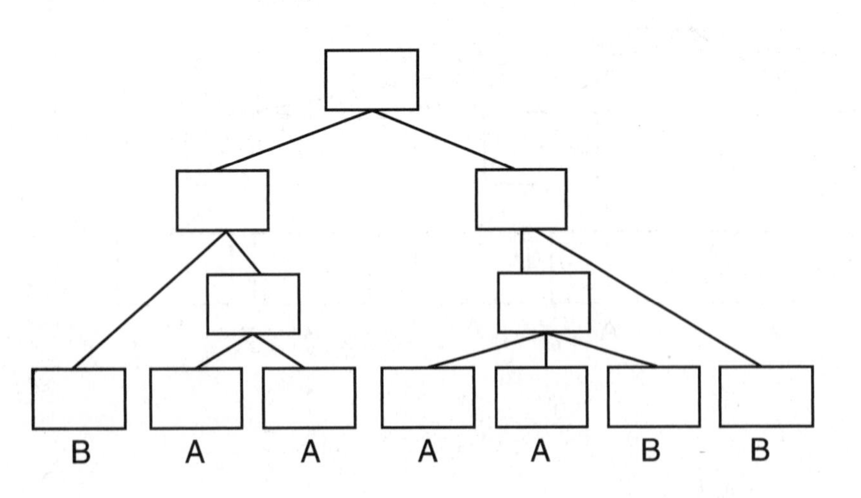

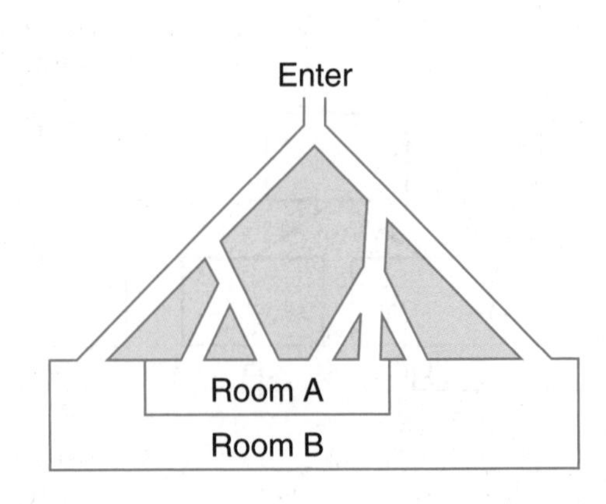

Date Time

LESSON 7•4

Math Boxes

1. Suppose one of the numbered marbles is chosen at random.

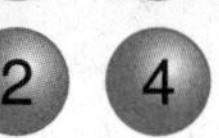 

What is the probability of choosing

a. a number ≤ 5? ______ fraction ______ percent

b. a 3 or an 8? ______ fraction ______ percent

c. a prime or a composite number? ______ fraction ______ percent

2. Complete.

a. $\frac{3}{4}$ of 80 = ______

b. ______ = $\frac{2}{9}$ of 27

c. ______ = $\frac{4}{5}$ of 60

d. $\frac{7}{11}$ of 88 = ______

3. Complete the table for the formula below. Then plot the points to make a graph.

Formula: $4h = g$

h	*g*
1	
2	
3	
	20
	26

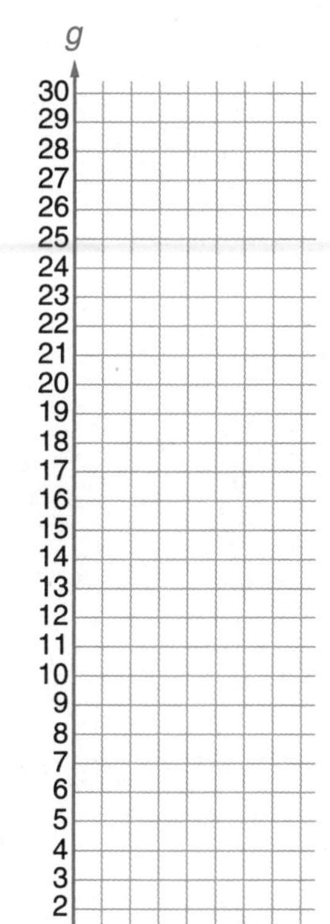

4. Name two solutions for each inequality.

Solutions

a. $n \geq 3\frac{11}{16} + 4\frac{1}{2}$ ______

b. $k < 6\frac{2}{5} * 15$ ______

5. Write each ratio in simplest form.

a. 20 to 12 ______

b. 64 out of 78 ______

c. 14 boys to 16 girls ______

d. 440 miles in 8 hours ______

SRB 117–119

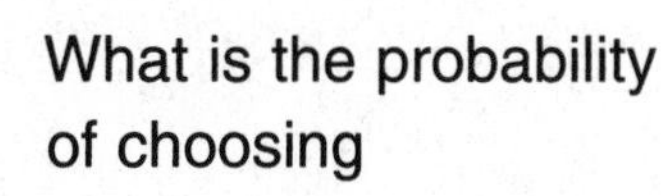

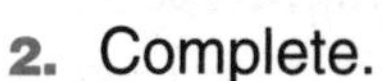

Date Time

LESSON 7•5

Probability Tree Diagrams

Complete the tree diagram for each maze.

Write a fraction next to each branch to show the probability of selecting that branch. Then calculate the probability of reaching each endpoint. Record your answers in the blank spaces beneath the endpoints.

1. What is the probability of entering Room A? ______

   What is the probability of entering Room B? ______

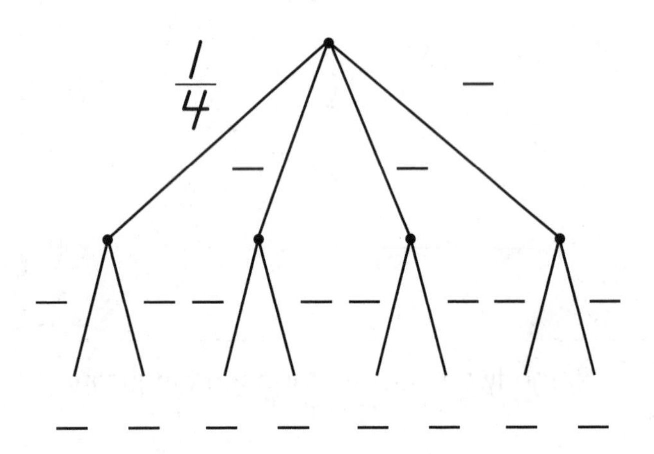

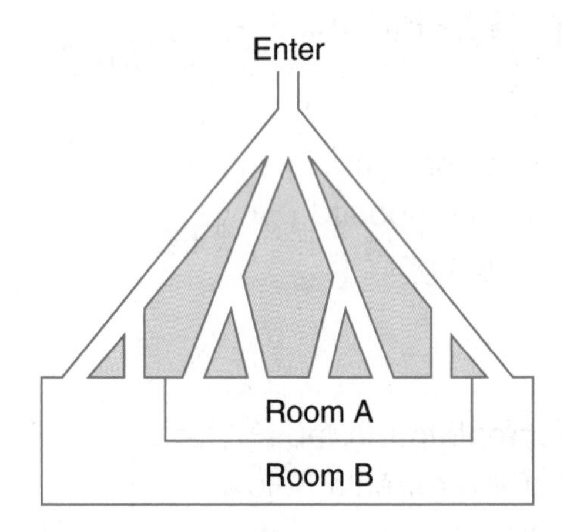

2. What is the probability of entering Room A? ______

   What is the probability of entering Room B? ______

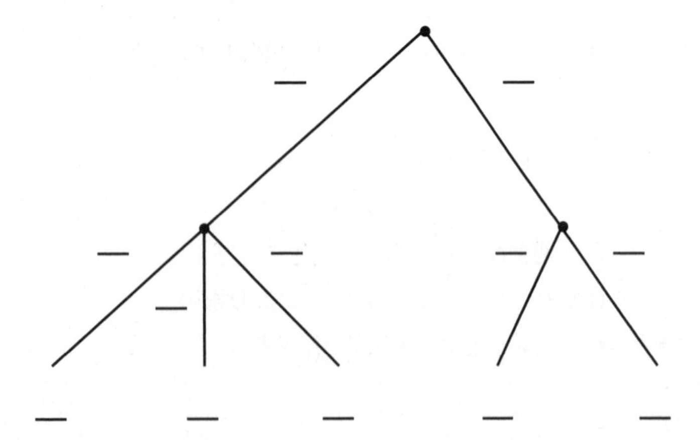

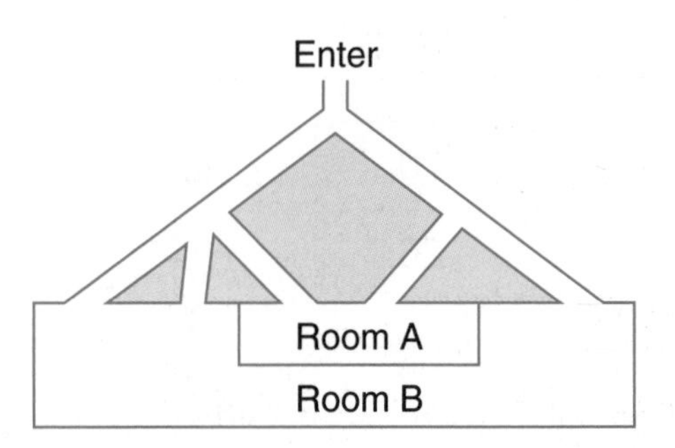

LESSON 7•5

Probability Tree Diagrams *continued*

3. Josh has 3 clean shirts (red, blue, and green) and 2 clean pairs of pants (tan and black). He randomly selects one shirt. Then he randomly selects a pair of pants.

a. Complete the tree diagram by writing a fraction next to each branch to show the probability of selecting that branch. Then calculate the probability of selecting each combination.

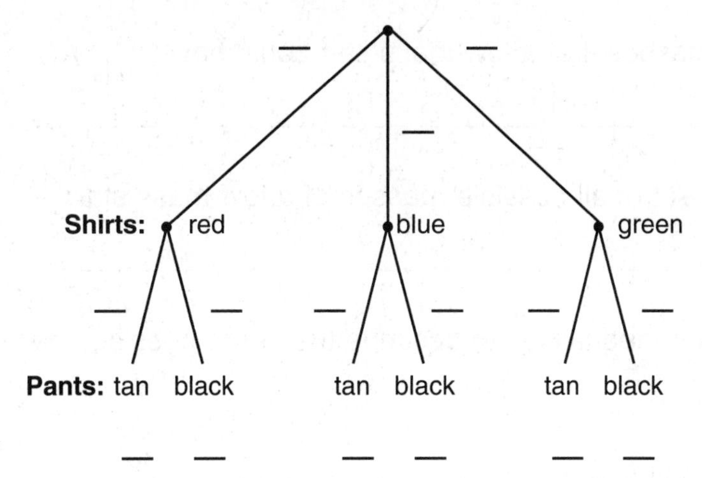

b. List all possible shirt-pants combinations. One has been done for you.

red–tan ______________________________

c. How many different shirt-pants combinations are there? ______________

d. Do all the shirt-pants combinations have the same chance of being selected? ________

e. What is the probability that Josh will select

the blue shirt? ______ the blue shirt and the tan pants? ______

the tan pants? ______ a shirt that is not red? ______

the black pants and a shirt that is not red? ______

Try This

Suppose Josh has 4 clean shirts and 3 clean pairs of pants. Explain how to calculate the number of different shirt-pants combinations without drawing a tree diagram.

____

____

Date Time

LESSON 7•5 Going to Space

Star masses are measured in a unit called a *solar mass,* which is equal to the mass of the Sun. One solar mass is equal to $2 * 10^{33}$ grams.

1. Low-mass stars use hydrogen fuel so slowly that they may shine for over 100 billion years. A low-mass star has a mass of at least 0.1 solar mass but less than 0.5 solar mass.

   a. List several possible masses that a low-mass star could have.

   ____

   b. Describe in words the set of all possible masses of a low-mass star.

   ____

   c. Use an inequality or two inequalities to describe the set of possible masses.

   ____

   d. Graph the set of all possible masses on the number line below.

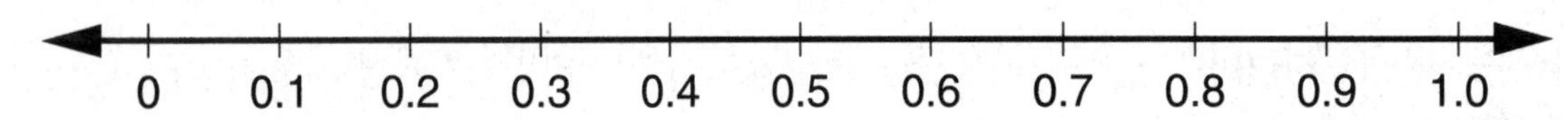

   e. Do the values represented on the graph make sense in the situation? Explain your answer.

   ____

   ____

2. High-mass stars have short lives and sometimes become black holes. When a high-mass star dies, an explosion occurs, leaving behind a stellar core. If the stellar core has a mass of at least 3 solar masses, the star becomes a black hole.

   a. List several possible masses of a stellar core that will become a black hole.

   ____

   b. Describe the set of possible masses in words.

   ____

LESSON 7•5

Going to Space *continued*

2. c. Use an inequality to describe the set of possible masses. ____________________

d. Graph the set of all possible masses on the number line below.

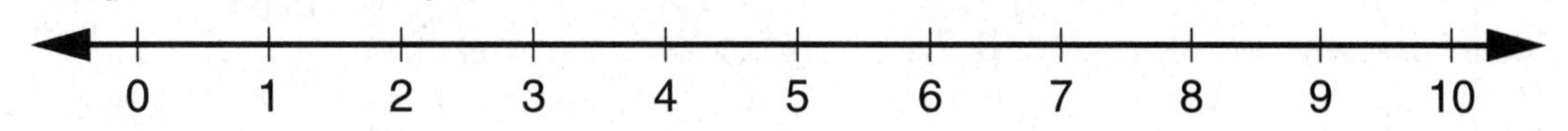

e. Do the values represented on the graph make sense in the situation? Explain your answer.

____

____

____

____

In the United States, people who pilot spacecrafts or work in space are called astronauts. In Russia and other former republics of the Soviet Union, these people are called cosmonauts.

3. Cosmonauts began flying the Soyuz series of spacecraft in 1967. These vehicles can transport up to 3 cosmonauts.

a. List the possible numbers of cosmonauts that can go on a mission in a Soyuz spacecraft.

____

b. Describe the possible numbers of cosmonauts in words.

____

c. Write an inequality or inequalities to represent the possible numbers of cosmonauts.

____

d. Graph the possible numbers of cosmonauts on the number line below.

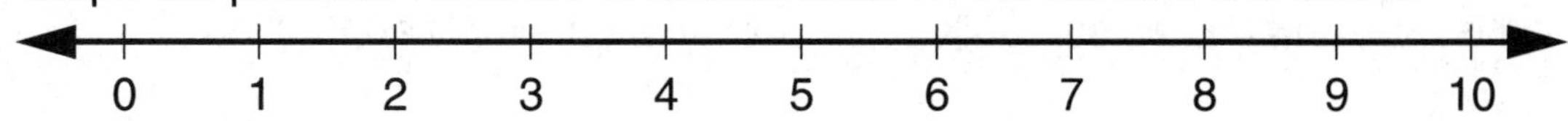

e. Do the values shown on the graph make sense in the situation? Explain your answer.

____

____

Date Time

LESSON 7·5 Math Boxes

1. Darnell has 3 jackets and 4 baseball hats. Complete the tree diagram.

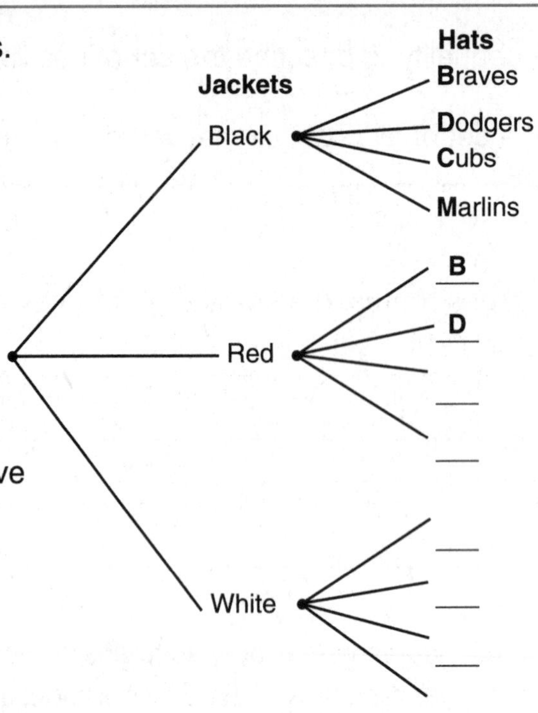

a. How many jacket-hat combinations are possible?

______

b. Do all the jacket-hat combinations have the same chance of being selected?

______

2. Which set of numbers is represented by the graph below? Choose the best answer.

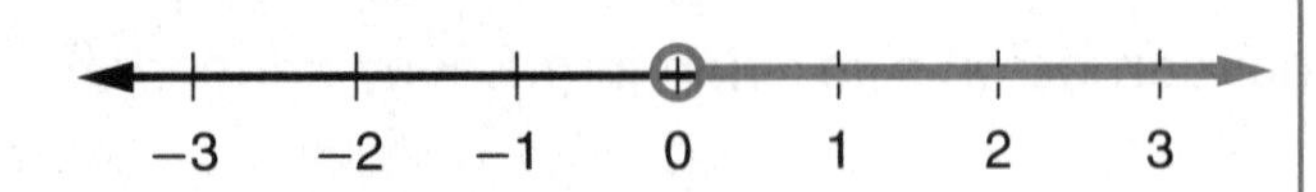

- ⬭ positive real numbers
- ⬭ positive integers
- ⬭ whole numbers
- ⬭ counting numbers

3. Write each number in standard notation.

a. 72 billion ______

b. 42.78 million ______

c. 89.6 billion ______

d. 0.5 million ______

4. Janella walks at a speed of 6.9 kilometers per hour. At this rate, how far can she walk

a. in 2 hours? ______ kilometers

b. in 20 minutes? ______ kilometers

c. in 1 hour 40 minutes?

______ kilometers

5. Express the length of $\overline{JK}$ to the length of $\overline{JL}$ as a simplified fraction.

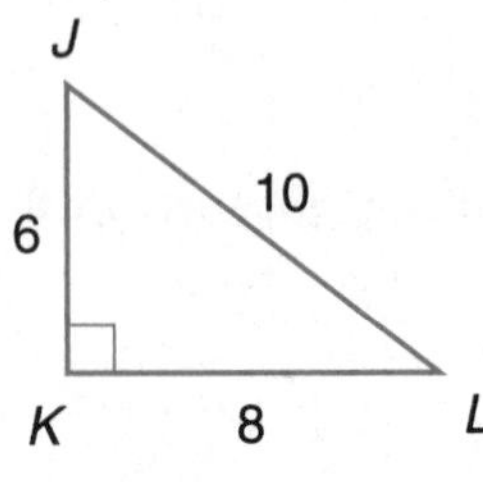

$\frac{JK}{JL} =$ ______

Date Time

LESSON 7•6

Math Boxes

1. The Venn diagram shows the relationships between sets of numbers within the set of real numbers. Use the diagram to answer the following questions.

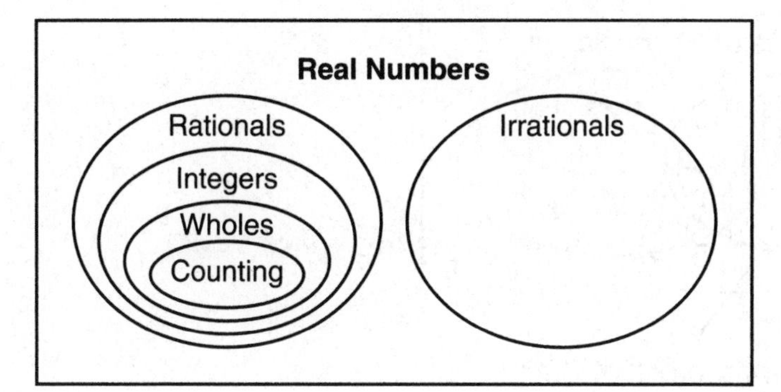

a. Can a number be rational and irrational?

____

b. Is a whole number a rational number?

____

2. Complete.

a. $\frac{1}{10}$ of 268 = __________

b. $\frac{1}{100}$ of 21,509 = __________

c. $\frac{1}{1,000}$ of 7,834 = __________

d. $\frac{1}{100}$ of 72 = __________

3. Fill in the missing numbers.

a. 947 ∗ 23 ∗ 16 = 16 ∗ 23 ∗ __________

b. 18 ∗ 7 ∗ 3 = 21 ∗ __________

c. __________ ∗ 51 ∗ 97 = 51 ∗ 97 ∗ 82

d. __________ ∗ 14 ∗ 182 = 28 ∗ 182

e. __________ ∗ 29 ∗ 30 = 150 ∗ 29

4. Use division to rename the fraction as a decimal rounded to the nearest hundredth. Show your work.

$\frac{14}{15}$

$\frac{14}{15}$ = __________

5. Estimate the percent equivalents for the following fractions.

a. $\frac{3}{26}$ Estimate __________

b. $\frac{7}{9}$ Estimate __________

c. $1\frac{3}{7}$ Estimate __________

Date Time

LESSON 7·6 Venn Diagrams

Math Message

The Venn diagram below shows the factors of 20 and 30.

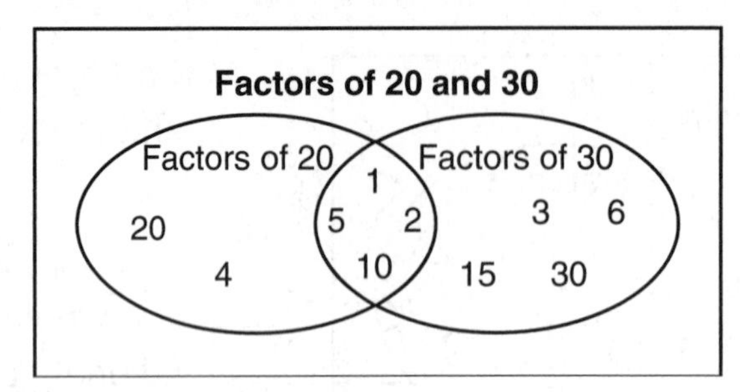

1. Which numbers are factors of 20 but are *not* factors of 30? ____________

2. Which numbers are factors of both 20 and 30? ____________

3. List the factors of 30. ____________

4. What is the greatest common factor of 20 and 30? ____________

---

5. Ms. Barrie teaches math and science. The Venn diagram below shows the number of students in her classes.

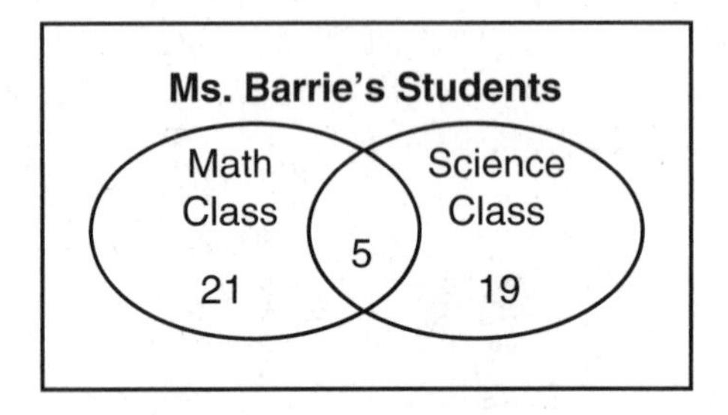

   a. How many students are in Ms. Barrie's math class? ____________ students

   b. How many students are in her science class? ____________ students

   c. How many students have Ms. Barrie for math and science? ____________ students

   d. How many students have Ms. Barrie as a teacher in at least one class? ____________ students

LESSON 7•6

Venn Diagrams *continued*

6. The sixth graders at Lincoln Middle School were asked whether they write with their left or right hand. A small number of students reported that they were *ambidextrous,* which means they write equally well with either hand.

The survey results are shown in the Venn diagram at the right.

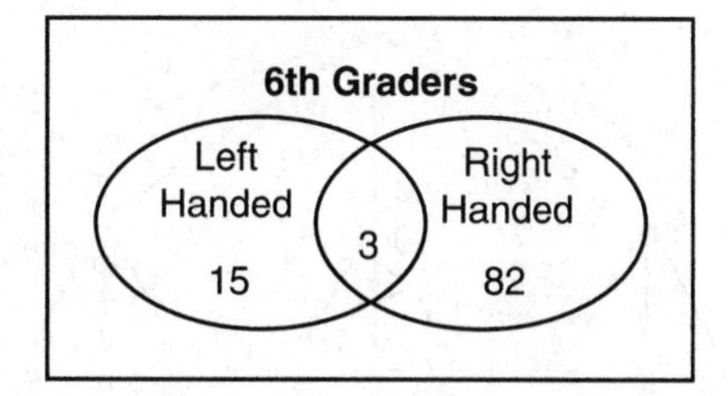

a. How many students were surveyed? ____________ students

b. How many students are ambidextrous (can write with either hand)? ____________ students

c. How many students always write with their left hand? ____________ students

d. How many students never write with their left hand? ____________ students

7. Mr. Wu has 32 students in his sixth-grade homeroom. He identified all students who scored 90% or above on each test. Mr. Wu then drew the Venn diagram at the right.

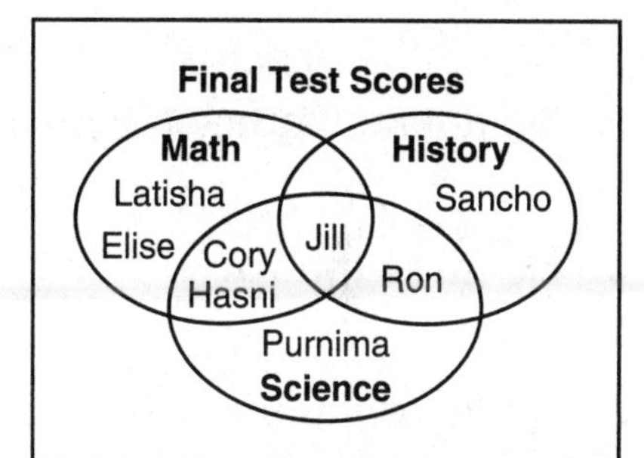

a. Which student(s) scored 90% or above on *all* 3 tests?

____

b. Which student(s) scored 90% or above on *exactly* 1 test?

____

c. Which student(s) scored 90% or above in *both* math and science?

____

Try This

d. What percent of the students in Mr. Wu's homeroom had a score of 90% or above on *at least* 2 tests? ____________

Date Time

LESSON 7•6 Probability Tree Diagrams

Mr. Gulliver travels to and from work by train. Trains to work leave at 6, 7, 8, 9, and 10 A.M. Trains from work leave at 3, 4, and 5 P.M. Suppose Mr. Gulliver selects a morning train at random and then selects an afternoon train at random.

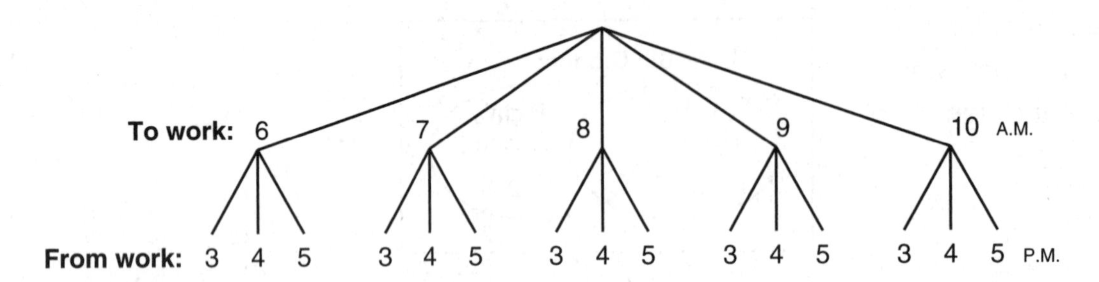

1. How many different combinations of trains to and from work can Mr. Gulliver take?

   ________ combinations

Calculate the probability of each of the following.

2. Mr. Gulliver takes the 7 A.M. train to work. ________
3. He returns home on the 4 P.M. train. ________
4. He takes the 7 A.M. train to work and returns home on the 4 P.M. train. ________
5. He takes the 9 A.M. train to work and returns home on the 5 P.M. train. ________
6. He leaves for work *before* 9 A.M. ________
7. He leaves for work at 6 A.M. or 7 A.M. and returns home at 3 P.M. ________
8. He returns home, but not on the 5 P.M. train. ________
9. He boards the return train 9 hours after leaving for work. ________

Date Time

LESSON 7•7

Math Boxes

1. The coin is flipped, and the spinner is spun. Complete the tree diagram. Write a fraction next to each branch to show the probability of selecting that branch.

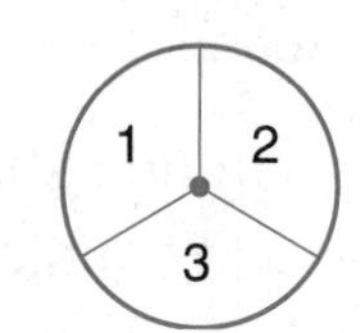

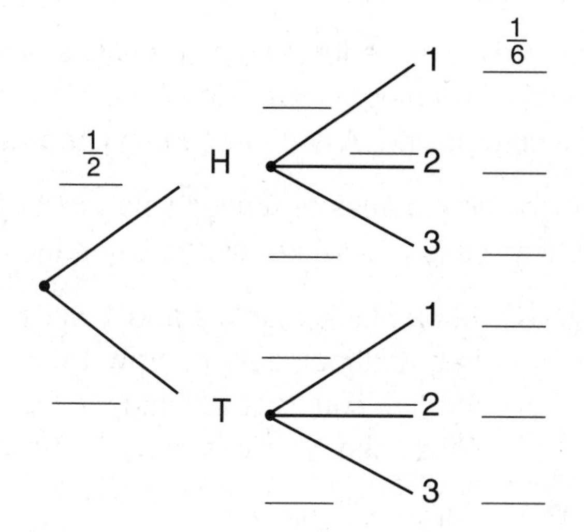

How many different outcomes are possible? ________

2. Describe the solution set for the graph below.

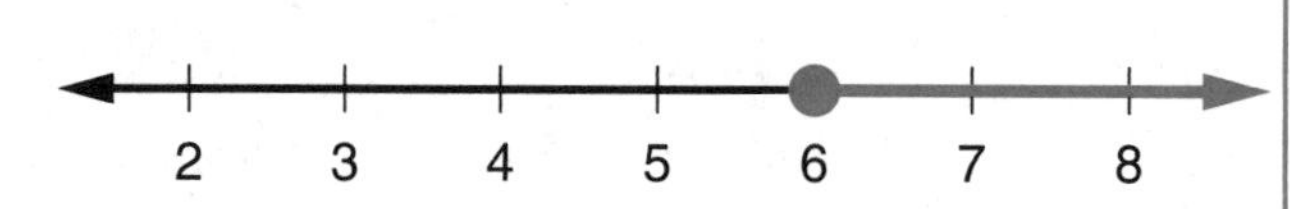

Solution set:

____

3. Write each number in standard notation.

a. 14.05 billion ________________

b. 2.01 trillion ________________

c. 0.25 million ________________

d. 0.75 thousand ________________

4. Suppose the average dalmation dog runs 1,668 feet in 1 minute. At this rate, how far can this dog run

a. in 3 minutes? ________ feet

b. in 1 minute 30 seconds?

________ feet

c. in 45 seconds? ________ feet

5. Express the ratio of the perimeter of square *ABCD* to the perimeter of square *EFGH* as a fraction in simplest form.

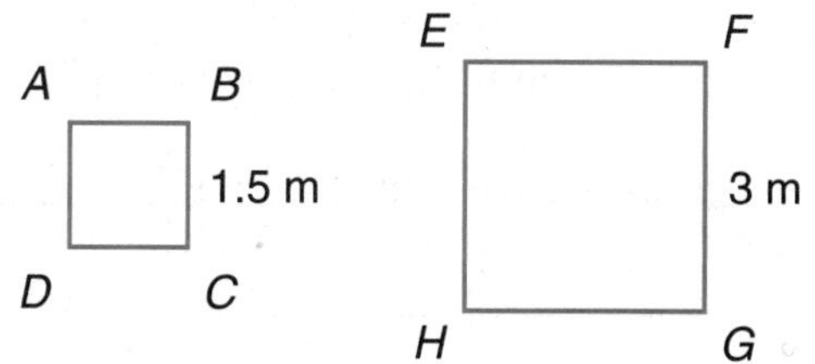

$\frac{\text{Perimeter } ABCD}{\text{Perimeter } EFGH}$ = ________

Date Time

LESSON 7·7

Fair and Unfair Games

Math Message

A game of chance for 2 or more players is a **fair game** if each player has the same chance of winning. A game for 1 player is fair if the player has an equal chance of winning or losing. Any other game is an **unfair game.**

Each of the 4 games described below is for 1 player. Play Games 1, 2, and 3 a total of 6 times each. Tally the results. Later, the class will combine results for each game.

Game 1 Put 2 black counters and 1 white counter into a paper bag and shake the bag. Without looking, draw 1 counter. Then draw a second counter without putting the first counter back into the bag. If the 2 counters are the same color, you win. Otherwise, you lose. Play 6 games and tally your results.

● ●
○

Tally for 6 games: Win ________ Lose ________

Do you think the game is fair? ________

Combined class data: Win ________ Lose ________

Game 2 Use 2 black counters and 2 white counters. The rules are the same.

● ●
○ ○

Tally for 6 games: Win ________ Lose ________

Do you think the game is fair? ________

Combined class data: Win ________ Lose ________

Game 3 Use 3 black counters and 1 white counter. The rules are the same.

● ●
● ○

Tally for 6 games: Win ________ Lose ________

Do you think the game is fair? ________

Combined class data: Win ________ Lose ________

Game 4 Suppose you use 4 black counters. The rules are the same.

● ●
● ●

Do you think the game is fair? ________

Explain your answer. ____

____

____

Date Time

LESSON 7·7

Fair Games and Probability

You can use a tree diagram to decide whether a game is fair or unfair. This tree diagram represents Game 1 on page 268.

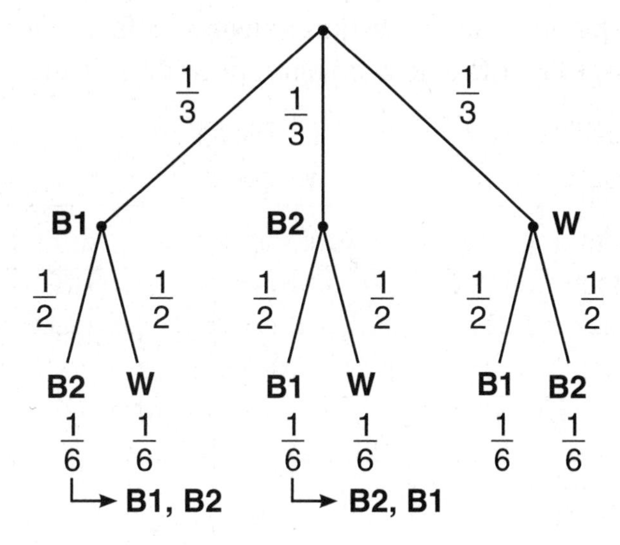

Before you draw the first counter, there are 3 counters in the bag. Although the 2 black counters look alike, they are not the same. To tell them apart, they are labeled B1 and B2. The probability of drawing either B1, B2, or W is $\frac{1}{3}$.

After the first draw, there are 2 counters left in the bag. The probability of drawing either of the 2 remaining counters is $\frac{1}{2}$.

There are 6 possible ways to draw the 2 remaining counters. The probability of each outcome is $\frac{1}{3} * \frac{1}{2}$, or $\frac{1}{6}$. There are 2 ways to draw the same color counters:

Draw B1 on the first draw and B2 on the second draw.
Draw B2 on the first draw and B1 on the second draw.

The chance of drawing 2 black counters is $\frac{1}{6} + \frac{1}{6}$, which is $\frac{2}{6}$, or $\frac{1}{3}$. Therefore, Game 1 is not a fair game.

Make tree diagrams to help you answer these questions.

1. What is the probability of winning Game 2? ____________

2. Is Game 2 a fair game? ____________

3. What is the probability of winning Game 3? ____________

4. Is Game 3 a fair game? ____________

Date Time

LESSON 7·8

Probabilities and Outcomes

Math Message

1. A game is played using the spinner at the right. The spinner will land on white $\frac{4}{5}$ of the time. Each time the spinner lands on white, Player A gets 1 point. Each time the spinner lands on blue, Player B gets 4 points. The winner is the player with more points after 20 spins.

   Is this a fair game? Explain. ______________________________

   ____

   ____

   ____

   ____

   ____

   ____

   ____

2. Suppose you are taking a multiple-choice test. Four possible answers are given for each question. There are 20 questions for which you don't know the correct answer. You decide to guess the answer for each of these questions.

   a. What is the probability of answering a question correctly? ______

   b. How many of the 20 questions would you expect to answer correctly by guessing? ______ questions

      Explain. ______________________________

3. In scoring the test, each correct answer is worth 1 point. To discourage guessing, there is a penalty of $\frac{1}{3}$ point for each incorrect answer. Do you think this is a fair penalty?

   Explain. ______________________________

   ____

   ____

   ____

   ____

Date Time

LESSON 7·8

Math Boxes

1. The Venn diagram shows the relationships between the family of quadrangles that includes squares, rhombuses, rectangles, trapezoids, and parallelograms. Use the diagram to indicate whether each statement is true or false.

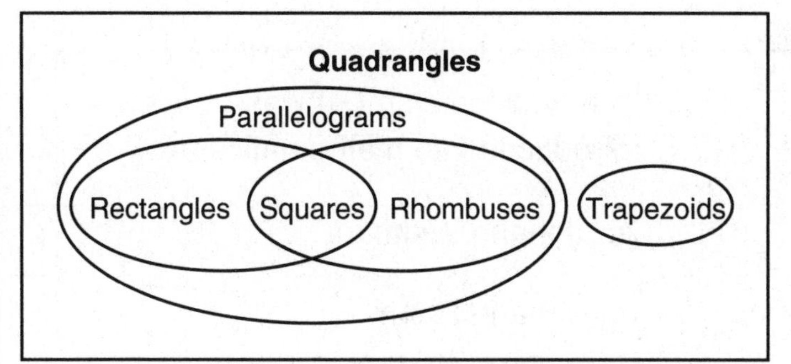

a. A square is a rectangle. ________

b. A rhombus is always a square. ________

c. A trapezoid is a quadrangle. ________

2. Write each fraction as a decimal.

a. $\frac{709}{10}$ = ________

b. $\frac{83,261}{1,000}$ = ________

c. $\frac{352}{10,000}$ = ________

d. $\frac{247.8}{100}$ = ________

3. Fill in the missing numbers.

a. 58 * 91 * 27 = 27 * 58 * ________

b. 24 * 16 * 10 = 240 * ________

c. ________ * 500 = 25 * 20 * 153

d. ________ * 426 * 81 = 81 * 945 * 426

e. ________ * 35 * 94 = 35 * 94 * 87

4. Use division to rename the fraction as a decimal rounded to the nearest hundredth. Show your work.

$\frac{31}{36}$

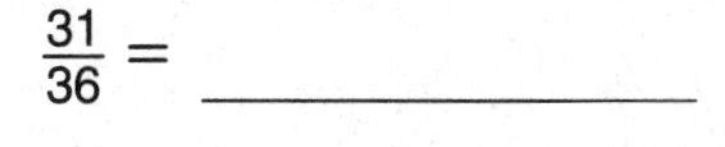
$\frac{31}{36}$ = ________

5. Which is the best estimate for the percent equivalent of $\frac{3}{40}$?

Circle the best answer.

A. 3%

B. 8%

C. 10%

D. 13%

Date Time

LESSON 7·8

Guessing on Multiple-Choice Tests

1. For each question on the following test, first draw a line through each answer that you know is not correct. Then circle one answer for each question. If you do not know the correct answer, guess.

Number correct ______________

1. A nautical mile is equal to

A. 1 foot.

B. 1 yard.

C. 1,832 meters.

D. 1,852 meters.

2. In 2000, the population of Nevada was

A. 1 billion.

B. 2,810,828.

C. 2,018,828.

D. 14,628.

3. Which region receives the greatest average annual rainfall?

A. Atlanta, Georgia

B. New Orleans, Louisiana

C. Mojave Desert, California

D. Sahara Desert, Africa

4. The addition sign (+) was introduced into mathematics by

A. Johann Widman.

B. Johann Rahn.

C. Abraham Lincoln.

D. Martin Luther King, Jr.

5. How many diagonals does a 13-sided polygon (13-gon) have?

A. 1

B. 54

C. 65

D. 13,000

6. The leading cause of death in the United States is

A. bungee jumping.

B. cancer.

C. drowning.

D. heart disease.

Date Time

LESSON 7·8

Guessing on Multiple-Choice Tests *continued*

2. When you can narrow the choices for a question to 2 possible answers, what is the chance of guessing the correct answer? ____________

3. How many of the 6 questions on page 272 do you think you answered correctly? _____ questions

4. Is it likely that you got all 6 correct? ________

5. Is it likely that you got all 6 incorrect? ________

6. Suppose each correct answer is worth 1 point and each incorrect answer carries a penalty of $\frac{1}{3}$ point. Complete the Total Points column of the table. You will complete the Class Tally column later.

Number Correct	Number Incorrect	Total Points	Class Tally
6	0	6	
5	1		
4	2		
3	3		
2	4	$2 - \frac{4}{3} = \frac{2}{3}$	
1	5		
0	6		

Date Time

LESSON 7•8

Guessing on Multiple-Choice Tests *continued*

7. For each question on the following test, first draw a line through each answer that you know is not correct. Then circle one answer for each question. If you do not know the correct answer, guess.

Number correct ______________

1. The neck of a 152-pound person weighs about

A. 100 pounds.

B. $12\frac{1}{2}$ pounds.

C. $11\frac{1}{2}$ pounds.

D. $10\frac{1}{2}$ pounds.

2. The average height of a full-grown weeping willlow tree is

A. 50 feet.

B. 45 feet.

C. 35 feet.

D. 2 feet.

3. The normal daily high temperature for July in Cleveland, Ohio, is

A. 84°F.

B. 82°F.

C. 80°F.

D. 0°F.

4. The circumference of Earth at the equator is about

A. 24,901.6 miles.

B. 24,801.6 miles.

C. 24,701.6 miles.

D. 2,000 miles.

5. In 2001, the average U.S. resident consumed about 138 pounds of which food?

A. sugar

B. spinach

C. potatoes

D. rice

6. A slice of white bread has about how many calories?

A. 3

B. 65

C. 70

D. 75

LESSON 7·8

Guessing on Multiple-Choice Tests *continued*

8. When you can narrow the choices for a question to 3 possible answers, what is the chance of guessing the correct answer? ________

9. How many of the 6 questions do you think you answered correctly?

_____ questions

10. Is it likely that you got all 6 correct? ________

11. Is it likely that you got all 6 incorrect? ________

12. Suppose each correct answer is worth 1 point and each incorrect answer carries a penalty of $\frac{1}{3}$ point. Complete the Total Points column of the table below.

Number Correct	Number Incorrect	Total Points	Class Tally
6	0	6	
5	1		
4	2		
3	3		
2	4	$2 - \frac{4}{3} = \frac{2}{3}$	
1	5		
0	6		

Date Time

LESSON 7·8

Algebraic Expressions

1. Write an algebraic expression for each situation. Use the suggested variable.

   a. Dakota is 14 inches taller than Gerry. If Gerry is h inches tall, how many inches tall is Dakota? ____________ (unit)

   b. Sam ran for $\frac{4}{5}$ the length of time that Justin ran. If Justin ran r minutes, how long did Sam run? ____________ (unit)

   c. Beyonce has x CDs in her collection. If Ann has 9 fewer CDs than Beyonce, how many CDs does Ann have? ____________ (unit)

   d. Charlie has d dollars. Leanna has 6 times as much money as Charlie. How much money does Leanna have? ____________ (unit)

   e. Erica has been a lifeguard for y years. That is 3 times as many years as Tom. How long has Tom worked as a lifeguard? ____________ (unit)

2. Write the rule for the numbers in the table.

x	y
1.5	6.75
2	9
–3	–13.5
0.25	1.125

Rule ____________

3. Write the rule for the numbers in the table.

a	b
$\frac{2}{3}$	$\frac{4}{15}$
$\frac{1}{6}$	$\frac{1}{15}$
$\frac{3}{4}$	$\frac{3}{10}$
$\frac{4}{5}$	$\frac{8}{25}$

Rule ____________

Translate each situation from words into an algebraic expression. Then solve the problem that follows.

4. Calida has 7 more crayons than 3 times the number of crayons Royce has. If Royce has c crayons, how many does Calida have? ____________

   If Royce has 12 crayons, how many does Calida have? ____________ (unit)

5. Alinda has seen 4 fewer than $\frac{1}{2}$ the number of movies that her sister has seen. If her sister has seen m movies, how many has Alinda seen? ____________

   If her sister has seen 20 movies, how many has Alinda seen? ____________ (unit)

Date Time

LESSON 7•9

Math Boxes

1. Solve each equation.

Solution

a. $\frac{28}{c} = 14$ __________

b. $\frac{15}{33} = \frac{5}{x}$ __________

c. $\frac{500}{10,000} = \frac{p}{100}$ __________

d. $\frac{25}{75} = \frac{d}{12}$ __________

2. Complete.

a. 20% of 80 = __________

b. 75% of 48 = __________

c. 55% of 1,000 = __________

d. 30% of 250 = __________

3. Use the graph to answer the following questions.

a. Suppose a black rhino runs at its top speed for 6 minutes.

How far would the rhino travel? __________ (unit)

b. According to the graph, how far would a grizzly bear run in 15 minutes? __________ (unit)

c. Which animal's speed is about 20% the speed of the cheetah? __________

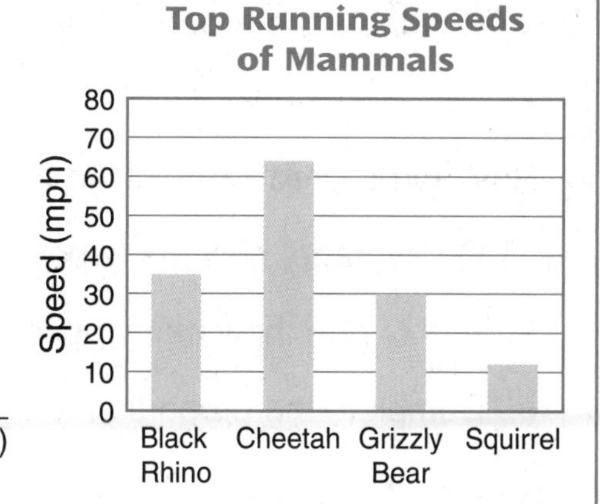

4. Divide. Express the remainder as a fraction in simplest form.

$75\overline{)7,698}$

7,698 ÷ 75 = __________

5. Express each ratio as a fraction in simplest form.

a. 100 to 60 __________

b. 80 out of 56 __________

c. 1,440 calories in 8 ounces __________

d. 112 males for every 98 females __________

LESSON 8•1

Solving Rate Problems

Math Message

1. A computer printer prints 70 pages in 2 minutes. How many pages will it print in 5 minutes? ____________

2. Anevay trains at an indoor track during the winter. She can run 24 laps in 8 minutes. At this rate, how many laps can Anevay run in 12 minutes? ____________

---

3. Ms. Marquez is reading stories that her students wrote. She has read 5 stories in 40 minutes.

   a. At this rate, how long would it take her to read 1 story? ____________

   b. How long will it take her to read all 30 of her students' stories? ____________

   c. Complete the proportion to show your solution. $\frac{\text{5 stories}}{\text{40 minutes}} = \frac{\text{30 stories}}{\square \text{ minutes}}$

4. Roshaun scored 75 points in the first 5 basketball games.

   a. On average, how many points did he score per game? ____________

   b. At this rate, how many points might he score in a 15-game season? ____________

   c. Complete the proportion to show your solution. $\frac{\text{75 points}}{\text{5 games}} = \frac{\square \text{ points}}{\text{15 games}}$

5. Last year, 55 students sold $1,210 worth of candy for their band's fund-raiser.

   a. On average, how many dollars' worth of candy did each student sell? ____________

   b. This year, 67 students will be selling candy. If they sell at the same rate as last year, how much money can they expect to raise? ____________

   c. Complete the proportion to show your solution. $\frac{\square \text{ students}}{\$\square} = \frac{\square \text{ students}}{\$\square}$

6. Anoki worked at the checkout counter from 5:30 P.M. to 11 P.M. He earned $33.

   a. How much did he earn per hour? ____________

   b. Anoki works $27\frac{1}{2}$ hours per week. How much will he earn in 1 week? ____________

   c. Complete the proportion to show your solution. $\frac{\$\square}{\square \text{ hours}} = \frac{\$\square}{\square \text{ hours}}$

Date Time

LESSON 8•1

Solving Rate Problems *continued*

7. The furlong is a unit of distance commonly used in horse racing. There are 40 furlongs in 5 miles.

   a. Fill in the rate table.

miles	1	2		5	8	10
furlongs			24	40		

   b. How many furlongs are in 8 miles? _______________

   Complete the proportion to show your solution.

   $\frac{\square \text{ miles}}{\square \text{ furlongs}} = \frac{\square \text{ miles}}{\square \text{ furlongs}}$

   c. How many miles are in 8 furlongs?

   ____

   Complete the proportion to show your solution.

   $\frac{\square \text{ miles}}{\square \text{ furlongs}} = \frac{\square \text{ miles}}{\square \text{ furlongs}}$

8. Nico's grandfather read a 240-page book in 6 hours.

   a. Fill in the rate table.

pages		120		240	80	
hours	1		4	6		8

   b. At this rate, how long would it take him to read a 160-page book? _______________

   Complete the proportion to show your solution.

   $\frac{\square \text{ pages}}{\square \text{ hours}} = \frac{\square \text{ pages}}{\square \text{ hours}}$

   c. How many hours did it take him to read 80 pages? _______________

   Complete the proportion to show your solution.

   $\frac{\square \text{ pages}}{\square \text{ hours}} = \frac{\square \text{ pages}}{\square \text{ hours}}$

Use any method you wish to solve the following problems. Write a proportion to show your solution.

9. A recipe for a 2-pound loaf of bread calls for 4 cups of flour. How many 2-pound loaves can you make with 12 cups of flour? _______________

   $\frac{\square \text{ loaves}}{\square \text{ cups}} = \frac{\square \text{ loaves}}{\square \text{ cups}}$

10. Two inches of rain fell between 7 A.M. and 3 P.M. The rain continued at the same rate until 7 P.M.

    How many inches of rain fell between 7 A.M. and 7 P.M.? _______________

    $\frac{\square \text{ inches}}{\square \text{ hours}} = \frac{\square \text{ inches}}{\square \text{ hours}}$

Date Time

LESSON 8•1

More Division Practice

Divide. Show your work in the space below.
For Problems 1–3, round answers to the nearest tenth.

1. $571 \div 8 =$ ________
2. $2,723 / 94 =$ ________
3. $815 \div 46 =$ ________

For Problems 4–6, round answers to the nearest hundredth.

4. $89 / 6 =$ ________
5. $3,714 / 42 =$ ________
6. $217 \div 18 =$ ________

7. Write a number story for Problem 6.

____

____

____

____

____

____

Date Time

LESSON 8•1

Math Boxes

1. Express each rate as a per-unit rate.

a. 108 words in 4 minutes

$\frac{\square \text{ words}}{1 \text{ minute}}$

b. 300 miles in 15 hours

☐ miles/hour

c. $102.50 for 10 hours

$ ☐ /hour

2. Tamika earns $500 for 40 hours of work.

Fill in the rate table.

hours	5	10	15
amount ($)	62.50		

At this rate, how much will Tamika earn for 55 hours of work?

____

SRB 110 111

3. For lunch, the school cafeteria offers a main course and a beverage. For the main course, students can choose spaghetti, hamburgers, or hot dogs. The beverage choices are milk, soda, and juice. Draw a tree diagram to show all the possible meal combinations.

If you choose a meal at random, what is the probability of getting a

a. hot dog?

____

b. hot dog and juice? _______

4. Add or subtract.

a. $14 + (-72) =$ _________

b. _________ $= 27 - (-28)$

c. _________ $= -63 + (-87)$

d. $-33 - (-89) =$ _________

5. Insert parentheses to make each number sentence true.

a. $5 * 10^2 + 10^2 * 2 = 2,000$

b. $9 * 3 + 4 = 63$

c. $16 + 2^2 - 5 + 3 = 12$

d. $7 * 4 - 2 + 1 = 21$

SRB 247

Date Time

LESSON 8·2

Rate Problems and Proportions

Math Message

Solve the problems below. Then write a proportion for each problem.

1. Robin rode her bike at an average speed of 8 miles per hour. At this rate, how far would she travel in 3 hours? ______

$\frac{\square \text{ miles}}{\square \text{ hours}} = \frac{\square \text{ miles}}{\square \text{ hours}}$

2. A high-speed copier makes 90 copies per minute. How long will it take to make 270 copies? ______

$\frac{\square \text{ copies}}{\square \text{ minutes}} = \frac{\square \text{ copies}}{\square \text{ minutes}}$

3. Talia rode her bike at an average speed of 10 miles per hour. At this rate, how long would it take Talia to ride 15 miles? ______

$\frac{\square \text{ miles}}{\square \text{ hours}} = \frac{\square \text{ miles}}{\square \text{ hours}}$

For each of the following problems, first complete the rate table. Use the table to write an open proportion. Solve the proportion. Then write the answer to the problem. Study the first problem.

4. Angela earns $6 per hour babysitting. How long must she work to earn $72?

dollars	6	72
hours	1	t

$\frac{6 * 12}{1 * 12} = \frac{72}{t}$

Answer: Angela must work ______ hours to earn $72.

5. There are 9 calories per gram of fat. How many grams of fat are in 63 calories?

calories		
grams (of fat)		

$\frac{\square}{\square} = \frac{\square}{\square}$

Answer: There are ______ grams of fat in 63 calories.

Date Time

Rate Problems and Proportions *continued*

6. If carpet costs $22.95 per square yard, how much will 12 square yards of carpet cost?

dollars		
square yards		

Answer: 12 square yards of carpet will cost ________.

7. Jasmine used 3 gallons of water to make lemonade. How many cups of water did she use?

gallons		
cups		

Answer: Jasmine used ________ cups of water.

8. A car goes 480 miles on a 12-gallon tank of gas. How many miles is this per gallon?

miles		
gallons		

Answer: The car will go ________ miles on 1 gallon of gas.

9. Corey's door is 1.2 meters wide. How many centimeters wide is Corey's door?

meters		
centimeters		

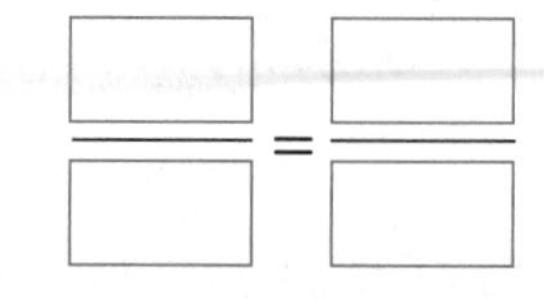

Answer: Corey's door is ________ centimeters wide.

10. A TV station runs 42 minutes of commercials in 7 half-hour programs. How many minutes of commercials does it run per hour?

commercial minutes		
hours		

Answer: The station runs ________ minutes of commercials per hour.

Date Time

LESSON 8•2

Dividing Decimals by Decimals: Part 2

In general, you can think of a fraction as a division problem.

The fraction $\frac{3}{8}$ equals $3 \div 8$ or $8\overline{)3}$

$\frac{3}{8}$ 3 ← numerator → 8 ← denominator → $8\overline{)3}$

As you do with a fraction, you can multiply the dividend and divisor by the same number without changing the value of the quotient.

$$\frac{17.48}{0.076} * \frac{1,000}{1,000} = \frac{17,480}{76}$$

original problem — equivalent problem

$0.076\overline{)17.480}$ $76\overline{)17,480}$

Rename each division problem to make the divisor a whole number. Then solve the equivalent problem using the method of your choice.

Equivalent Problem

1. $3.5\overline{)27.3}$ = ______________

Quotient = ________

Equivalent Problem

2. $1.9\overline{)7.03}$ = ______________

Quotient = ________

Equivalent Problem

3. $5.06\overline{)47.058}$ = ______________

Quotient = ________

Equivalent Problem

4. $0.56\overline{)21.28}$ = ______________

Quotient = ________

Date Time

LESSON 8•2

Math Boxes

1. The formula $d = r * t$ gives the distance d traveled at speed r in time t. Use this formula to solve the problem below.

The distance from San Francisco to Los Angeles is about 420 miles. About how many hours would it take to drive from San Francisco to Los Angeles at a constant rate of 55 miles per hour? Round your answer to the nearest tenth.

$t =$ ________

2. Solve.

Solution

a. $\frac{42}{18} = \frac{x}{3}$ ________

b. $\frac{125}{n} = \frac{5}{1}$ ________

c. $\frac{36}{w} = \frac{18}{2}$ ________

d. $\frac{k}{150} = \frac{3}{5}$ ________

e. $\frac{90}{3} = \frac{9}{u}$ ________

3. Thirty-three sixth graders at Maple Middle School belong to the drama club. Thirty-seven sixth graders are in the school choir.

Use this information to complete the Venn diagram below.

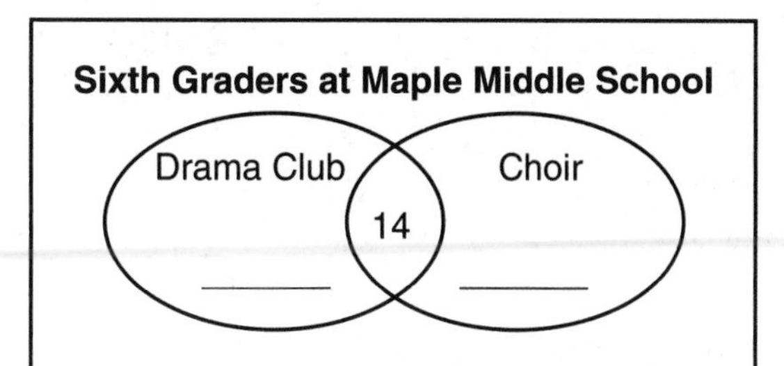

4. Solve.

a. $\frac{-9 * 5}{3} =$ ________

b. ________ $= (-1)(-7)^2$

c. $\frac{-6 + (-3) + (-7)}{4} =$ ________

5. Fill in the blanks.

a. $-8 +$ ________ $= 0$

b. $\frac{2}{3} +$ ________ $= 0$

c. $-5 -$ ________ $= 0$

Date Time

LESSON 8·3

Equivalent Fractions and Cross Products

Math Message

SRB 114 115

For Part a of each problem, write = or ≠ in the answer box.
For Part b, calculate the cross products.

1. a. $\frac{3}{5}$ ☐ $\frac{6}{10}$

b. $10 * 3 =$ ______ ______ $= 5 * 6$

$\frac{3}{5}$ $\frac{6}{10}$

2. a. $\frac{7}{8}$ ☐ $\frac{2}{3}$

b. $3 * 7 =$ ______ ______ $= 8 * 2$

3. a. $\frac{2}{3}$ ☐ $\frac{6}{9}$

b. $9 * 2 =$ ______ ______ $= 3 * 6$

$\frac{2}{3}$ $\frac{6}{9}$

4. a. $\frac{6}{9}$ ☐ $\frac{8}{12}$

b. $12 * 6 =$ ______ ______ $= 9 * 8$

$\frac{6}{9}$ $\frac{8}{12}$

5. a. $\frac{2}{8}$ ☐ $\frac{4}{10}$

b. $10 * 2 =$ ______ ______ $= 8 * 4$

$\frac{2}{8}$ $\frac{4}{10}$

6. a. $\frac{10}{12}$ ☐ $\frac{5}{8}$

b. $8 * 10 =$ ______ ______ $= 12 * 5$

$\frac{10}{12}$ $\frac{5}{8}$

7. a. $\frac{1}{4}$ ☐ $\frac{5}{20}$

b. $20 * 1 =$ ______ ______ $= 4 * 5$

$\frac{1}{4}$ $\frac{5}{20}$

8. a. $\frac{5}{7}$ ☐ $\frac{15}{21}$

b. $21 * 5 =$ ______ ______ $= 7 * 15$

$\frac{5}{7}$ $\frac{15}{21}$

9. a. $\frac{10}{16}$ ☐ $\frac{4}{8}$

b. $8 * 10 =$ ______ ______ $= 16 * 4$

$\frac{10}{16}$ $\frac{4}{8}$

10. a. $\frac{3}{5}$ ☐ $\frac{10}{15}$

b. $15 * 3 =$ ______ ______ $= 5 * 10$

$\frac{3}{5}$ $\frac{10}{15}$

11. What pattern can you find in Parts a and b in the problems above?

____

Date Time

LESSON 8·3 Math Boxes

1. Which rate is equivalent to 70 km in 2 hr 30 min? Fill in the circle next to the best answer.

   ○ **A.** 35 km in 75 min

   ○ **B.** 70,000 m in 230 min

   ○ **C.** 140 km in 4 hr 30 min

   ○ **D.** 1,400 m in 300 min

2. A boat traveled 128 kilometers in 4 hours.

   Fill in the rate table.

distance (km)	24		72		144
hours	$\frac{3}{4}$	$1\frac{1}{2}$		3	

   At this rate, how far did the boat travel in 2 hours 15 minutes?

   ____

3. A bag contains 1 red counter, 2 blue counters, and 1 white counter. You pick 1 counter at random. Then you pick a second counter without replacing the first counter.

   a. Draw a tree diagram to show all possible counter combinations.

   b. What is the probability of picking 1 red counter and 1 white counter (in either order)? ________

4. Add or subtract.

   a. $-303 + (-28) =$ __________

   b. __________ $= 245 - 518$

   c. __________ $= -73 + 89$

   d. $280 - (-31) =$ __________

5. Insert parentheses to make each number sentence true.

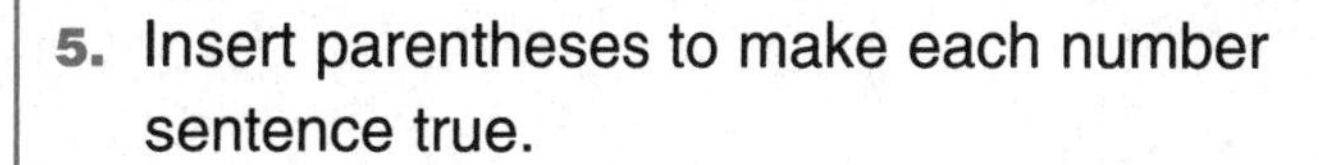

   a. $0.01 * 7 + 9 / 4 = 0.04$

   b. $\frac{4}{5} * 25 - 10 / 2 = 15$

   c. $\sqrt{64} / 5 + 3 * 3 = 3$

   d. $5 * 10^2 + 10^2 * 2 = 700$

Date Time

LESSON 8·3

Solving Proportions with Cross Products

Use cross multiplication to solve these proportions.

SRB 114 115

Example: $\frac{4}{6} = \frac{p}{15}$

$15 * 4 =$ ____ ____ $= 6 * p$

$\frac{4}{6} = \frac{p}{15}$

$15 * 4 = 6 * p$

$60 = 6p$

$\frac{60}{6} = p$

$10 = p$

1. $\frac{3}{6} = \frac{y}{10}$ ____________

2. $\frac{7}{21} = \frac{3}{c}$ ____________

3. $\frac{m}{20} = \frac{2}{8}$ ____________

4. $\frac{2}{10} = \frac{5}{z}$ ____________

5. $\frac{9}{15} = \frac{12}{k}$ ____________

6. $\frac{10}{12} = \frac{d}{9}$ ____________

7. $\frac{2}{9} = \frac{t}{54}$ ____________

8. $\frac{4}{10} = \frac{26}{z}$ ____________

9. $\frac{3}{4} = \frac{r}{28}$ ____________

10. $\frac{16}{p} = \frac{128}{40}$ ____________

11. $\frac{51}{102} = \frac{6}{h}$ ____________

12. $\frac{j}{8} = \frac{72}{192}$ ____________

Solving Proportions with Cross Products *continued*

For Problems 13–16, set up a proportion and solve it using cross multiplication. Show how the units cancel. Then write the answer.

Example: Jessie swam 6 lengths of the pool in 4 minutes. At this rate, how many lengths will she swim in 10 minutes?

Proportion: $\frac{\text{6 lengths}}{\text{4 minutes}} = \frac{\text{n lengths}}{\text{10 minutes}}$

Solution: $\frac{6}{4} = \frac{n}{10}$

$10 * 6 =$ ___ ___ $= 4 * n$

$\frac{6}{4} = \frac{n}{10}$

10 minutes $*$ 6 lengths $=$ 4 minutes $*$ n lengths

60 minutes $*$ lengths $=$ 4 minutes $*$ n lengths

$\frac{\text{60 minutes * lengths}}{\text{4 minutes}} = n$ lengths

15 lengths $= n$ lengths

Answer: Jessie will swim _________ lengths in 10 minutes.

13. Belle bought 8 yards of ribbon for $6. How many yards could she buy for $9?

Solution:

$\frac{\boxed{}}{\boxed{}} = \frac{\boxed{}}{\boxed{}}$

Answer: Belle could buy _________ yards of ribbon for $9.

LESSON 8·3

Solving Proportions with Cross Products *continued*

14. Before going to France, Maurice exchanged $25 for 20 euros. At that exchange rate, how many euros could he get for $80?

Solution:

$\frac{\square}{\square} = \frac{\square}{\square}$

Answer: Maurice could get ________ euros for $80.

15. One gloomy day, 4 inches of rain fell in 6 hours. At this rate, how many inches of rain had fallen after 4 hours?

Solution:

$\frac{\square}{\square} = \frac{\square}{\square}$

Answer: ________ inches of rain had fallen in 4 hours.

16. Adelio's apartment building has 9 flights of stairs. To climb to the top floor, he must go up 144 steps. How many steps must he go up to climb 5 flights?

Solution:

$\frac{\square}{\square} = \frac{\square}{\square}$

Answer: Adelio must climb ________ steps.

Date Time

LESSON 8·3 Solving Proportions with Cross Products *continued*

Set up a proportion for each problem and solve it using cross multiplication.

17. Sarah uses 5 scoops of coffee beans to brew 8 cups of coffee. How many scoops of beans does Sarah use per cup?

Solution:

$\frac{\square}{\square} = \frac{\square}{\square}$

Answer: Sarah uses ________ scoop(s) of beans per cup of coffee.

18. Jeremiah ran $1\frac{1}{4}$ miles in 12 minutes. At this pace, how long would it take him to run 5 miles?

Solution:

$\frac{\square}{\square} = \frac{\square}{\square}$

Answer: It would take Jeremiah ________ minutes to run 5 miles.

19. It took Zach 12 days to read a book that was 186 pages long. If he read the same amount each day, how many pages did he read in one week?

Solution:

$\frac{\square}{\square} = \frac{\square}{\square}$

Answer: Zach read ________ pages in one week.

20. At sea level, sound travels 0.62 mile in 3 seconds. What is the speed of sound in miles per hour? (*Hint:* First find the number of seconds in 1 hour.)

Solution:

$\frac{\square}{\square} = \frac{\square}{\square}$

Answer: Sound travels at the rate of ________ miles per hour.

Date Time

LESSON 8•4 Rate * Time = Distance

Math Message

For each problem, make a rate table. Then write a number model and solve it.

1. Grandma Riley drove her car at 60 miles per hour for 4 hours. How far did she travel?

   Number model ______________________

   Answer: She traveled ____________ miles in 4 hours.

2. A bamboo plant grows 8 inches per day. How tall will it be after 7 days?

   Number model ______________________

   Answer: The plant will be ____________ inches tall.

3. A rocket is traveling at 40,000 miles per hour. How far will it travel in 168 hours?

   Number model ______________________

   Answer: The rocket will travel ____________ miles in 168 hours.

4. Amora can ride her bicycle at 9 miles per hour. At this rate, how long will it take her to ride 30 miles?

   Number model ______________________

   Answer: It will take her ____________ hours to ride 30 miles.

5. Australia is moving about 3 inches per year with respect to the southern Pacific Ocean. How many *feet* will it move in 50 years?

   Number model ______________________

   Answer: Australia will move ____________ feet in 50 years.

LESSON 8•4

Using Unit Fractions to Find a Whole

Example 1:

Alex collects sports cards. Seventy of the cards feature basketball players. These 70 cards are $\frac{2}{3}$ Alex's collection. How many sports cards does Alex have?

- If $\frac{2}{3}$ the collection is 70 cards, then $\frac{1}{3}$ is 35 cards.
- Alex has all the cards—that's $\frac{3}{3}$ the cards.
- Therefore, Alex has 3 * 35, or 105 cards.

Example 2:

Barb's grandmother baked cookies. She gave Barb 12 cookies, which was $\frac{2}{5}$ the total number she baked. How many cookies did Barb's mother bake?

- If $\frac{2}{5}$ the total is 12 cookies, then $\frac{1}{5}$ is 6 cookies.
- Barb's mother baked all the cookies—that's $\frac{5}{5}$ the cookies.
- She baked 5 * 6, or 30 cookies.

1. Six jars are filled with cookies. The number of cookies in each jar is not known. For each clue given in the table, find the number of cookies in the jar.

Clue	Number of Cookies in Jar
$\frac{1}{2}$ jar contains 31 cookies.	
$\frac{2}{8}$ jar contains 10 cookies.	
$\frac{3}{5}$ jar contains 36 cookies.	
$\frac{3}{8}$ jar contains 21 cookies.	
$\frac{4}{7}$ jar contains 64 cookies.	
$\frac{3}{11}$ jar contains 45 cookies.	

2. Jin is walking to a friend's house. He has gone $\frac{6}{10}$ the distance in 48 minutes. If he continues at the same speed, about how long will the entire walk take? __________

3. A candle burned $\frac{3}{8}$ the way down in 36 minutes. If it continues to burn at the same rate, about how many more minutes will the candle burn before it is used up? __________

LESSON 8•4

How Many Calories Do You Use Per Day?

Your body needs food. It uses the materials in food to produce energy—energy to keep your body warm and moving, to live and grow, and to build and repair muscles and tissues.

The amount of energy a food will produce when it is digested by the body is measured in a unit called the **calorie.** A calorie is not a substance in food.

1. The following table shows the number of calories used per minute and per hour by the average sixth grader for various activities. Complete the table. Round your answers for calories per minute to the nearest tenth and calories per hour to the nearest ten.

Calorie Use by Average Sixth Graders

Activity	Calories/Minute (to nearest 0.1)	Calories/Hour (to nearest 10)
Sleeping	0.7	40
Studying, writing, sitting	1.2	70
Eating, talking, sitting in class	1.2	70
Standing	1.3	80
Dressing, undressing		90
Watching TV	1.0	60
Walking (briskly, at 3.5 mph)	3.0	180
Doing housework, gardening	2.0	
Playing the piano	2.7	160
Raking leaves	3.7	220
Shoveling snow	5.0	300
Bicycling (6 mph)		170
Bicycling (13 mph)	4.5	
Bicycling (20 mph)	8.3	
Running (5 mph)	6.0	360
Running (7.5 mph)		560
Swimming (20 yd/min)	3.3	200
Swimming (40 yd/min)	5.8	350
Basketball, soccer (vigorous)	9.7	580
Volleyball	4.0	240
Aerobic dancing (vigorous)	6.0	360
Bowling	3.4	200

LESSON 8•4

How Many Calories Do You Use Per Day? *continued*

2. Think of all the things you do during a typical 24-hour day during which you go to school.

   a. List your activities in the table below.

   b. Record your estimate of the time you spend on each activity (to the nearest 15 minutes). Be sure the times add up to 24 hours.

   c. For each activity, record the number of calories used per minute or per hour. Then calculate the number of calories you use for the activity.

Example:

Suppose you spend 8 hours and 15 minutes sleeping.
Choose the per-hour rate: Sleeping uses 40 calories per hour.
Multiply: 8.25 hours * 40 calories per hour = 330 calories

My Activities during a Typical School Day (24 hr)

Activity	Time Spent on Activity	Calorie Rate (cal/min or cal/hr)	Calories Used for Activity

3. After you complete the table, find the total number of calories you use in 24 hours.

   In a typical 24-hour day during which I go to school, I use about __________ calories.

Date Time

LESSON 8•4 Math Boxes

1. The formula $d = r * t$ gives the distance d traveled at speed r in time t. Use this formula to solve the problem below.

   Which formula is equivalent to $d = r * t$? Choose the best answer.

   ◯ $r = \frac{d}{t}$

   ◯ $r = d - t$

   ◯ $r = \frac{t}{d}$

   ◯ $r = t + d$

2. Solve.

   Solution

   a. $\frac{6}{p} = \frac{2}{7}$ ____________

   b. $\frac{f}{2} = \frac{3}{12}$ ____________

   c. $\frac{4}{15} = \frac{12}{x}$ ____________

   d. $\frac{24}{36} = \frac{6}{w}$ ____________

   e. $\frac{7}{8} = \frac{y}{2}$ ____________

3. Of the 330 students at Pascal Junior High, 45 run track and 67 play basketball. Twenty-two students participate in both sports.

   Use this information to complete the Venn diagram below. Label each set (ring). Write the number of students belonging to each individual set and the intersection of the sets.

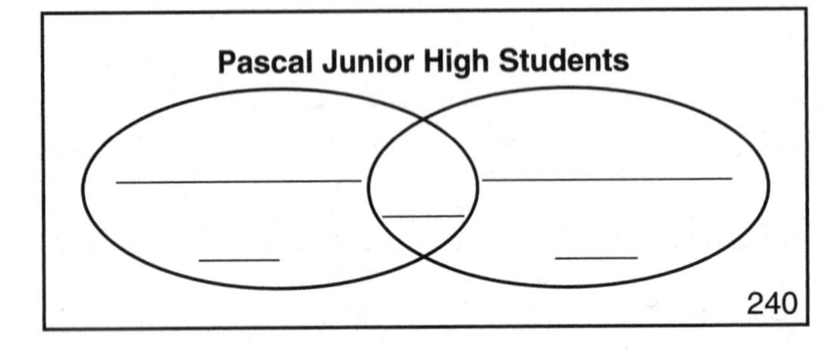

4. Solve.

   a. $-\frac{32}{2} + \frac{-75}{-15} =$ ____________

   b. $(-9)^2 (-1)^5 =$ ____________

   c. $\frac{-68 - 112}{-10} =$ ____________

5. Fill in the blanks. (*Hint*: For decimals, think fractions.)

   a. $\frac{7}{3} *$ ____________ $= 1$

   b. $0.01 *$ ____________ $= 1$

   c. $0.5 *$ ____________ $= 1$

Date Time

LESSON 8•5

Food Nutrition Labels

Use the information from the food label at the right to complete the following statements.

Low-fat yogurt

Nutrition Facts	
Serving Size 1 container (227 g)	
Amount Per Serving	
Calories 240 Calories from Fat 27	
	% Daily Value
Total Fat 3 g	**5%**
Saturated Fat 1.5 g	**8%**
Cholesterol 15 mg	**5%**
Sodium 150 mg	**6%**
Potassium 450 mg	**13%**
Total Carbohydrate 44 g	**15%**
Dietary Fiber 1 g	**4%**
Sugars 43 g	
Protein 9 g	
Vitamin A 2% • Vitamin C 10%	
Calcium 35% • Iron 0%	
Calories per gram: Fat 9 • Carbohydrate 4 • Protein 4	

1. There are 240 calories per serving. Of these 240 calories,

   _________ calories come from fat.

2. There are _________ grams of total carbohydrate per serving.

   Complete the proportion.

   $$\frac{1 \text{ g of carbohydrate}}{4 \text{ calories}} = \frac{\square \text{ g of carbohydrate}}{\square \text{ calories}}$$

   Solve the proportion. How many calories come from carbohydrate?

   _________ calories

3. Write the unit rate you can use to calculate calories from protein.

   ____

For each food label below, record the number of calories from fat. Then calculate the numbers of calories from carbohydrate and from protein. Add to find the total calories per serving.

White bread

Nutrition Facts	
Serving Size 1 slice (23 g)	
Servings Per Container 20	
Amount Per Serving	
Calories 65 Calories from Fat 9	
	% Daily Value
Total Fat 1 g	2%
Total Carbohydrate 12 g	4%
Protein 2 g	

Hot dog

Nutrition Facts	
Serving Size 1 link (45 g)	
Servings Per Container 10	
Amount Per Serving	
Calories 150 Calories from Fat 120	
	% Daily Value
Total Fat 13 g	**20%**
Total Carbohydrate 1 g	**<1%**
Protein 7 g	

4. Calories

   From fat _________

   From carbohydrate _________

   From protein _________

   Total calories _________

5. Calories

   From fat _________

   From carbohydrate _________

   From protein _________

   Total calories _________

LESSON 8•5

Plan Your Lunch

1. Choose 5 items you would like to have for lunch from the following menu. Choose your favorite foods—pay no attention to calories. Make a check mark next to each item.

Food	Total Calories	Calories from Fat	Calories from Carbohydrate	Calories from Protein
Ham sandwich	265	110	110	45
Turkey sandwich	325	70	155	100
Hamburger	330	135	120	75
Cheeseburger	400	200	110	90
Double burger, cheese, sauce	500	225	175	100
Grilled cheese sandwich	380	220	100	60
Peanut butter and jelly sandwich	380	160	170	50
Chicken nuggets (6)	250	125	65	60
Bagel	165	20	120	25
Bagel with cream cheese	265	105	125	35
Hard-boiled egg	80	55	0	25
French fries (small bag)	250	120	115	15
Apple	100	10	90	0
Carrot	30	0	25	5
Orange	75	0	70	5
Cake (slice)	235	65	160	10
Cashews (1 oz)	165	115	30	20
Doughnut	200	100	75	25
Blueberry muffin	110	30	70	10
Apple pie (slice)	250	125	115	10
Frozen-yogurt cone	100	10	75	15
Orange juice (8 fl oz)	110	0	104	8
2% milk (8 fl oz)	145	45	60	40
Skim milk (8 fl oz)	85	0	50	35
Soft drink (8 fl oz)	140	0	140	0
Diet soft drink (8 fl oz)	0	0	0	0

Date Time

LESSON 8•5 Plan Your Lunch *continued*

2. In the table below, record the 5 items you chose. Fill in the rest of the table and write the total number of calories for each column.

Food	Total Calories	Calories from Fat	Calories from Carbohydrate	Calories from Protein
Total				

What percent of the total number of calories in your lunch comes from fat? ________

From carbohydrate? ________ From protein? ________

3. Suppose a nutritionist recommends that, at most, 20–35% of the total number of calories should come from fat, about 35% from protein, and no more than 45–65% from carbohydrate.

Does the lunch you chose meet these recommendations? ________

4. Plan another lunch. This time use only 4 foods and try to limit the percent of calories from fat to between 20 and 35%, from protein to between 25 and 36%, and from carbohydrate to between 45 and 65%.

Food	Total Calories	Calories from Fat	Calories from Carbohydrate	Calories from Protein
Total				

What percent of the total number of calories in your lunch comes from fat? ________

From carbohydrate? ________ From protein? ________

Date Time

LESSON 8•5

Unit Percents to Find a Whole

Example 1:

The sale price of a CD player is $120, which is 60% of its list price. What is the list price?

- If 60% of the list price is $120, then 1% is $2. ($120 / 60 = 2$)
- The list price (the whole, or 100%) is $200. ($100 * 2 = 200$)

Example 2:

A toaster is on sale for $40, which is 80% of its list price. What is the list price?

- If 80% of the list price is $40, then 1% is $0.50. ($40 / 80 = 0.5$)
- The list price (100%) is $50. ($100 * 0.5 = 50$)

Use your percent sense to estimate the list price for each item below. Then calculate the list price. (*Hint:* First use your calculator to find what 1% is worth.)

Sale Price	Percent of List Price	Estimated List Price	Calculated List Price
$120	60%	$180	$200
$100	50%		
$8	32%		
$255	85%		
$77	55%		
$80	40%		
$9	60%		
$112.50	75%		
$450	90%		

Date Time

LESSON 8·5

Math Boxes

1. If there are 975 calories in 3 oz of ice cream, how many calories are in 4 oz of ice cream?

Set up a proportion below.

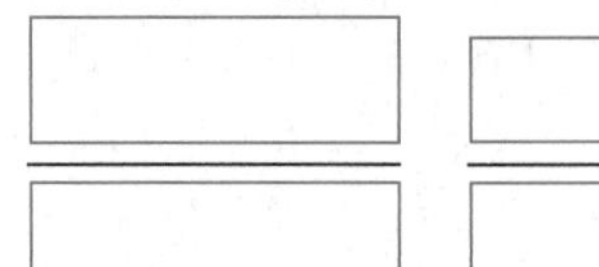

Solve the proportion in the space below.
Solution:

There are ____________ calories in 4 oz of ice cream.

2. Solve.

Solution

a. $\frac{c}{15} = \frac{21}{9}$ ____________

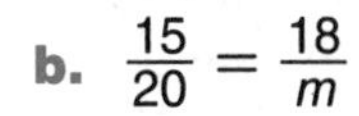

b. $\frac{15}{20} = \frac{18}{m}$ ____________

c. $\frac{w}{15} = \frac{14}{10}$ ____________

3. Use order of operations to evaluate each expression.

a. $9 * 5 / 10 + 3 - 2 =$ ______

b. $8 - 6 * 4 + 8 / 2 =$ ______

c. ______ $= 2 + 2 * 12 + 3^2 - 5$

d. ______ $= 15 / (2 + 3) - 8 * 2$

e. ______ $= 5^2 * 2 + 9 * 2$

4. Which equation describes the relationship between the numbers in the table? Circle the best answer.

x	*y*
0.55	$\frac{1}{2}$
0.6	1
1	5
1.5	10

A. $(y + 0.1) * \frac{1}{2} = x$

B. $(y * 0.1) + \frac{1}{2} = x$

C. $\frac{0.1}{2}y = x$

D. $(y + \frac{1}{2}) * 0.1 = x$

5. The area of square *CAMP* is 25 cm^2. Squares *CAMP* and *DAME* are congruent.

What is the area of triangle *PAE?*

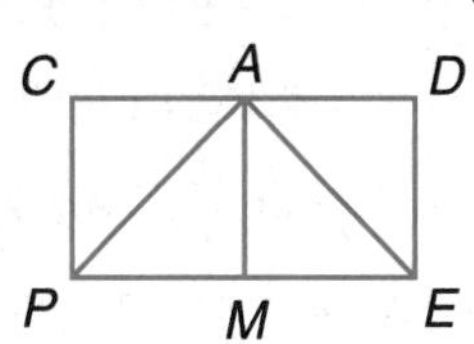

Area = ______ cm^2

Date Time

LESSON 8·6

Ratios

Math Message

Work with a partner. You may use a deck of cards to help you with these problems.

1. There are 2 facedown cards for every faceup card. If 6 of the cards are faceup, how many cards are facedown? ________ cards

2. You have 12 cards. One out of every 4 cards is faceup. The rest are facedown. How many cards are faceup? ________ cards

3. There are 4 facedown cards for every 3 faceup cards. If 8 of the cards are facedown, how many cards are faceup? ________ cards

4. Three out of every 5 cards are faceup. If 12 cards are faceup, how many cards are there in all? ________ cards

5. There are 2 faceup cards for every 5 facedown. If there are 21 cards in all, how many cards are faceup? ________ cards

6. The table at the right shows the average number of wet days in selected cities for the month of October.

City	Wet Days
Beijing, China	3
Boston, United States	9
Frankfurt, Germany	14
Mexico City, Mexico	13
Moscow, Russia	15
Sydney, Australia	12

   a. How many more wet days does Moscow have than Beijing? ________

   b. Moscow has how many times as many wet days as Beijing? ________

   c. The number of wet days in Beijing is what fraction of the number of wet days in Sydney? ________

Try This

7. You have 5 faceup cards and no facedown cards. You add some facedown cards so 1 in every 3 cards is faceup. How many cards are there now? ________ cards

8. You have 5 faceup cards and 12 facedown cards. You add some faceup cards so 2 out of every 5 cards are faceup. How many cards are there now? ________ cards

9. You have 8 faceup cards and 12 facedown cards. You add some faceup cards so $\frac{2}{3}$ of the cards are faceup.

   How many cards are faceup? ________ cards Facedown? ________ cards

Date Time

LESSON 8•6

Math Boxes

1. Solve.

   a. If 9 counters are $\frac{1}{6}$ of a set, how many counters are in the set?

   ____

   b. If $\frac{5}{7}$ of a mystery number is 80, then $\frac{1}{7}$ of the mystery number is ________.

   The mystery number is ________.

2. An artist mixes yellow paint with blue paint to make a certain shade of green. The ratio of yellow to blue is 3 to 5. How much yellow paint should the artist mix with 20 ounces of blue paint?

   Write a proportion. Then solve.

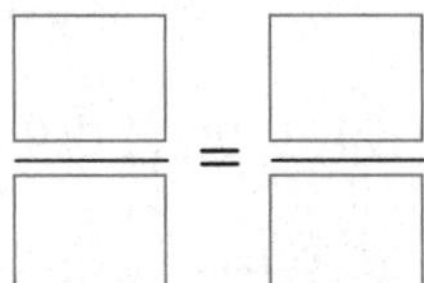

   ______ ounces of yellow paint

3. Janine watches about 12 hours of television per week. Complete the table. Then use your protractor to make a circle graph of the information.

Type of Show	Number of Hours	Percent of Hours	Degrees
Comedy	4		
Educational	1		
News	2		
Sports	3		
Cartoon	2		
Total			

____

(title)

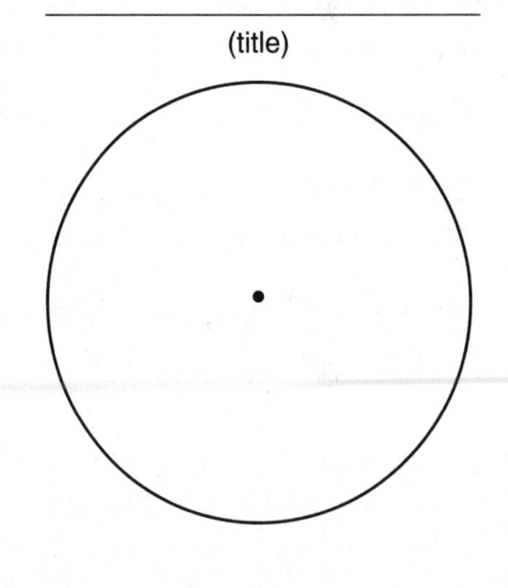

4. Use the general pattern

   $a(b + c) = (a * b) + (a * c)$

   to compute the following products mentally. One has been done for you.

   $3 * 1\frac{2}{3} = (3 * 1) + (3 * \frac{2}{3}) = 3 + 2 = 5$

   a. $9 * 2\frac{1}{9} =$ ____________________

   b. $5 * 2\frac{3}{10} =$ ____________________

5. The formula $F = \frac{9}{5}C + 32$ can be used to convert temperatures from Celsius to Fahrenheit.

   F is the temperature in degrees Fahrenheit.
   C is the temperature in degrees Celsius.

   Calculate the temperature in degrees Fahrenheit for the following Celsius temperatures.

   a. 20°C = ______°F

   b. −25°C = ______°F

Date Time

LESSON 8•6

Ratio Number Stories

For each problem, write and solve a proportion.

1. A spinner lands on blue 4 times for every 6 times it lands on green. How many times does it land on green if it lands on blue 12 times?

   a. $\frac{\text{(lands on blue)}}{\text{(lands on green)}}$ $\frac{\square}{\square} = \frac{\square}{\square}$

   b. It lands on green ______ times.

2. It rained 2 out of 5 days in the month of April. On how many days did it rain that month?

   a. $\frac{\text{(days with rain)}}{\text{(days)}}$ $\frac{\square}{\square} = \frac{\square}{\square}$

   b. It rained ______ days in April.

3. Of the 42 animals in the Children's Zoo, 3 out of 7 are mammals. How many mammals are in the Children's Zoo?

   a. $\frac{\text{(mammals)}}{\text{(animals)}}$ $\frac{\square}{\square} = \frac{\square}{\square}$

   b. Answer: ______

4. Last week, Mikasi spent 2 hours doing homework for every 3 hours he watched TV. If he spent 6 hours doing homework, how many hours did he spend watching TV?

   a. $\frac{\square}{\square}$ $\frac{\square}{\square} = \frac{\square}{\square}$

   b. Answer: ______

5. Five out of 8 students at Lane Junior High School play a musical instrument. If 140 students play an instrument, how many students attend Lane School?

   a. $\frac{\square}{\square}$ $\frac{\square}{\square} = \frac{\square}{\square}$

   b. Answer: ______

6. A choir has 50 members. Twenty members are sopranos. How many sopranos are there for every 5 members of the choir?

   a. $\frac{\square}{\square}$ $\frac{\square}{\square} = \frac{\square}{\square}$

   b. Answer: ______

Date Time

LESSON 8·6

Ratio Number Stories *continued*

7. Mr. Dexter sells subscriptions to a magazine for $18 each. For each subscription he sells, he earns $8. One month, he sold $900 in subscriptions. How much did he earn?

   a. $\frac{\square}{\square}$ $\frac{\square}{\square} = \frac{\square}{\square}$

   b. Answer: ______

8. At Kozminski School, the ratio of weeks of school to weeks of vacation is 9 to 4. How many weeks of vacation do students at the school get in 1 year?

   a. Complete the table.

Weeks of school	9	18	27		
Weeks of vacation	4				
Total weeks	13				

   b. Write a proportion.

   $\frac{\square}{\square}$ $\frac{\square}{\square} = \frac{\square}{\square}$

   c. Answer: ______

9. The class library has 3 fiction books for every 4 nonfiction books. If the library has a total of 63 books, how many fiction books does it have?

   a. $\frac{\square}{\square}$ $\frac{\square}{\square} = \frac{\square}{\square}$

   b. Answer: ______

Try This

10. There are 48 students in the sixth grade at Robert's school. Three out of 8 sixth graders read 2 books last month. One out of 3 students read only 1 book. The rest of the students read no books at all. How many books in all did the sixth graders read last month? ______

    Tell what you did to solve the problem. ______

Date Time

Using Proportions to Solve Problems

Math Message

1. In a recent game, the Mansfield School basketball team took 15 three-point shots and made 6 of them. What percent of its shots did the team make? _________%

2. The team also took 20 two-point shots and made 45% of them. How many two-point shots did the players make? _________ two-point shots

3. The team made 80% of its free throws (one-point shots). If players made 16 free throws, how many free throws did they attempt? _________ free throws

4. How many shots did the team take in all? _________ shots

   How many points did the team score in all? _________ points

---

For each problem, use a variable to represent each part, whole, or percent that is unknown. Complete and solve each proportion.

5. Nigel's dog had a litter of puppies. Three of the puppies were male and 5 were female. What percent of the puppies were male?

   a. $\frac{\text{(male puppies)}}{\text{(total)}} \quad \frac{\square}{\square} = \frac{\square}{100}$

   b. _________% of the puppies were male.

6. The 12 boys in Mr. Stiller's class make up 40% of the class. How many students are in Mr. Stiller's class?

   a. $\frac{\text{(boys)}}{\text{(total)}} \quad \frac{\square}{\square} = \frac{\square}{100}$

   b. There are _________ students in Mr. Stiller's class.

7. Apples are about 85% water. What is the weight of the water in 5 pounds of apples?

   a. $\frac{\text{(weight of water)}}{\text{(weight of apples)}} \quad \frac{\square}{\square} = \frac{\square}{100}$

   b. There are _________ pounds of water in 5 pounds of apples.

Date Time

LESSON 8·7 Using Proportions to Solve Problems *cont.*

For each problem, use a variable to represent each part, whole, or percent that is unknown. Complete and solve each proportion.

8. 24 is what percent of 60?

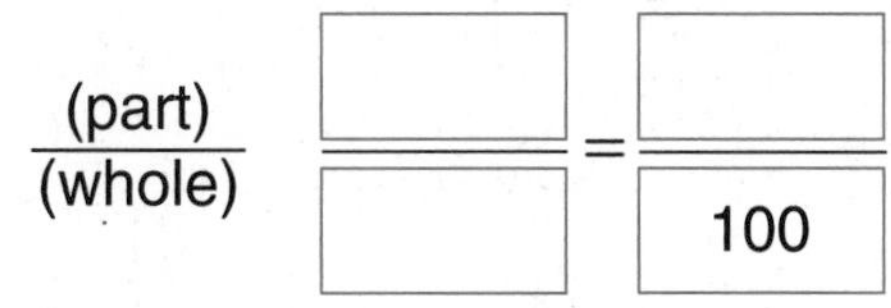

24 is ________% of 60.

9. 54 is what percent of 75?

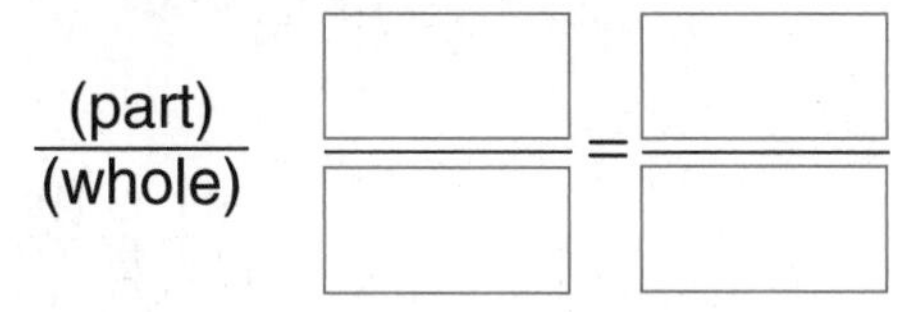

54 is ________% of 75.

10. 1 is what percent of 200?

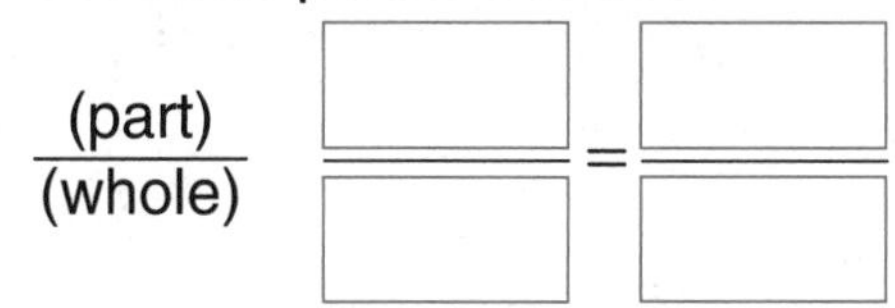

1 is ________% of 200.

11. 75% of what number is 24?

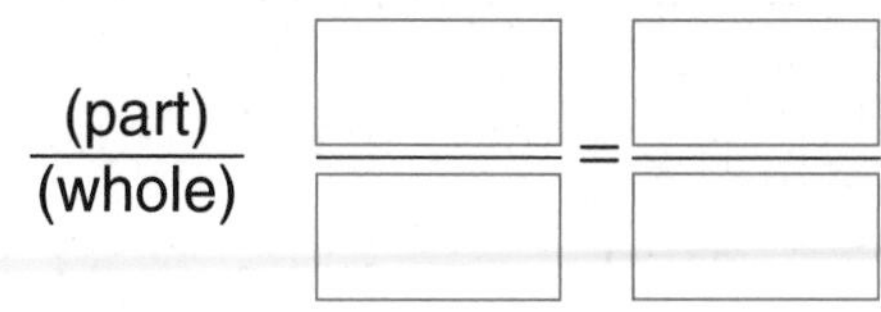

75% of ________ is 24.

12. 20 is 4% of what number?

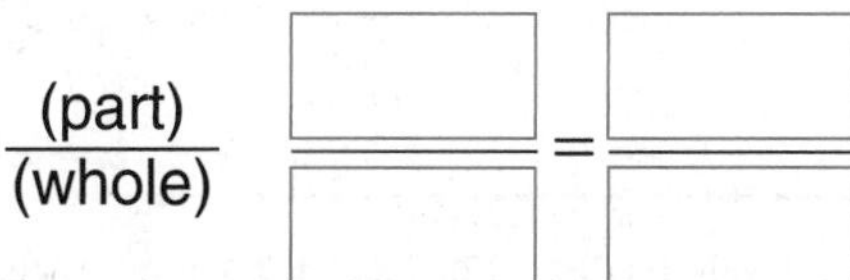

20 is 4% of ________.

Try This

Translate each question into an equation.

13. 75% of what number is 24?

Equation: ________________

14. 20 is 4% of what number?

Equation: ________________

15. Compare each equation you wrote to the cross products in Problems 11 and 12.

What do you notice? ________________________________

____

LESSON 8·7

Math Boxes

1. A kilometer is about $\frac{5}{8}$ mile. About how many miles are in $4\frac{2}{5}$ kilometers? Set up a proportion below.

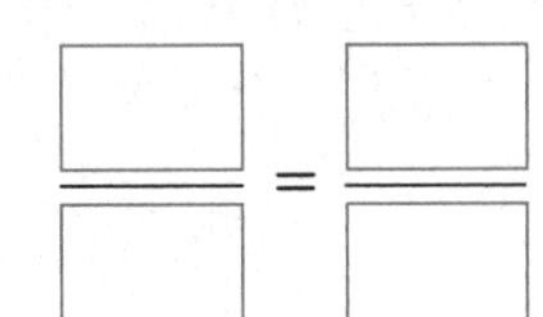

Solve the proportion in the space below.
Solution:

There are about ______ miles in $4\frac{2}{5}$ kilometers.

2. Solve.

Solution

a. $\frac{6}{24} = \frac{5}{n}$ ______

b. $\frac{p}{3.9} = \frac{100}{30}$ ______

c. $\frac{f}{\frac{7}{8}} = \frac{1\frac{1}{7}}{5}$ ______

3. Use order of operations to evaluate each expression.

a. $3 * 8 / 4 + 7 =$ ______

b. $9 + 3 * 5 - 7 =$ ______

c. ______ $= 6 * 5 + 7 * 3$

d. ______ $= 80 / (2 + 8) * 3^3 + 5$

e. ______ $= 28 - 7 * 4 * 0 + 2$

4. Which equation describes the relationship between the numbers in the table? Circle the best answer.

x	y
1	$\frac{3}{8}$
2	$\frac{3}{4}$
8	3
24	9

A. $\frac{y}{8} * 3 = x$

B. $(3 * y) + 8 = x$

C. $\frac{x}{8} * 3 = y$

D. $(3 * x) + 8 = y$

5. The area of triangle *FOG* is 12 cm^2. What is the perimeter of rectangle *FROG?*

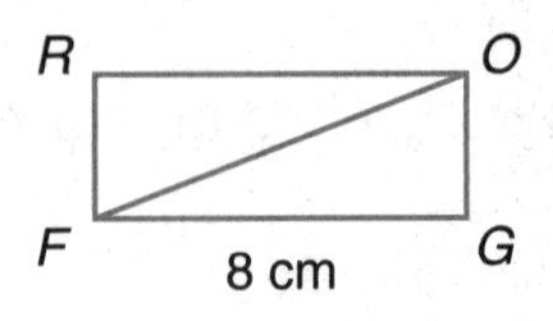

Perimeter = ______ cm

Date Time

LESSON 8·8 Math Boxes

1. Solve.

a. If \$933 is $\frac{1}{3}$ the original price, how much is the original price?

______

b. If $\frac{4}{9}$ of a box is 36 cookies, how many cookies are in the whole box?

______ cookies

2. The ratio of managers to workers in a company is 3 to 11. If there are 42 managers, how many workers are there?

Write a proportion. Then solve.

$$\frac{\square}{\square} \quad \frac{\square}{\square} = \frac{\square}{\square}$$

There are ______ workers.

3. Peabody's Bookstore had a sale. Complete the table. Then use your protractor to make a circle graph of the information.

Book Category	Number Sold	Percent of Total	Degrees
Fiction	280		
Sports	283		
Children's	125		
Travel	212		
Computer	100		
Total			

______
(title)

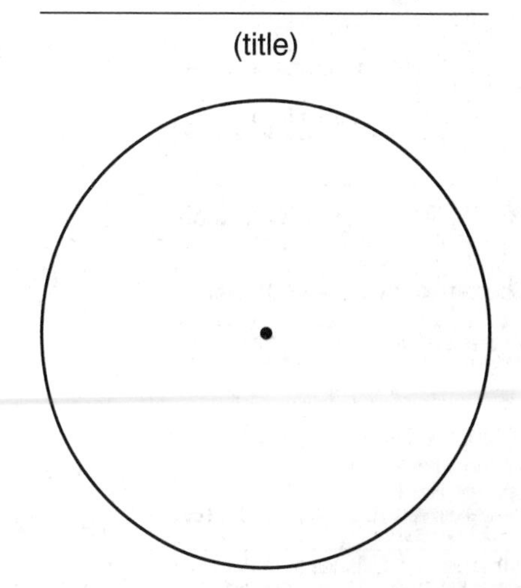

4. Use the general pattern

$x(y - z) = (x * y) - (x * z)$

to compute the following products mentally. One has been done for you.

$8 * 99 = 8(100 - 1) = (8 * 100) - (8 * 1) = 792$

a. $5 * 97 =$ ______

b. $12 * 18 =$ ______

5. The area A of a circle is given by the formula $A = \pi * r^2$, where r is the radius of the circle. Calculate the area of the circle below. Use $\pi = 3.14$.
Circle the best answer.

3 cm

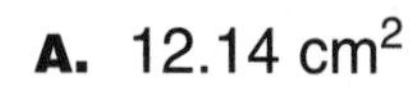
A. 12.14 cm^2

B. 18.84 cm^2

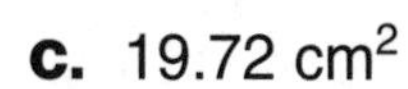
C. 19.72 cm^2

D. 28.26 cm^2

Date Time

LESSON 8·8

The Fat Content of Foods

1. Use the calorie information from the food labels on this page and the next.

   a. Write the ratio of calories that come from fat to the total number of calories as a fraction.

   b. Estimate the percent of total calories that come from fat. Do not use your calculator.

   c. Use your calculator to find the percent of calories that come from fat. (Round to the nearest whole percent.)

Food Label	Food	Calories from Fat / Total Calories	Estimated Fat Percent	Calculated Fat Percent
Nutrition Facts Serving Size 1 slice (28 g) Servings Per Container 12 **Amount Per Serving** **Calories** 90 Calories from Fat 80	Bologna	$\frac{80}{90}$	About 90	89%
Nutrition Facts Serving Size 2 waffles (72 g) Servings Per Container 4 **Amount Per Serving** **Calories** 190 Calories from Fat 50	Waffle			
Nutrition Facts Serving Size 2 tablespoons (32 g) Servings Per Container 15 **Amount Per Serving** **Calories** 190 Calories from Fat 140	Peanut butter			
Nutrition Facts Serving Size 1 slice (19 g) Servings Per Container 24 **Amount Per Serving** **Calories** 70 Calories from Fat 50	American cheese			
Nutrition Facts Serving Size 1 egg (50 g) Servings Per Container 12 **Amount Per Serving** **Calories** 70 Calories from Fat 40	Egg			

Date Time

LESSON 8·8

The Fat Content of Foods *continued*

Food Label	Food	$\frac{\text{Calories from Fat}}{\text{Total Calories}}$	Estimated Fat Percent	Calculated Fat Percent
Nutrition Facts Serving Size 1 cup (60 mL) Servings Per Container 6 **Amount Per Serving** **Calories** 110 Calories from Fat 0	Orange juice			
Nutrition Facts Serving Size 1/2 cup (125 g) Servings Per Container About 3 1/2 **Amount Per Serving** **Calories** 90 Calories from Fat 5	Corn			
Nutrition Facts Serving Size 1 package (255 g) Servings Per Container 1 **Amount Per Serving** **Calories** 280 Calories from Fat 90	Macaroni and cheese			
Nutrition Facts Serving Size 1/2 cup (106 g) Servings Per Container 4 **Amount Per Serving** **Calories** 270 Calories from Fat 160	Ice cream			

2. Compare whole milk to skim (nonfat) milk.

Type of Milk	Total Calories	Calories from Fat	Calories from Carbohydrate	Calories from Protein
1 cup whole milk	160	75	50	35
1 cup skim milk	85	trace	50	35

a. For whole milk, what percent of the total calories come from

fat? _______% carbohydrate? _______% protein? _______%

b. For skim milk, what percent of the total calories come from

fat? _______% carbohydrate? _______% protein? _______%

3. Find the missing percents.

a. 25% + 30% + _______% = 100% **b.** 82% + _______% + 9% = 100%

Date Time

LESSON 8•9 Enlargements

A copy machine was used to make 2X enlargements of figures on the Geometry Template.

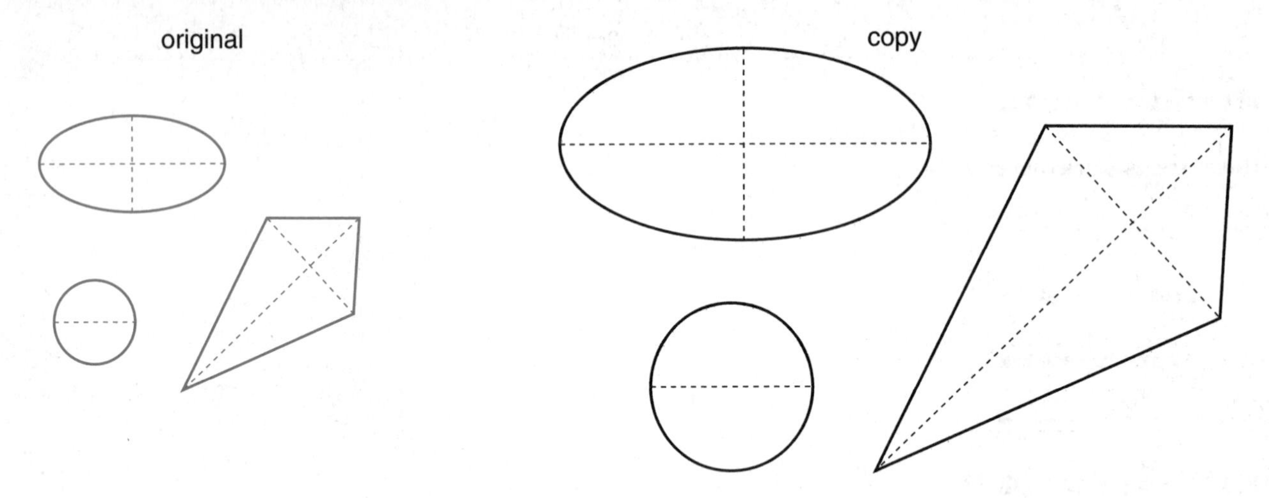

1. Use your ruler to measure the line segments shown in the figures above to the nearest $\frac{1}{16}$ inch. Then fill in the table below.

Line Segment	Length of Original	Length of Enlargement	Ratio of Enlargement to Original
Diameter of circle			
Longer axis of ellipse			
Shorter axis of ellipse			
Longer side of kite			
Shorter side of kite			
Longer diagonal of kite			
Shorter diagonal of kite			

2. Are the figures in the enlargements similar to the original figures? ______________

3. What does a 3.5X enlargement mean? ______________________________

____

____

____

LESSON 8•9

Map Scale

This map shows the downtown area of the city of Chicago. The shaded area shows the part of Chicago that was destroyed in the Great Chicago Fire of 1871.

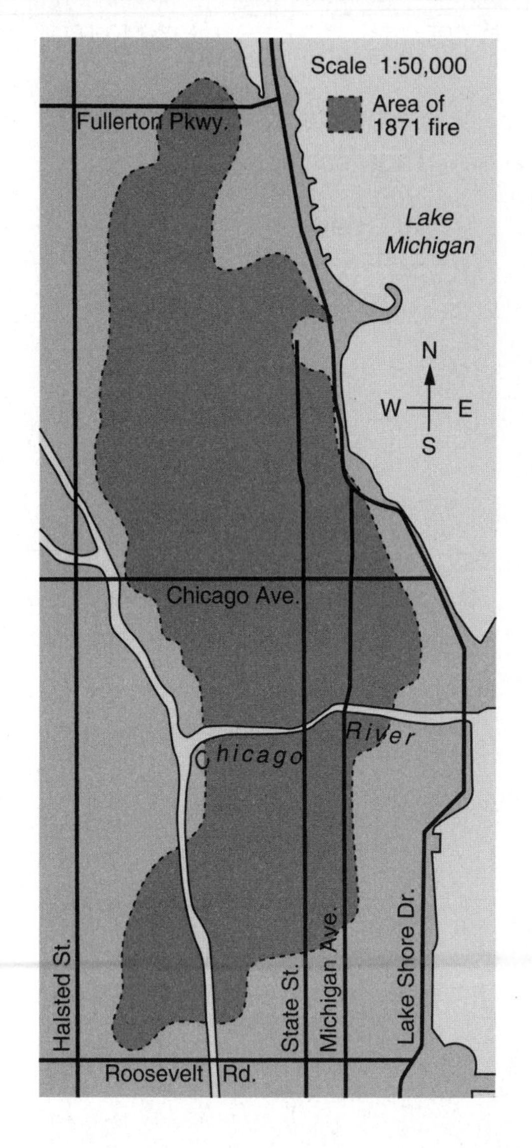

The map was drawn to a scale of 1:50,000. This means that each 1-inch length on the map represents 50,000 inches (about $\frac{3}{4}$ mile) of actual distance.

$$\frac{\text{map distance}}{\text{actual distance}} = \frac{1}{50{,}000}$$

1. Measure the distance on the map between Fullerton Parkway and Roosevelt Road, to the nearest $\frac{1}{4}$ inch. This is the approximate north–south length of the part that burned.

   Burn length on map = ______ inches

2. Measure the width of the part that burned along Chicago Avenue, to the nearest $\frac{1}{4}$ inch. This is the approximate east–west length of the part that burned.

   Burn width on map = ______ inches

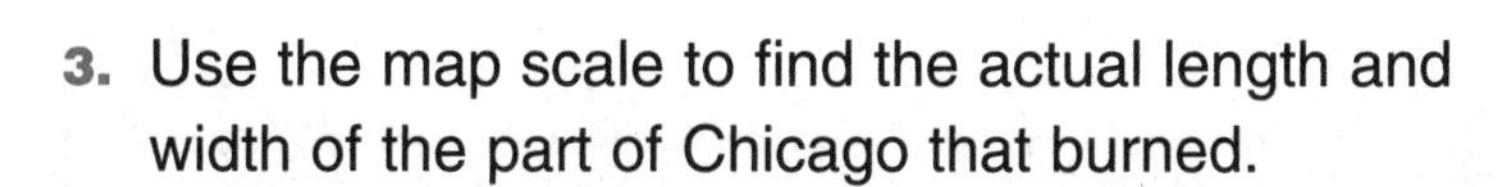

3. Use the map scale to find the actual length and width of the part of Chicago that burned.

   a. Actual burn length = ________________ inches

   b. Actual burn width = ________________ inches

4. Convert the answers in Problem 3 from inches to miles, to the nearest tenth of a mile.

   a. Actual burn length = ______ miles

   b. Actual burn width = ______ miles

5. Estimate the area of the part of Chicago that burned, to the nearest square mile.

   About ______________ square miles

LESSON 8•9

Math Boxes

1. 15 is what percent of 25?

Complete the proportion.
Then solve.

$\frac{\text{(part)}}{\text{(whole)}}$ $\frac{\square}{\square} = \frac{\square}{\square}$

15 is ________% of 25.

2. Find the value of x so each ratio is expressed in terms of a common unit.

a. 4 inches:5 feet = 4 inches:x inches

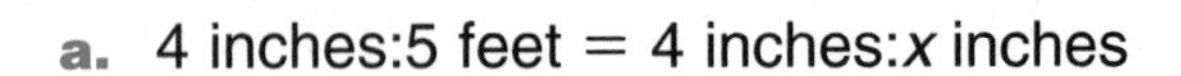

$x =$ ________

b. $\frac{2.4 \text{ m}}{80 \text{ cm}} = \frac{2.4 \text{ m}}{x \text{ m}}$ $x =$ ________

c. 840 mm to 7 cm = x cm to 7 cm

$x =$ ________

3. Subtract. Write your answer as a fraction or a mixed number in simplest form.

a. $7\frac{3}{4} - 3\frac{3}{8} =$ ________

b. ________ $= \frac{5}{2} - 1\frac{5}{6}$

c. ________ $= 5\frac{1}{3} - 2\frac{5}{9}$

d. ________ $= 17 - 13\frac{4}{5}$

4. Write 5 names for the number 10 in the name-collection box. Each name should include the number (−2) and involve subtraction.

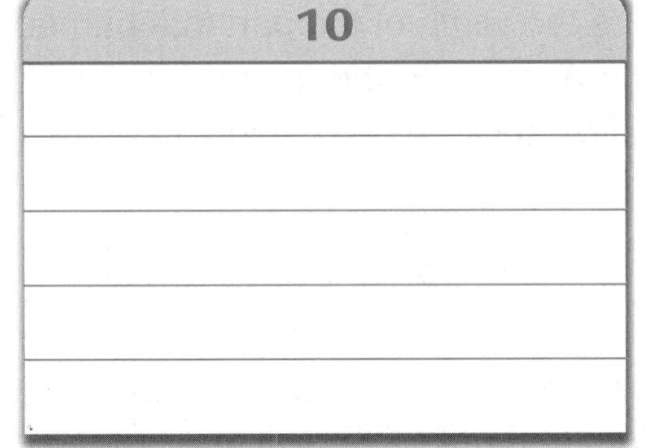

5. The spreadsheet shows how Jonas spent his money for the first quarter of the year.

a. In which cell is the largest amount that Jonas spent?

____

b. Calculate the values for cells E2, E3, and E4 and enter them in the spreadsheet.

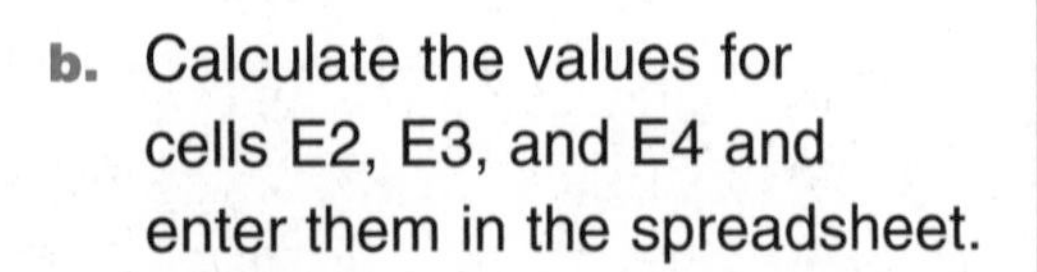

	A	B	C	D	E
1	Month	Food	Movies	Music	Total
2	January	\$38.50	\$34.00	\$62.50	
3	February	\$29.45	\$28.70	\$26.89	
4	March	\$34.90	\$41.86	\$48.30	

c. Circle the correct formula for calculating the amount of money Jonas spent in February.

D1 + D2 + D3 D3 − C2 + C3 B3 + C3 + D3

Date Time

LESSON 8·10

Math Boxes

1. The owner of a restaurant knows that only about 75% of those who make a dinner reservation actually show up. At this rate, how many reservations should the owner take to fill his restaurant of 180 seats?

Write a proportion. Then solve.

Proportion: ______________________

Solution: ________

2. Triangles JKL and PQR are similar.

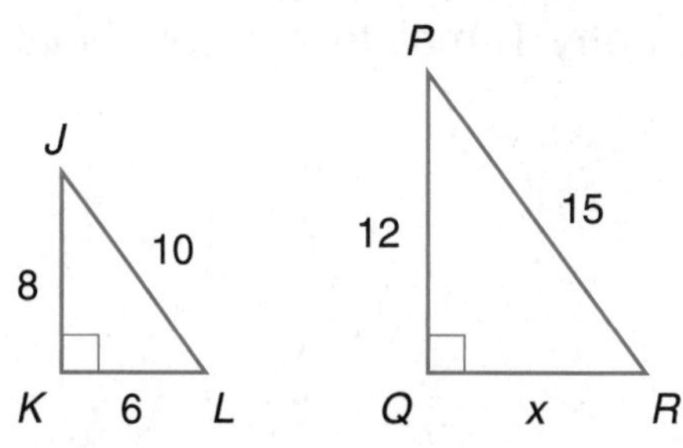

a. Find the ratio $JL:PR$. __________

b. The length of $\overline{QR}$ = ______

3. Rename the fraction to 2 decimal places.

$\frac{5}{7}$

$\frac{5}{7}$ rounded to the nearest hundredth = ________

4. Find the value of $x + y + z$.

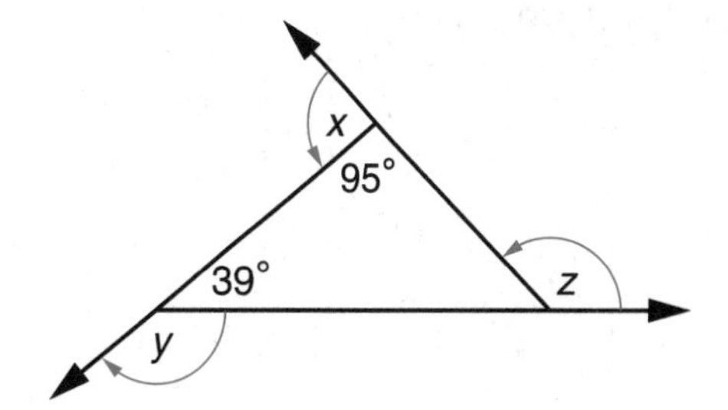

$x + y + z =$ ________

5. Which phrase is represented by the algebraic expression $4 - x$? Fill in the circle next to the best answer.

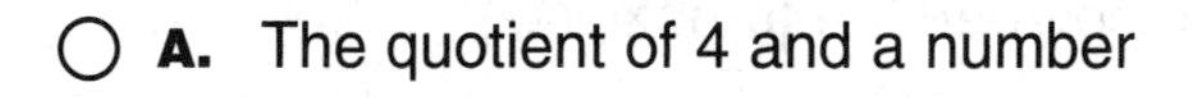

○ **A.** The quotient of 4 and a number

○ **B.** 4 less than a number

○ **C.** The difference of a number and 4

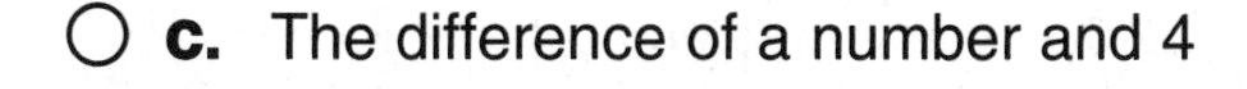

○ **D.** 4 decreased by a number

6. Write an equation to represent the congruent line segments shown below.

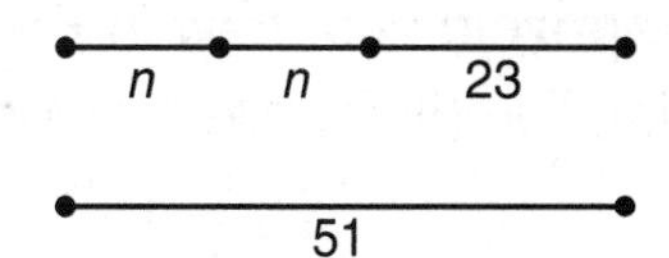

Equation: ______________________

Solve for n.

SRB 251 252

$n =$ ________

LESSON 8·10

Similar Polygons

1. Use pattern-block trapezoids to construct a trapezoid whose sides are twice the length of the corresponding sides of a single pattern-block trapezoid. Then use your Geometry Template to record what you did.

2. Draw a trapezoid whose sides are 3 times the length of a single pattern-block trapezoid. You may use any drawing or measuring tools you wish, such as a compass, a ruler, a protractor, the trapezoid on your Geometry Template, or a trapezoid pattern block.

Which tools did you use? ______________________________

Try This

3. Cover the trapezoid you drew in Problem 2 with pattern-block trapezoids from the Geometry Template. Then record the way you covered the trapezoid.

LESSON 8•10 Similar Polygons *continued*

4. Measure line segments *AB, CD,* and *EF* to the nearest millimeter. Draw a line segment *GH* so the ratio of the lengths of *AB* to *CD* is equal to the ratio of the lengths of *EF* to *GH*.

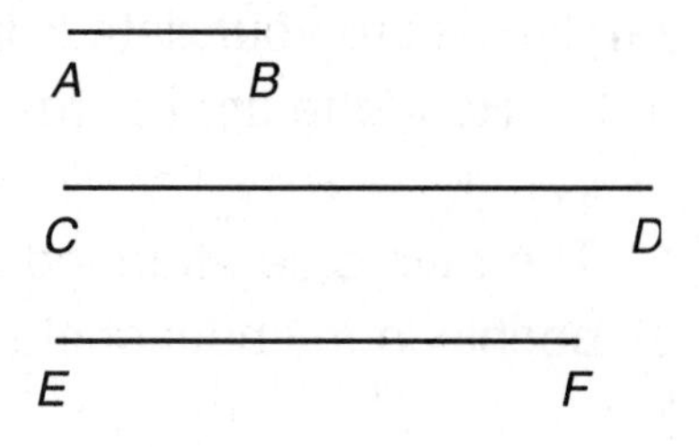

$$\frac{\text{length of } \overline{AB}}{\text{length of } \overline{CD}} = \frac{\text{length of } \overline{EF}}{\text{length of } \overline{GH}}$$

5. Pentagons *PAINT* and *MODEL* are similar polygons. Find the missing lengths of sides.

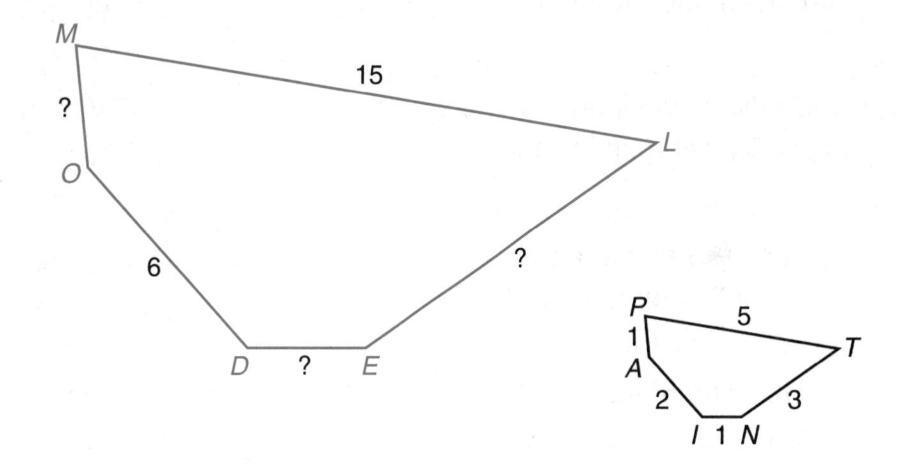

a. Length of side *MO* =

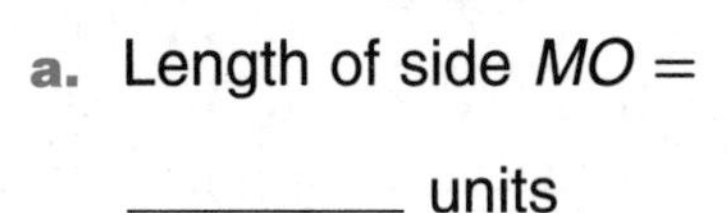

________ units

b. Length of side *EL* =

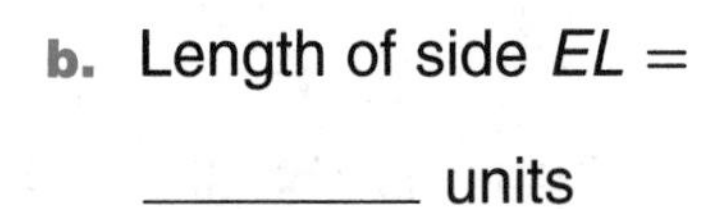

________ units

c. Length of side *DE* =

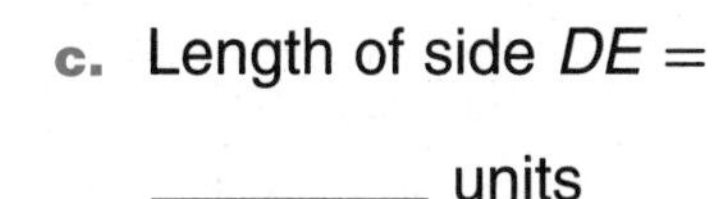

________ units

6. Triangles *PAL* and *CUT* are similar figures. Find the missing lengths of sides.

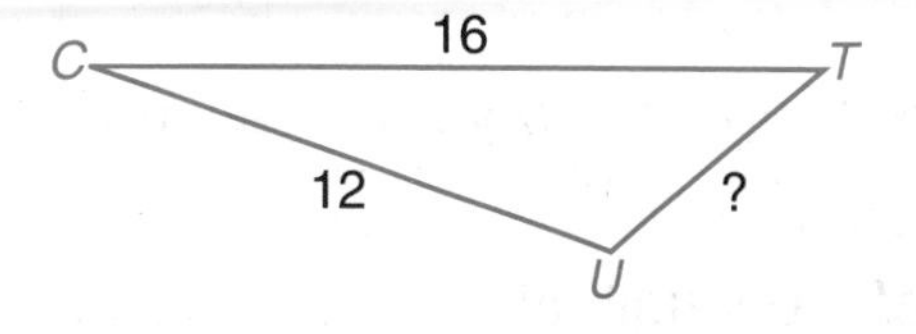

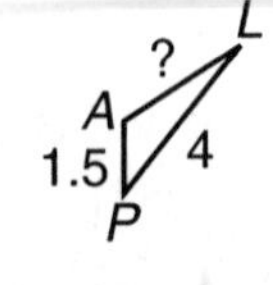

a. Length of side *AL* = ________ units

b. Length of side *UT* = ________ units

7. Alona is looking at a map of her town. The scale given on the map is 1 inch represents $\frac{1}{2}$ mile. Alona measures the distance from her home to school on the map. It's $3\frac{3}{4}$ inches. What is the actual distance from her home to school?

________ miles

8. For a school fair in the cafeteria, Jariah wants to construct a scale model of the 984-foot Eiffel Tower. She plans to use a scale of 1 to 6. Every length of the scale model will be $\frac{1}{6}$ the actual size of the Eiffel Tower. Does this scale seem reasonable? If yes, explain why. If not, suggest a more reasonable scale.

____

____

Date Time

LESSON 8·11

Renaming and Comparing Ratios

Part 1 Record your data from Study Link 8-8 as *n*-to-1 ratios. Use your calculator to divide. Round to the nearest tenth.

1. The ratio of left-handed to right-handed people in my household — $\frac{\text{(left-handed)}}{\text{(right-handed)}}$ $\frac{\square}{\square}$ is about $\frac{\square}{1}$

2. The ratio of the length of the American flag I found to its width — $\frac{\text{(flag length)}}{\text{(flag width)}}$ $\frac{\square}{\square}$ is about $\frac{\square}{1}$

3. The ratio of the length of the screen of my TV set to its width — $\frac{\text{(TV length)}}{\text{(TV width)}}$ $\frac{\square}{\square}$ is about $\frac{\square}{1}$

4. a. The ratio of the length of a small book to its width — $\frac{\text{(small book length)}}{\text{(small book width)}}$ $\frac{\square}{\square}$ is about $\frac{\square}{1}$

   b. The ratio of the length of a medium book to its width — $\frac{\text{(medium book length)}}{\text{(medium book width)}}$ $\frac{\square}{\square}$ is about $\frac{\square}{1}$

   c. The ratio of the length of a large book to its width — $\frac{\text{(large book length)}}{\text{(large book width)}}$ $\frac{\square}{\square}$ is about $\frac{\square}{1}$

   d. What is the shape of a book with a length-to-width ratio of 1 to 1? ____________________

5. a. The ratio of the length of a postcard to its width — $\frac{\text{(postcard length)}}{\text{(postcard width)}}$ $\frac{\square}{\square}$ is about $\frac{\square}{1}$

   b. The ratio of the length of an index card to its width — $\frac{\text{(index card length)}}{\text{(index card width)}}$ $\frac{\square}{\square}$ is about $\frac{\square}{1}$

   c. The ratio of the length of a regular-size envelope to its width — $\frac{\text{(envelope length)}}{\text{(envelope width)}}$ $\frac{\square}{\square}$ is about $\frac{\square}{1}$

   d. The ratio of the length of a business envelope to its width — $\frac{\text{(business envelope length)}}{\text{(business envelope width)}}$ $\frac{\square}{\square}$ is about $\frac{\square}{1}$

   e. The ratio of the length of a sheet of notebook paper to its width — $\frac{\text{(notebook paper length)}}{\text{(notebook paper width)}}$ $\frac{\square}{\square}$ is about $\frac{\square}{1}$

Date Time

LESSON 8•11 Renaming and Comparing Ratios *cont.*

6. Measure the length and width of each rectangle in Problem 6 on Study Link 8-8 to the nearest millimeter. Find the ratio of length to width for each rectangle.

a. $\frac{\text{(length of A)}}{\text{(width of A)}}$

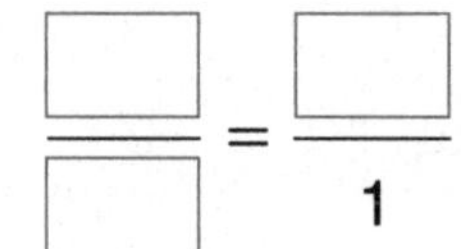

b. $\frac{\text{(length of B)}}{\text{(width of B)}}$

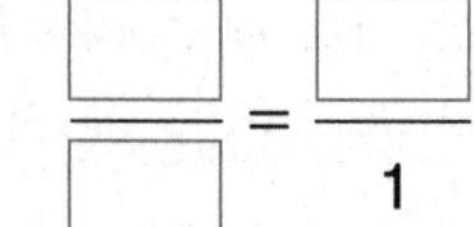

c. $\frac{\text{(length of C)}}{\text{(width of C)}}$

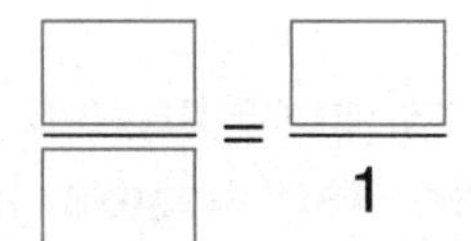

d. $\frac{\text{(length of D)}}{\text{(width of D)}}$

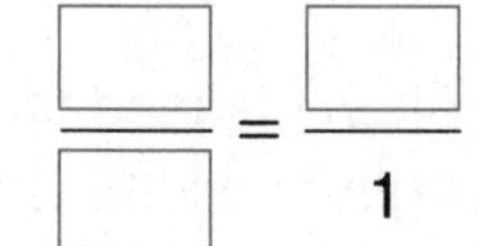

e. Which of the four rectangles was the most popular? ______________

7. The ratio of the rise to the run of my stairs is $\frac{\text{(rise)}}{\text{(run)}}$ $\frac{\square}{\square} = \frac{\square}{1}$

Part 2 Share the data you recorded in Problems 1–7 with the other members of your group. Use these data to answer the following questions.

8. Which group member has the largest ratio of left-handed people to right-handed people at home? ______________ What is this ratio? ______________

9. By law, the length of an official American flag must be 1.9 times its width.

a. Did the flag you measured meet this standard? ______________

b. What percent of the flags measured by your group meet this standard? ______________

c. One of the largest American flags was displayed at the J. L. Hudson store in Detroit, Michigan. The flag was 235 feet by 104 feet. Does this flag meet the legal requirements? ______________

How can you tell? ______________

____

10. For standard television sets, the ratio of length to width of the screen is about 4 to 3. For widescreen TVs, this ratio is about 16:9.

a. Is this true for the television sets in your group? ______________

b. Which television screen, standard or widescreen, is closest to having an *n*-to-1 ratio of 1.6?

____

LESSON 8•11

Renaming and Comparing Ratios *cont.*

11. Compare the ratios for the small, medium, and large books from Problem 4 that were measured by your group.

Which size of books tends to have the largest ratio of length to width? ______________

Which size tends to have the smallest ratio? ______________

12. It is often claimed that the nicest looking rectangular shapes have a special ratio of length to width. Such rectangles are called **Golden Rectangles.** In a Golden Rectangle, the ratio of length to width is about 8 to 5.

A B C D

a. Which of the four rectangles—A, B, C, or D—is closest to having the shape of a Golden Rectangle? ________

b. Did most people in your family choose the Golden Rectangle? ______________

c. Draw a Golden Rectangle in the space at the right whose shorter sides are 2 centimeters long.

d. Which TV screen in Problem 10 on page 317 is closest to a Golden Rectangle? ______________

13. Most stairs in homes have a rise of about 7 inches and a run of about $10\frac{1}{2}$ inches. Therefore, the rise is about $\frac{2}{3}$ the run.

a. Is this true of your stairs? ______________

b. Which stairs would be steeper, stairs with a rise-to-run ratio of 2:3 or 3:2?

____

c. Which member of your group has the steepest stairs? ______________

What is the ratio of rise to run? ______________

d. On the grid at the right, draw stairs whose rise is $\frac{2}{5}$ the run.

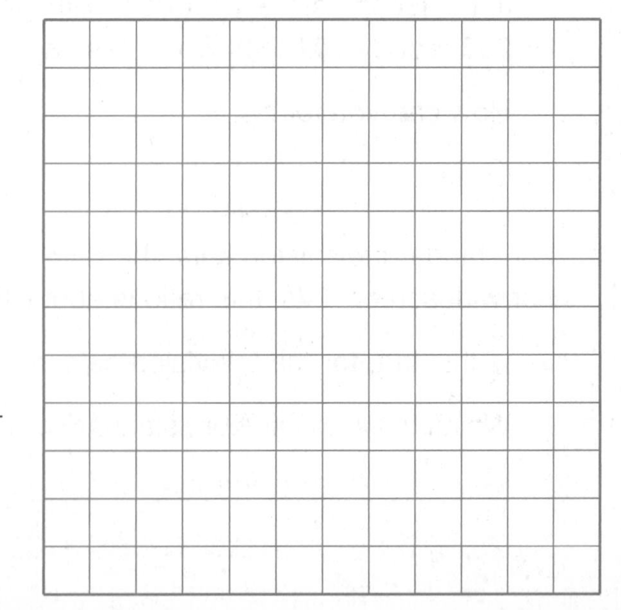

Date Time

LESSON 8·11

Math Boxes

1. What percent of 56 is 14?

Complete the proportion. Then solve.

$\frac{\text{(part)}}{\text{(whole)}} \quad \frac{\square}{\square} = \frac{\square}{\square}$

14 is ________% of 56.

2. Find the value of x so each ratio is expressed in terms of a common unit.

a. 9 inches:4 yards = x yards:4 yards

$x =$ ________

b. $\frac{6 \text{ hours}}{3 \text{ days}} = \frac{x \text{ days}}{3 \text{ days}}$ $x =$ ________

c. 140 quarts to 560 pints = 140 quarts to x quarts

$x =$ ________

3. Subtract. Write your answer as a fraction or a mixed number in simplest form.

a. $\frac{3}{2} - \frac{5}{8} =$ ________

b. ________ $= 4\frac{2}{3} - 1\frac{1}{2}$

c. ________ $= 3\frac{1}{4} - 1\frac{5}{6}$

d. $5\frac{8}{9} - \frac{25}{25} =$ ________

4. Write 5 names for the number 8 in the name-collection box. Each name should include the fraction $\frac{1}{3}$ and involve multiplication.

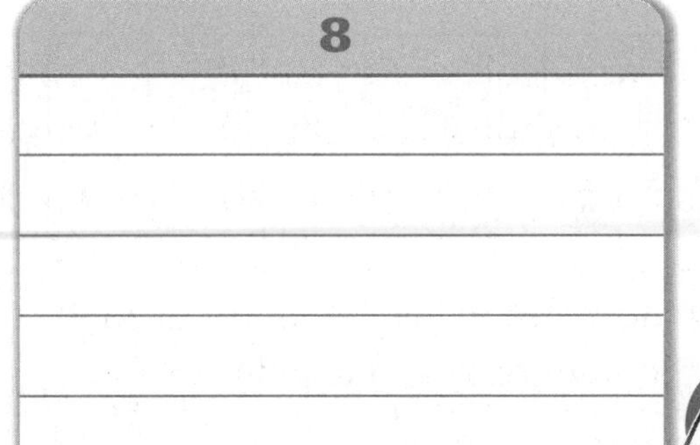

5. The spreadsheet shows Hiroshi's utility bills for 2 months.

a. If Hiroshi entered the wrong amount for the January electric bill, which cell should he correct?

____

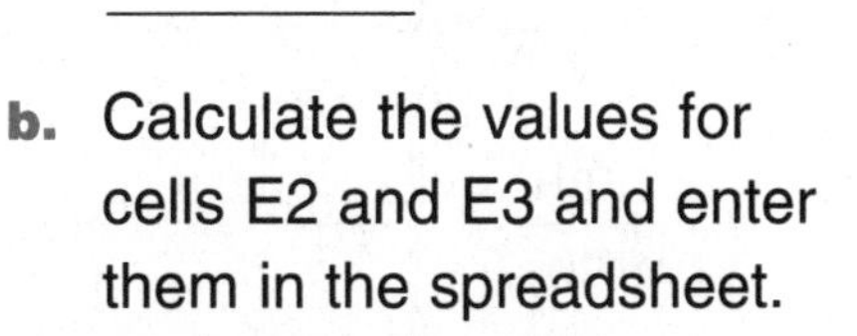

b. Calculate the values for cells E2 and E3 and enter them in the spreadsheet.

	A	B	C	D	E
1	Month	Phone	Electric	Gas	Total
2	January	$17.95	$38.50	$120.50	
3	February	$34.70	$35.60	$148.96	

c. Circle the correct formula for calculating the total cost of the utilities for February.

A2 + B2 + C2 + D2   B3 + C3 + D3   (B3 + C2 + D2) / 3

LESSON 8·12

Rectangle Length-to-Width Ratios

Math Message

1. Draw a large rectangle.

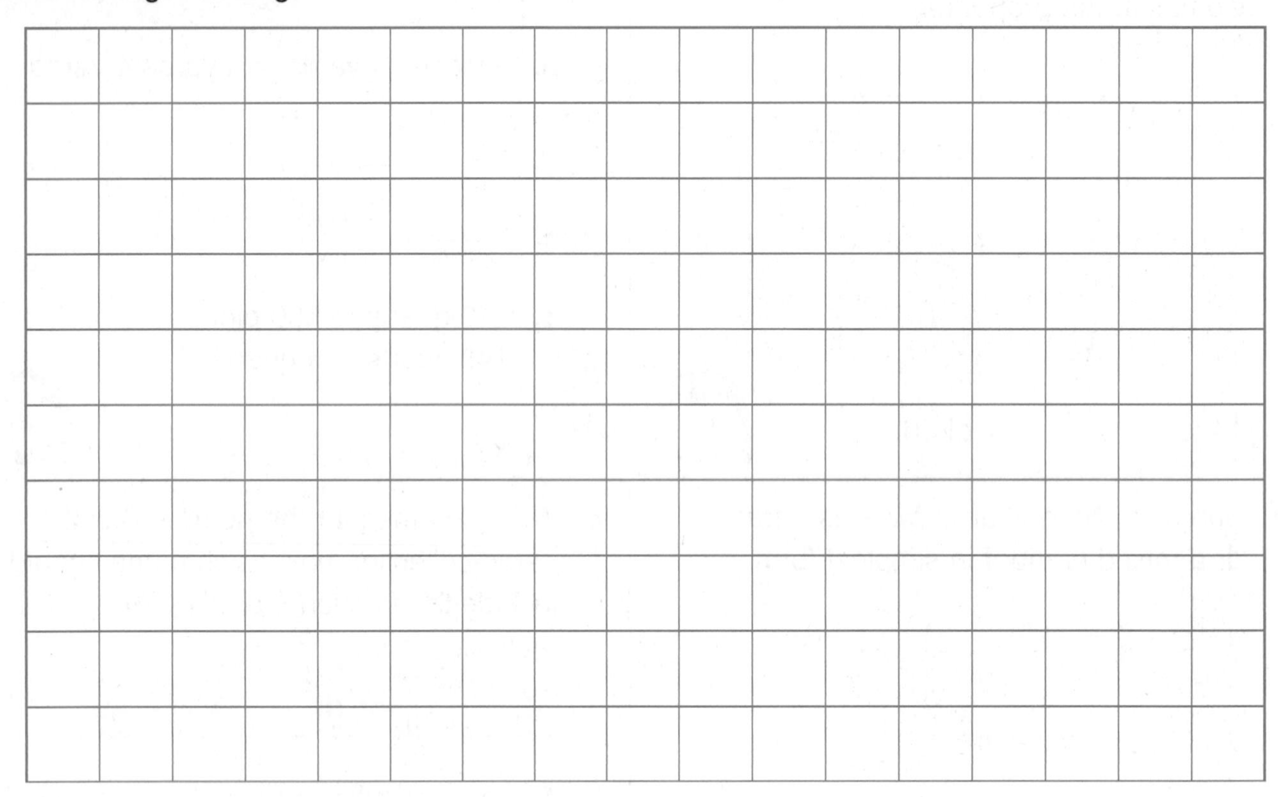

2. Measure the length and width of your rectangle to the nearest millimeter. Calculate the ratio of length to width. (Call the longer side the length and the shorter side the width.)

$\frac{\text{(length)}}{\text{(width)}} \quad \frac{\square}{\square} = \frac{\square}{1}$

3. Using a compass, draw 2 arcs on your rectangle as shown at the right. The arcs are drawn with the compass point at vertices that are next to each other. The compass opening is the same for both arcs.

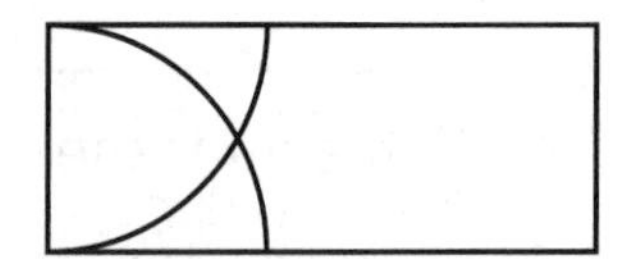

4. Connect the ends of your arcs to make a square. Shade the square. Your rectangle should now look something like this:

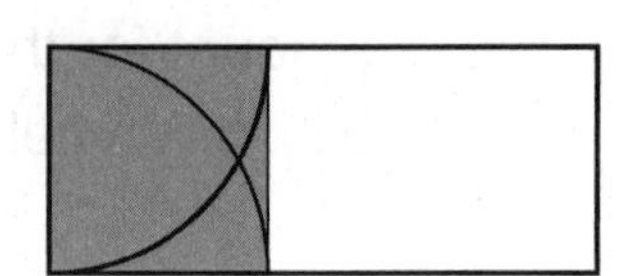

5. Measure the length and width of the unshaded part of your rectangle to the nearest millimeter. Calculate the ratio of the length of the rectangle to its width. (As before, call the longer side the length and the shorter side the width.)

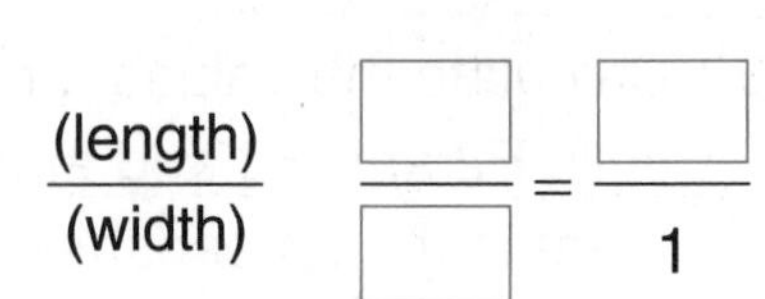

6. Are the ratios you calculated in Problems 2 and 5 equal? ________

LESSON 8•12 Ratios in a Golden Rectangle

1. Measure the length and width of rectangle $ABCD$ to the nearest millimeter. (Call the longer side the length and the shorter side the width.) Calculate the length-to-width ratio to the nearest tenth.

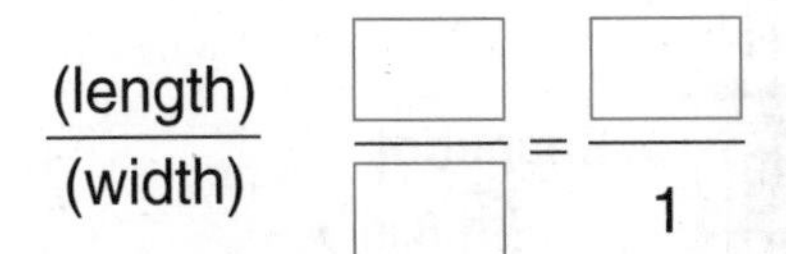

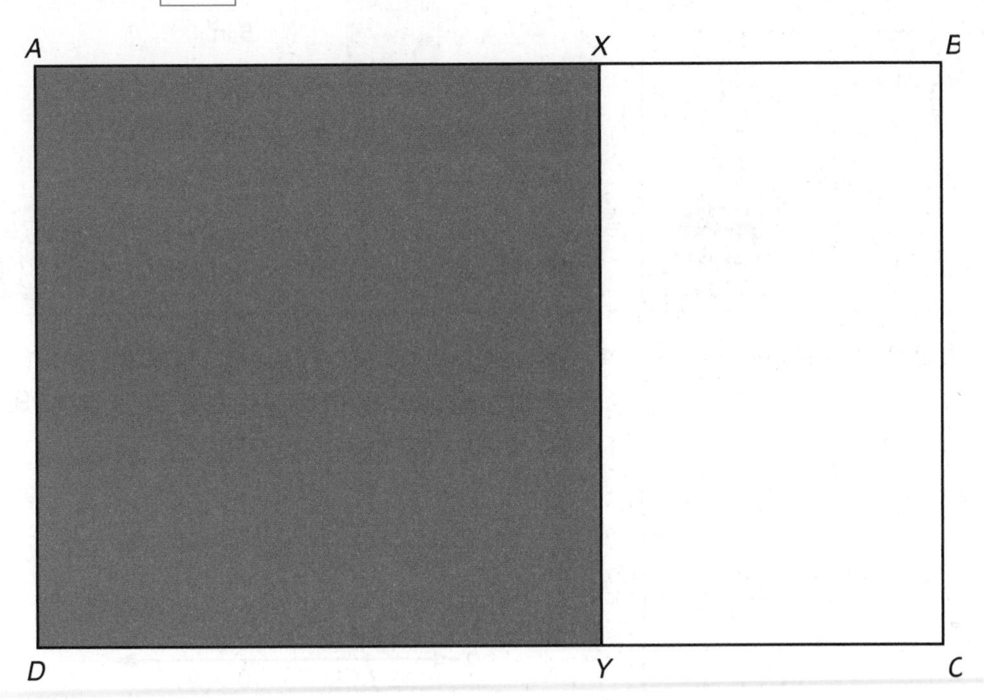

2. Measure the sides of rectangle $AXYD$. What kind of rectangle is $AXYD$?

____

3. Measure the length and width of rectangle $XBCY$. (Call the longer side the length and the shorter side the width.) Calculate the length-to-width ratio.

$\frac{\text{(length)}}{\text{(width)}}$ $\frac{\square}{\square} = \frac{\square}{1}$

4. What do you notice about the ratios you calculated in Problems 1 and 3?

____

____

LESSON 8•12

Math Boxes

1. A scale drawing of a giraffe is 2% of actual size. If the drawing is 12 cm high, what is the actual height of the giraffe?

Write a proportion. Then solve.

Proportion: ______________________

Solution: ____________

2. The snapshot is a reduction of the poster. Find the poster's width. Show your work.

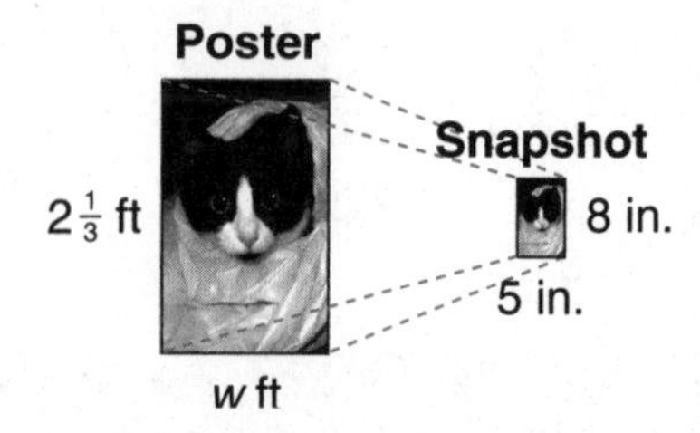

$w =$ ________

3. Rename the fraction to 2 decimal places.

$\frac{9}{13}$

$\frac{9}{13}$ rounded to the nearest hundredth = ________

4. Without using a protractor, find the sum of the angles numbered 1, 3, 5, 7, and 9. (Lines *a* and *b* are parallel.)

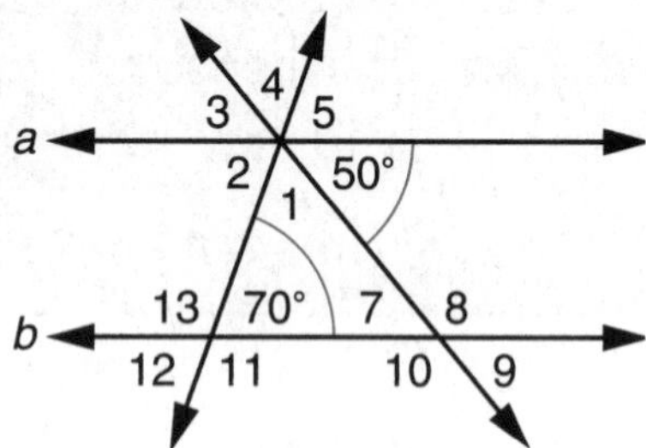

$m\angle 1 + m\angle 3 + m\angle 5 + m\angle 7 + m\angle 9 =$

____

5. Write an algebraic expression for the area of the rectangle below. Use $A = b * h$.

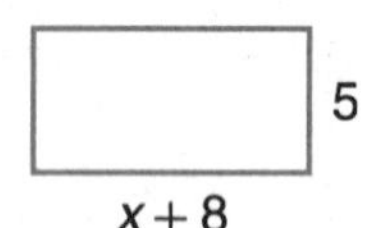

Expression: ______________________

Find the area of the rectangle when $x = 3$.

Area = __________ units^2

6. Write an equation to find the value of *m*.

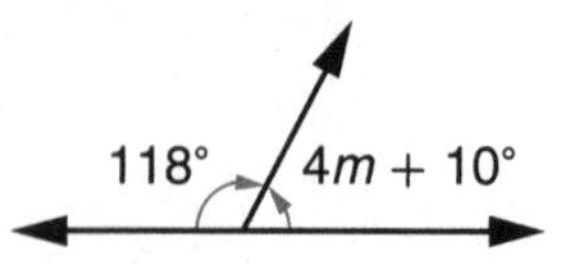

Equation: ______________________

Solve for *m*.

$m =$ ________

LESSON **8•13**

Math Boxes

1. Insert parentheses to make each equation true.

a. $\frac{1}{2} * 18 + 2 * 15 = 150$

b. $9 + 7 \div 4 * 2 = 8$

c. $5 / 3 + 3 * 5 = 4\frac{1}{6}$

d. $6 \div 17 - 11 * 14 = 14$

e. $4 / 9 + 3 * 6 = 2$

2. The circumference of a circle is given by the formula $C = \pi * d$, where C is the circumference and d is the diameter. Circle an equivalent formula.

$d = \frac{C}{\pi}$ $\quad$ $d = C * \pi$

$d = \frac{\pi}{C}$ $\quad$ $d = \pi + C$

Find the circumference for a circle with a diameter of 5 cm. Use 3.14 for π.

$C =$ ________ cm

3. Mr. and Mrs. Gauss keep a record of their expenses on a spreadsheet.

	A	B	C
1	Total of Expense	June	July
2	Rent and Utilities	\$755	\$723
3	Food	\$125	\$189
4	Car Expenses	\$179	\$25
5	Clothing	\$65	\$0
6	Miscellaneous	\$45	\$23
7	Total		

a. If the Gausses entered the wrong amount for car expenses in July, which cell should they correct? ________

b. In which month were the total expenses greater? ________

How much greater? ________

c. Circle the correct formula for the total for June.

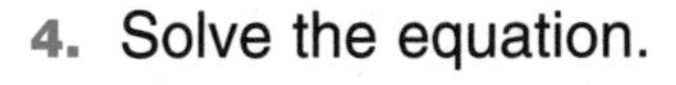
A7 + B7 + C7

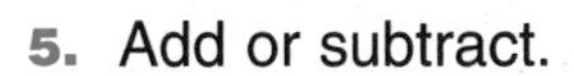
B2 + B3 + B4 + B5 + B6 $\quad$ (B2 + B3 + B4 + B5 + B6) / 5

4. Solve the equation.

$-47 + 6k = 3k - 2$

$k =$ ________

5. Add or subtract.

a. $235 + (-150) =$ ________

b. $-76 - 24 =$ ________

c. ________ $= 143 - 258$

d. ________ $= -99 + 167$

e. ________ $= 380 - (-59)$

Date Time

LESSON 9•1

Two Methods for Finding Areas of Rectangles

Math Message

Rectangle A

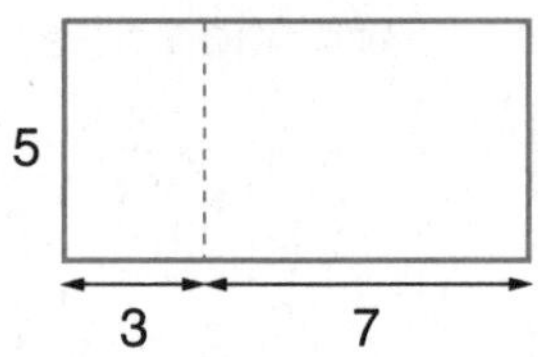

1. What is the area of Rectangle A? ________ square units

   We can express the area of Rectangle A in 4 ways.

   5 * (3 + 7) (5 * 3) + (5 * 7)

   (3 + 7) * 5 (3 * 5) + (7 * 5)

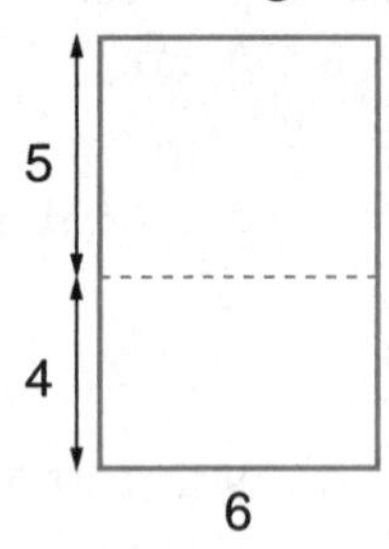

2. Write 2 number sentences to express the area of Rectangle B.

   ________ * (________ + ________) = ________

   (________ * ________) + (________ * ________) = ________

Rectangle C

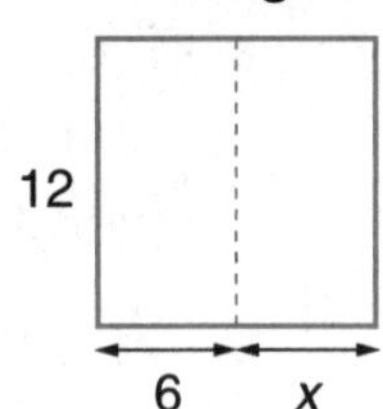

3. The area of Rectangle C is 144 square units.

   a. What is the value of x? ________

   b. Write 2 number sentences to express the area of Rectangle C.

   ________ * (________ + x) = 144

   (________ * ________) + (________ * x) = 144

4. Each of the following expressions describes the area of one of the rectangles below. Write the letter of the rectangle next to the correct expression. The first one has been done for you.

Rectangle D

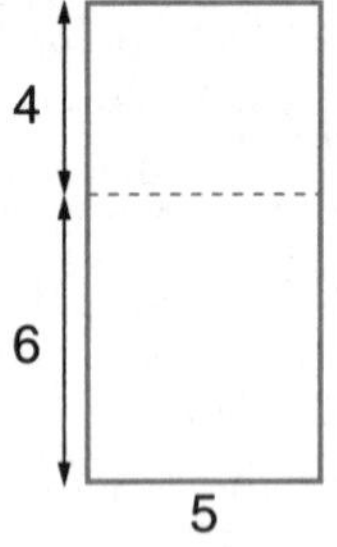

Rectangle E

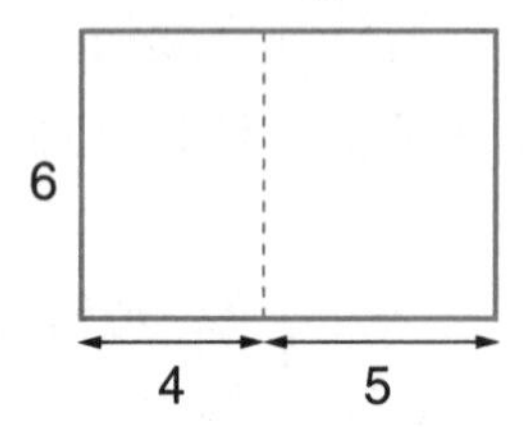

Rectangle F

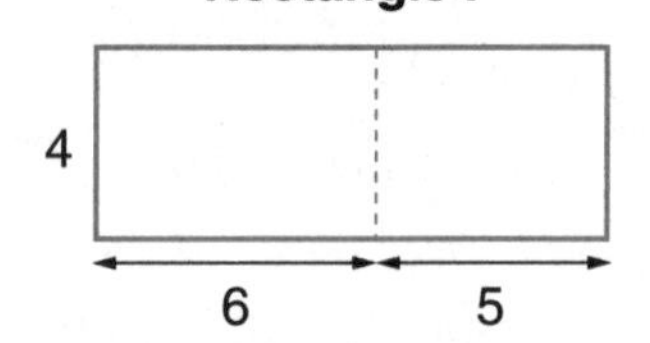

a. 6 * (5 + 4) E

b. (4 + 6) * 5 ________

c. 44 ________

d. 24 + 30 ________

e. (6 * 4) + (5 * 4) ________

f. 50 ________

g. (5 * 6) + (4 * 6) ________

h. 24 + 20 ________

i. (6 + 5) * 4 ________

j. (5 * 6) + (5 * 4) ________

LESSON 9•1

Two Methods for Finding Areas of Rectangles *cont.*

5. What is the area of the shaded part of Rectangle G?

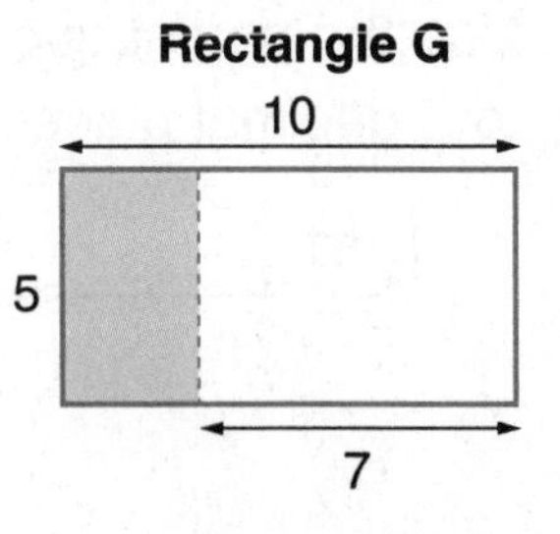

Area of shaded part = ________ square units

We can express the area of the shaded part of Rectangle G with a number sentence in 4 ways.

$5 * (10 - 7) = 15$ $(5 * 10) - (5 * 7) = 15$

$(10 - 7) * 5 = 15$ $(10 * 5) - (7 * 5) = 15$

6. Write 2 number sentences to express the area of the shaded part of Rectangle H.

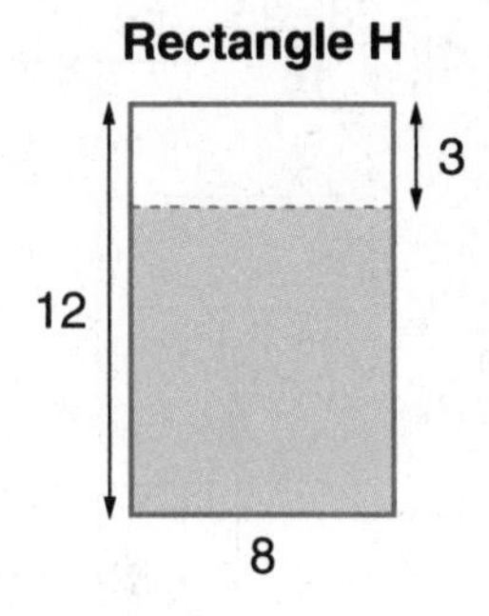

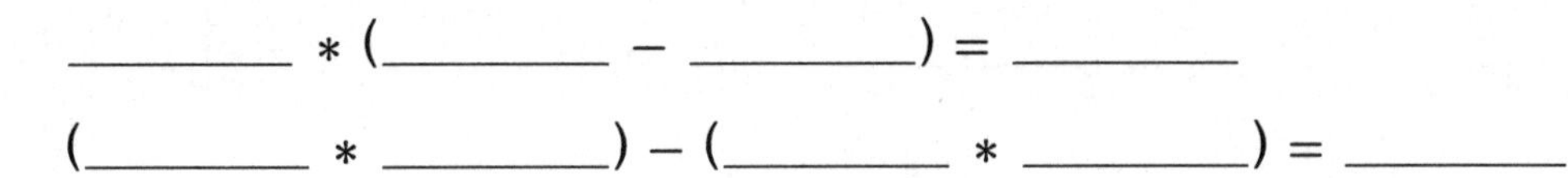

________ * (________ – ________) = ________

(________ * ________) – (________ * ________) = ________

7. The area of Rectangle I is 48 square units.

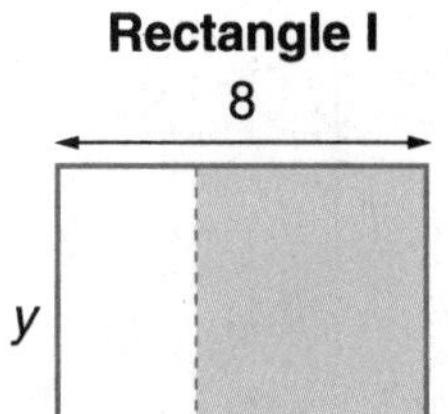

a. What is the value of *y*? ________

b. Write 2 number sentences to express the area of the shaded part of Rectangle I.

(________ – ________) $* y = 30$

(________ $* y$) – (________ $* y$) $= 30$

8. Each of the following expressions describes the area of the shaded part of one of the rectangles below. Write the letter of the rectangle next to the correct expression.

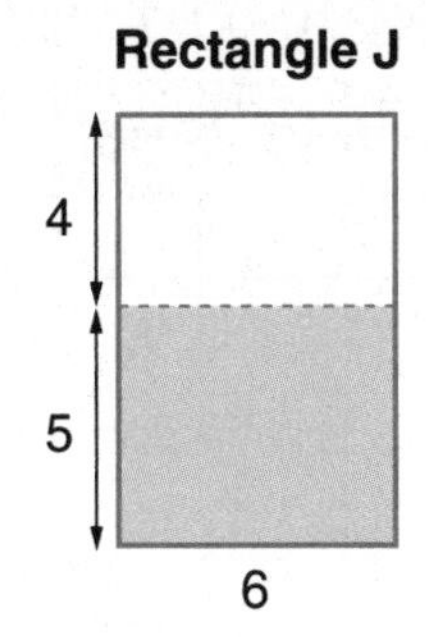

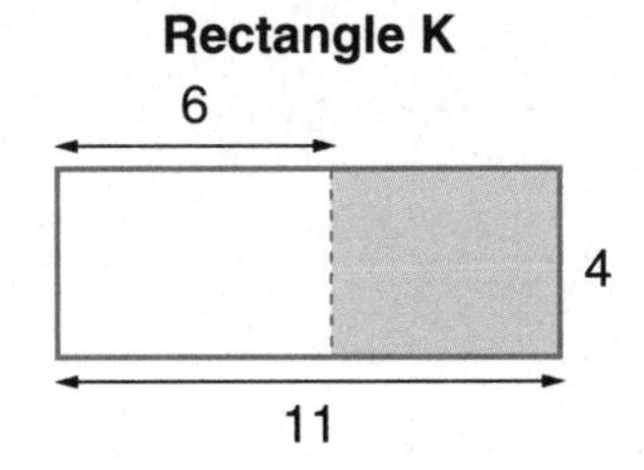

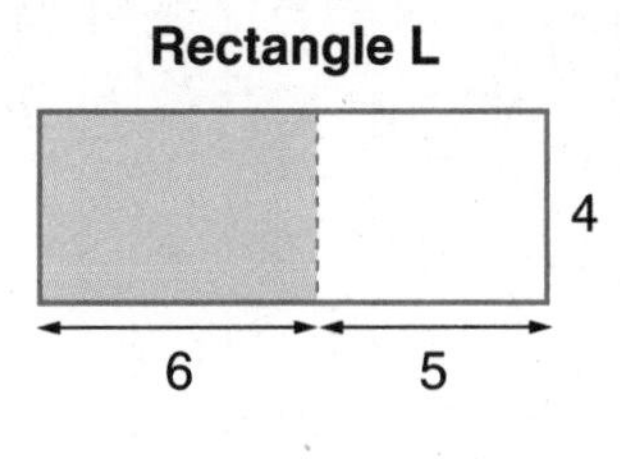

a. $4 * (11 - 6)$ ________

b. $44 - 20$ ________

c. 30 ________

d. $(6 * 9) - (6 * 4)$ ________

e. $(4 * 11) - (4 * 6)$ ________

f. $(11 - 5) * 4$ ________

g. $(11 * 4) - (5 * 4)$ ________

h. $6 * (9 - 4)$ ________

Date Time

LESSON 9·1

Partial-Quotients Division

Use the partial-quotients division algorithm to find quotients that are correct to 2 decimal places. Show your work on the computation grid below.

1. $\frac{1,285}{7}$ ____________

2. 3,709 ÷ 18 ____________

3. $42\overline{)7,956}$ ____________

4. $\frac{282.25}{16}$ ____________

5. 19.015 ÷ 38 ____________

6. $3.8\overline{)746.85}$ ____________

Date Time

LESSON 9•1 Math Boxes

1. Write 2 different number sentences for the area of the shaded part of the rectangle.

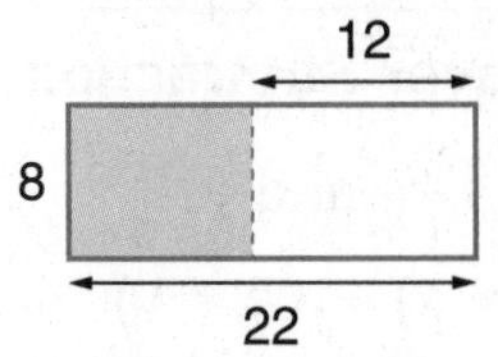

a. (______ − ______) ∗ ______ = 80

b. (______ ∗ ______) − (______ ∗ ______) = 80

2. Solve.

a. $15 = t - 6$

Solution ________

b. $y - (-12) = -7$

Solution ________

c. $6 + 10k = 256$

Solution ________

3. Triangles *QRS* and *XYZ* are similar.

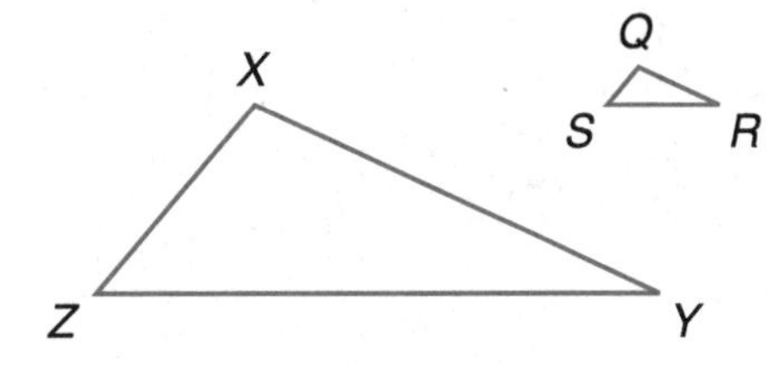

a. Find *SQ* when *SR* = 7, *ZX* = 15, and *ZY* = 35. *SQ* = ______ units

b. Find the size-change factor: $\frac{\text{triangle } XYZ}{\text{triangle } QRS}$ = ______:______

4. Circle all the regular polygons.

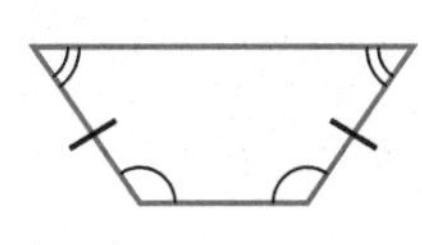
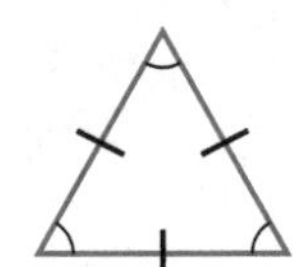
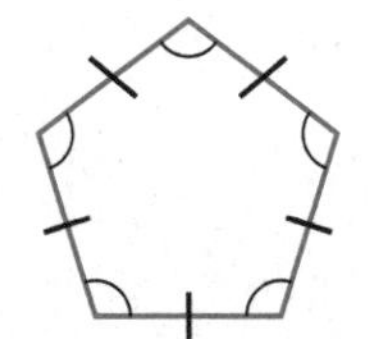
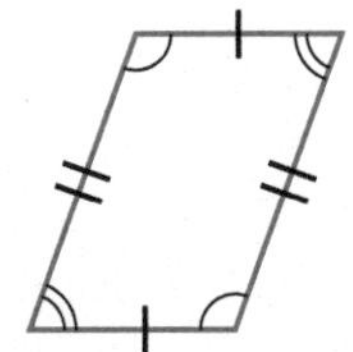

5. Use the diagonals to find the sum of the interior angle measures of the pentagon shown below.

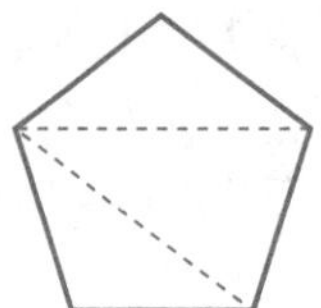

Sum of the interior angle measures = ______ °

Date Time

LESSON 9•2

The Distributive Property

The distributive property is a number property that combines multiplication with addition or multiplication with subtraction. The distributive property can be stated in 4 different ways.

Multiplication over Addition	Multiplication over Subtraction
For any numbers *a*, *x*, and *y*:	For any numbers *a*, *x*, and *y*:
$a * (x + y) = (a * x) + (a * y)$	$a * (x - y) = (a * x) - (a * y)$
$(x + y) * a = (x * a) + (y * a)$	$(x - y) * a = (x * a) - (y * a)$

Use the distributive property to fill in the blanks.

1. $4 * (70 + 8) = (4 * \_\_\_\_) + (4 * \_\_\_\_)$
2. $6 * 34 = (\_\_\_\_ * 30) + (\_\_\_\_ * 4)$
3. $(6 * 70) - (6 * 4) = \_\_\_\_ * (70 - \_\_\_\_)$
4. $(\_\_\_\_ + \_\_\_\_) * 8 = (40 * 8) + (6 * \_\_\_\_)$
5. $8 * (90 + 3) = (\_\_\_\_ * 90) + (8 * 3)$
6. $(50 * 7) + (8 * \_\_\_\_) = (\_\_\_\_ + \_\_\_\_) * 7$
7. $9 * (20 - 7) = (9 * \_\_\_\_) - (\_\_\_\_ * 7)$
8. $4 * (5 + 6) = (\_\_\_\_ * \_\_\_\_) + (\_\_\_\_ * \_\_\_\_)$
9. $(41 + 19) * 7 = (\_\_\_\_ * \_\_\_\_) + (\_\_\_\_ * \_\_\_\_)$
10. $(18 - 4) * r = (18 * \_\_\_\_) - (\_\_\_\_ * r)$
11. $7 * (w - \_\_\_\_) = (\_\_\_\_ * w) - (\_\_\_\_ * 6)$
12. $n * (13 - 27) = (\_\_\_\_ * \_\_\_\_) - (\_\_\_\_ * \_\_\_\_)$
13. $(f - 8) * 15 = (\_\_\_\_ * \_\_\_\_) - (\_\_\_\_ * \_\_\_\_)$
14. $(29 * x) + (12 * x) = (\_\_\_\_ + \_\_\_\_) * \_\_\_\_$
15. $6 * (d - 7) = \_\_\_\_\_\_\_\_\_\_\_\_\_\_\_\_$
16. $5 * (12 - h) = \_\_\_\_\_\_\_\_\_\_\_\_\_\_\_\_$

Date Time

LESSON 9·2

Factoring Sums

The Distributive Property of Multiplication over Addition

$a * (x + y) = a * x + a * y$ $(a * x) + (a * y) = a * (x + y)$

$(x + y) * a = x * a + y * a$ $(x * a) + (y * a) = (x + y) * a$

The equation $21 + 15 = 3 * (7 + 5)$ is a special case of the distributive property in which the addends on the left side are written as whole numbers instead of products. This equation tells you a lot about these numbers. Here are some true statements about the relationships among the numbers and expressions in this equation:

- 3 is a *factor* of both 21 and 15.
- 3 is a *factor* of the expression 21 + 15.
- The expression 21 + 15 is a *multiple* of the expression 7 + 5.

When you use the distributive property to write 21 + 15 as $3 * (7 + 5)$, we say you *factor 3 out of the sum 21 + 15.*

You can use the distributive property to factor a sum of any two whole numbers. In this activity, you will factor out the greatest common factor of two addends. You will be rewriting the original sum as a multiple of another sum whose addends have no common factors other than 1.

Complete the following steps for each sum.

a. Find the greatest common factor of the two addends.

b. Use the distributive property to factor the GCF out of the sum. Complete the number sentence to show the result.

c. Fill in the blanks to give an example of what your number sentence shows.

Example: 16 + 6

a. Greatest common factor: 2

b. Number sentence: 16 + 6 = 2 * (8 + 3)

c. Fill in the blanks: The expression 16 + 6 is a multiple of the expression 8 + 3.

LESSON 9•2

Factoring Sums *continued*

Follow the directions on journal page 328A.

1. 18 + 10

   a. Greatest common factor: ________

   b. Number sentence: 18 + 10 = ________ * (________ + ________)

   c. Fill in the blanks: ________ is a factor of the sum ________ + ________.

2. 30 + 25

   a. Greatest common factor: ________

   b. Number sentence: 30 + 25 = (________ + ________) * ________

   c. Fill in the blank: The expression 30 + 25 is a ______________ of the expression 6 + 5.

3. 48 + 28

   a. Greatest common factor: ________

   b. Number sentence: ________ + ________ = ________ * (________ + ________)

   c. Fill in the blanks: ________ is the greatest common factor of the numbers ________ and 28.

4. 27 + 63

   a. Greatest common factor: ________

   b. Number sentence: ______________________________

   c. Fill in the blanks: When the number ________ is factored out of the sum ______________, the result is the expression 9 * (3 + 7).

5. 36 + 60

   a. Greatest common factor: ________

   b. Number sentence: ______________________________

   c. Write your own sentence.

____

____

Date Time

LESSON 9•2

Math Boxes

1. Use the distributive property to fill in the blanks.

   a. $8 * (30 + 4) =$

   $(\_\_\_\_ * \_\_\_\_) + (\_\_\_\_ * \_\_\_\_)$

   b. $(\_\_\_\_ * 7) + (\_\_\_\_ * 6) = 9 * (7 + 6)$

   c. $(20 + 6) * 10 = (20 * 10) + (6 * \_\_\_\_)$

   d. $\_\_\_\_ (9 + 12) = (5)(9) + (5)(12)$

2. Circle the expressions that represent the area of the rectangle.

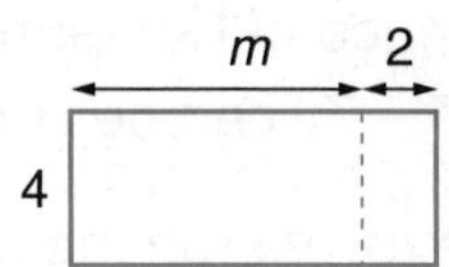

   a. $4m + 8$

   b. $4 * 2m$

   c. $8m$

   d. $(m + 2) * 4$

   e. $4(2 + m)$

   f. $8 + 2m$

3. Find the number.

   a. $\frac{1}{10}$ of what number is 17? ____

   b. $\frac{3}{4}$ of what number is 75? ____

   c. $\frac{2}{5}$ of what number is 14? ____

   d. $\frac{3}{20}$ of what number is 9? ____

   e. $\frac{7}{8}$ of what number is 84? ____

4. Write >, <, or =.

   a. $28 + (-15)$ ____ $36 \div (-2)$

   b. $\frac{1}{2} + (-\frac{3}{4})$ ____ $\frac{2}{3} * \frac{7}{8}$

   c. $-400 * -3$ ____ 20^2

   d. $2 + 15 / 3$ ____ $7 * 10^{-1}$

   e. $\frac{3}{7} + 6\frac{2}{3}$ ____ $\frac{12}{2} \div \frac{7}{9}$

5. Plot and label points on the coordinate grid as directed.

   a. Plot (4,−2). Label it *A*.

   b. Plot (−4,2). Label it *B*.

   c. Draw line segment $\overline{AB}$.

   d. Name the coordinates of the midpoint of $\overline{AB}$.

   (____, ____)

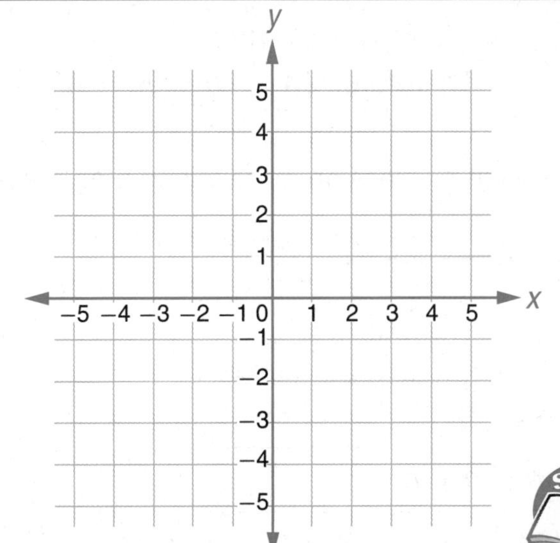

Date Time

LESSON 9·3 Combining Like Terms

Algebraic expressions contain **terms.** For example, the expression $4y + 2x - 7y$ contains the terms $4y$, $2x$, and $7y$. The terms $4y$ and $7y$ are called **like terms** because they are multiples of the same variable, y. To **combine like terms** means to rewrite the sum or difference of like terms as a single term. In the case of $4y + 2x - 7y$, the like terms $4y$ and $-7y$ can be combined and rewritten as $-3y$.

To **simplify an expression** means to write the expression in a simpler form. Combining like terms is one way to do that. *Reminder:* The multiplication symbol ($*$) is often not written. For example, $4 * y$ is often written as $4y$, and $(x + 3) * 5$ as $(x + 3)5$.

Example 1: Simplify the expression $5x - (-8)x$. Use the distributive property.

$$\begin{aligned} 5x - (-8)x &= (5 * x) - (-8 * x) \\ &= (5 - (-8)) * x \\ &= (5 + 8) * x \\ &= 13 * x, \text{ or } 13x \end{aligned}$$

Check your answer by substituting several values for the variable.

Check: Substitute 5 for the variable.

$$\begin{aligned} 5x - (-8)x &= 13x \\ (5 * 5) - (-8 * 5) &= 13 * 5 \\ 25 - (-40) &= 65 \\ 65 &= 65 \end{aligned}$$

Check: Substitute 2 for the variable.

$$\begin{aligned} 5x - (-8)x &= 13x \\ (5 * 2) - (-8 * 2) &= 13 * 2 \\ 10 - (-16) &= 26 \\ 26 &= 26 \end{aligned}$$

If there are more than 2 like terms, you can add or subtract the terms in the order in which they occur and keep a running total.

Example 2: Simplify the expression $2n - 7n + 3n - 4n$.

$2n - 7n = -5n$

$-5n + 3n = -2n$

$-2n - 4n = -6n$

Therefore, $2n - 7n + 3n - 4n = -6n$.

Simplify each expression by rewriting it as a single term.

1. $6y + 13y$ ________
2. $7g - 12g$ ________
3. ________ $5\frac{1}{2}x - 1\frac{1}{2}x$
4. $3c - (-5)c$ ________
5. $5y - 3y + 11y$ ________
6. $6g - 8g + 5g - 4g$ ________
7. $n + n + n + n + n$ ________
8. $n + 3n + 5n - 7n$ ________
9. $2x + 4x - (-9)x$ ________
10. $-7x + 2x + 3x$ ________

LESSON 9·3

Combining Like Terms *continued*

An expression such as $2y + 6 + 4y - 8 - 9y + (-3)$ is difficult to work with because it is made up of 6 different terms that are added and subtracted.

There are 2 sets of like terms in the expression. The terms $2y$, $4y$, and $9y$ are 1 set of like terms. The constant terms 6, 8, and (-3) are a second set of like terms. Each set of like terms can be combined into a single term. To simplify an expression that has more than one set of like terms, combine each set into a single term.

Example 3: Simplify $2y + 6 + 4y - 8 - 9y + (-3)$ by combining like terms.

Step 1 Combine the y terms. $2y + 4y - 9y = 6y - 9y = -3y$

Step 2 Combine the constant terms. $6 - 8 + (-3) = -2 + (-3) = -5$

Final result: $2y + 6 + 4y - 8 - 9y + (-3) = -3y + (-5) = -3y - 5$

Check: Substitute 2 for y in the original expression and the simplified expression.

$$2y + 6 + 4y - 8 - 9y + (-3) = -3y - 5$$
$$(2 * 2) + 6 + (4 * 2) - 8 - (9 * 2) + (-3) = (-3 * 2) - 5$$
$$4 + 6 + 8 - 8 - 18 + (-3) = -6 - 5$$
$$-11 = -11$$

Simplify each expression by combining like terms. Check each answer by substituting several values for the variable.

11. $4 + 7y + 20$ ____________________

12. $5x - 3x + 8$ ____________________

13. $5n + 6 - 8n - 2 - 3n$ ____________________

14. $n + \pi + 2n - \frac{1}{2}\pi$ ____________________

15. $-2.5x + 9 + 1.4x + 0.6$ ____________________

16. $9d + 2a - (-6a) + 3d - 15d$ ____________________

Date Time

LESSON 9•3

Estimating and Measuring in Millimeters

All measurements are approximations. The pencil shown below measures about 4 inches. A more precise measurement is about $3\frac{10}{16}$ inches. An even more precise measurement is about 93 millimeters.

The smaller the unit you use to measure an object, the more precise the measurement will be.

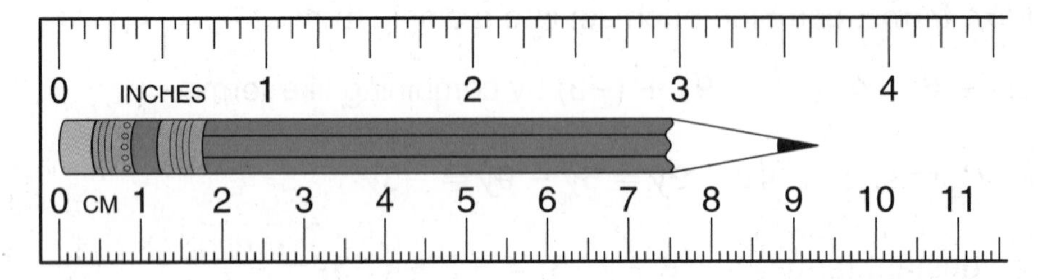

1. Explain why 93 millimeters is a more precise measurement than $3\frac{10}{16}$ inches for the length of the pencil.

____

____

Measure each line segment below to the nearest millimeter.

2. A ———— B Length of $\overline{AB}$ = ________ mm

3. D ———————————— E Length of $\overline{DE}$ = ________ mm

4. Measure $\overline{PQ}$. Then, in the space provided, draw a line segment that is $\frac{5}{8}$ the length of $\overline{PQ}$. Label the segment $\overline{RS}$.

P ——————————— Q Length of $\overline{PQ}$ = ________ mm

Length of $\overline{RS}$ = ________ mm

5. Measure $\overline{FG}$. Then, in the space provided, draw a line segment that is 125% the length of $\overline{FG}$. Label the segment $\overline{JK}$.

F ——————————— G Length of $\overline{FG}$ = ________ mm

Length of $\overline{JK}$ = ________ mm

Date Time

LESSON 9•3 Math Boxes

1. The area of the rectangle is 238 $units^2$.

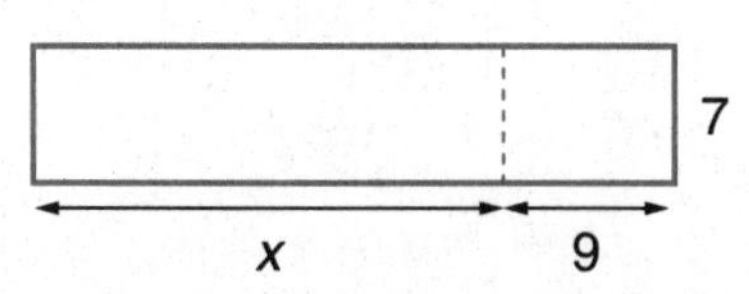

Write a number sentence to find the value of *x*.

Number sentence ____________

Solve for *x*.

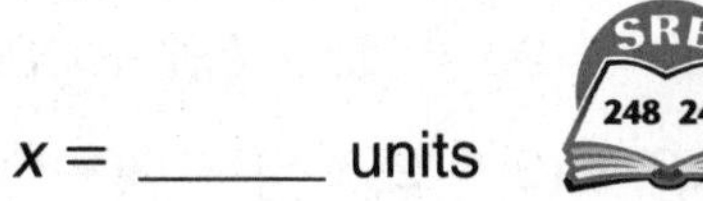

$x =$ ______ units

2. Solve.

a. $-100 = 12w - 4$

Solution ______

b. $-24 + p + 8 = 4$

Solution ______

c. $4j - 24 - 6j = 120$

Solution ______

3. Triangles *ABC* and *DEF* are similar.

a. Length of $\overline{AB}$ = ______ units

b. Length of $\overline{DF}$ = ______ units

c. $\frac{\text{Perimeter of triangle } ABC}{\text{Perimeter of triangle } DEF}$ = ______ units

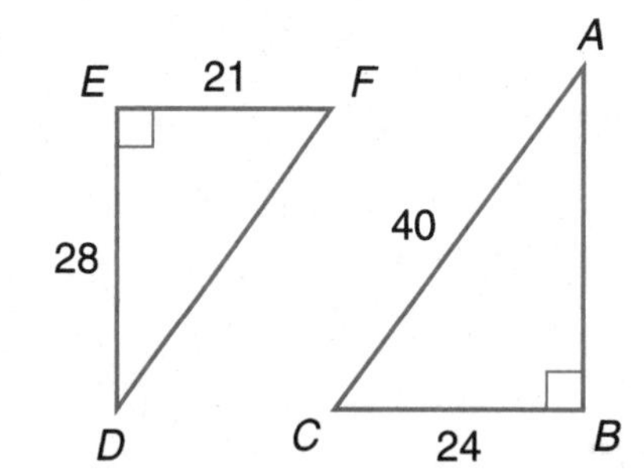

4. I am a regular polygon with all obtuse angles. I have the smallest number of sides of any polygon with obtuse angles. How many sides do I have?

______

Use your Geometry Template to draw this polygon below.

5. Use the diagonals to find the sum of the interior angle measures of the octagon below.

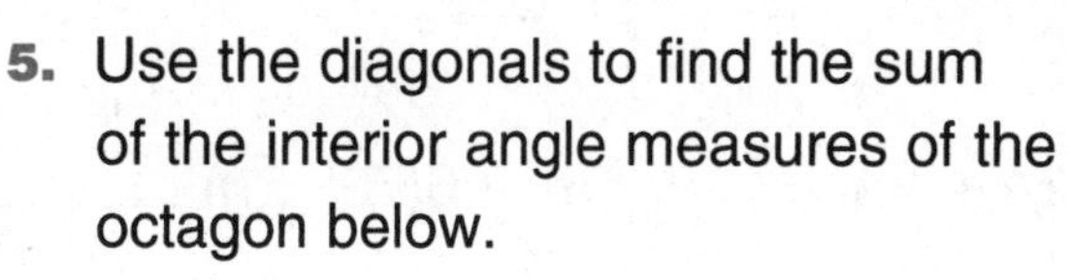

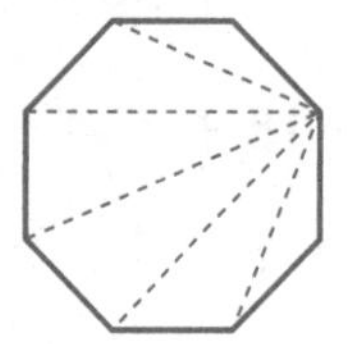

Sum of the interior angle measures = ______ °

Date Time

LESSON 9•4 Simplifying Algebraic Expressions

Simplify the following expressions. First use the distributive property to remove the parentheses. Then combine like terms. Check the answer by substituting a value for the variable.

Example: Simplify $20 * (3 + 2x) + 30x$.

Step 1 Use the distributive property to remove the parentheses.

$$20 * (3 + 2x) + 30x = (20 * 3) + (20 * 2x) + 30x$$
$$= 60 + 40x + 30x$$
$$= 60 + 70x$$

Step 2 Simplify the expression by combining like terms.

Therefore, $20 * (3 + 2x) + 30x = 60 + 70x$

Check by substituting 5 for x.

$$20 * (3 + 2 * 5) + 30 * 5 = 60 + 70 * 5$$
$$20 * (3 + 10) + 150 = 60 + 350$$
$$20 * 13 + 150 = 410$$
$$260 + 150 = 410$$
$$410 = 410$$

1. $7 + (5 - 3) * x + 1$ ________________

2. $2(g - 1) + 1 - 5g$ ________________

3. $\frac{1}{2}(2m + 1) + \frac{1}{2}$ ________________

4. $n + 2n + 3n + (4 + 5)n + 6(7 + 2n)$ ________________

Try This

5. $6(p - 7) - 5p + 15 + (3p + 2)4$ ________________

6. $12.4(2f - 5) - 11.6(3f - 2) + 3.4(0.5f + 2)$ ________________

Date Time

LESSON 9·4

Simplifying Equations

Simplify both sides of the following equations. Do NOT solve them.

Example: $2b + 5 + 3b = 8 - b + 21$

$(2b + 3b) + 5 = 8 - b + 21$

$5b + 5 = 8 - b + 21$

$5b + 5 = -b + 21 + 8$

$5b + 5 = -b + 29$

1. $5h + 13h = 20 - 2$

2. $2 + x + 2x + 4 = x + 16$

3. $2(y + 2) = 4(y + 3)$

4. $(4 - 1)m - m = (m - 1) * 4$

5. $4y + 6 = 8(1 + y)$

6. $5(x + 3) - 2x = 35 + x$

7. $3 * (3.2 - 2c) = 4.6 + 4c$

8. $4(2z - 5) = z + 1$

Date Time

LESSON 9•4

Math Boxes

1. Simplify each expression.

a. $8(2y - 3) - 6y$ ______

b. $18 - 4(8m + 5)$ ______

c. $-(6w - 1) + 3(w - 4)$

______

2. Circle the expressions that represent the area of the shaded rectangle.

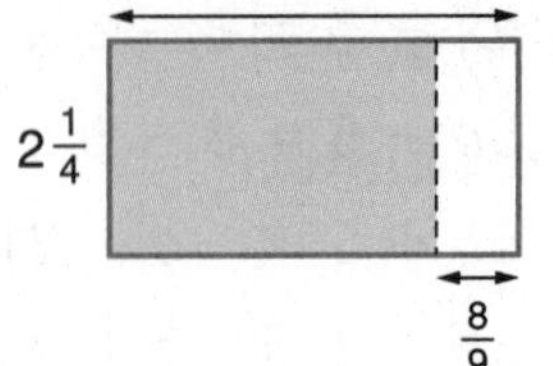

a. $(2 - b)$ b. $\frac{9}{4}(b - \frac{8}{9})$ c. $2b$

d. $\frac{9}{4}b$ e. $2\frac{1}{4}b - 2$ f. $2 - 2\frac{1}{4}b$

3. Find the number.

a. 75% of what number is 96? ______

b. 40% of what number is 30? ______

c. $33\frac{1}{3}$% of what number is 48? ______

d. 10% of what number is 11? ______

e. $12\frac{1}{2}$% of what number is 50? ______

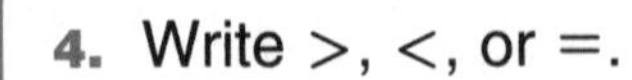

4. Write >, <, or =.

a. $12 - (-3)$ ______ $\frac{7}{8} \div \frac{1}{20}$

b. $5^2 + 3^3$ ______ $5\frac{20}{4} + 10\frac{50}{10}$

c. $3\frac{6}{7} + 2\frac{3}{5}$ ______ $\frac{100}{18}$

d. $0.48 * 2.5$ ______ $3 * 0.26$

e. $-4 * -8$ ______ $-(2^5)$

5. Plot and label points on the coordinate grid as directed.

a. Plot (−4,−3). Label it *L*.

b. Plot (4,−1). Label it *M*.

c. Draw line segment *LM*.

d. Name the coordinates of the midpoint of $\overline{LM}$.

(______,______)

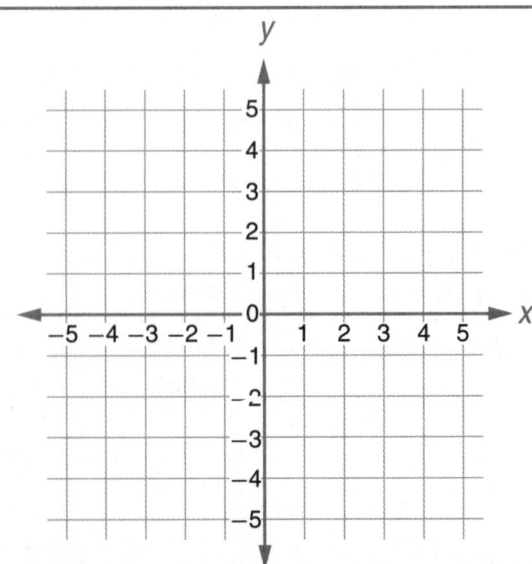

SRB 234

Date Time

LESSON 9•5

Number Stories and the Distributive Property

Solve each problem mentally. Then record the number model you used.

1. A carton of milk costs $0.60. John bought 3 cartons of milk one day and 4 cartons the next day.

   How much did he spend in all? ____________

   Number model ____________

2. During a typical week, Karen runs 16 miles and Jacob runs 14 miles.

   About how many miles in all do Karen and Jacob run in 8 weeks? ____________

   Number model ____________

3. Mark bought 6 CDs that cost $12 each. He returned 2 of them.

   How much did he spend in all? ____________

   Number model ____________

4. Max collects stamps. He had 9 envelopes, each containing 25 stamps. He sold 3 envelopes to another collector.

   How many stamps did he have left? ____________

   Number model ____________

5. Jean is sending party invitations to her friends. She has 8 boxes with 12 invitations in each box. She has already mailed 5 boxes of invitations.

   How many invitations are left? ____________

   Number model ____________

Date Time

LESSON 9·5 Simplifying and Solving Equations

Simplify each equation. Then solve it. Record the operations you used for each step.

1. $6y - 2y = 40$

Solution ________________

2. $5p + 28 = 88 - p$

Solution ________________

3. $8d - 3d = 65$

Solution ________________

4. $12e - 19 = 7 - e$

Solution ________________

5. $3n + \frac{1}{2}n = 42$

Solution ________________

6. $3m - 1 + m + 6 = 2 - 9$

Solution ________________

7. $3(1 + 2y) = y + 2y + 4y$

Solution ________________

8. $8 - 12x = 6 * (1 + x)$

Solution ________________

9. $-4.8 + b + 0.6b = 1.8 + 3.6b$

Solution ________________

10. $4t - 5 = t + 7$

Solution ________________

Date Time

LESSON 9·5 Simplifying and Solving Equations *continued*

11. $8v - 25 = v + 80$

Solution ____________

12. $3z + 6z = 60 - z$

Solution ____________

13. $g + 3g + 32 = 27 + 5g + 2$

Solution ____________

14. $16 + 3s - 2s = 24 + 2s - 20$

Solution ____________

15. Are the following 2 equations equivalent? ________

$5y + 3 = -6y + 4 + 12y$ $\quad\quad$ $5y + 3 = -6y + 4(1 + 3y)$

Explain your answer. ____

____

____

16. Are the following 2 equations equivalent? ________

$5(f - 2) + 6 = 16$ $\quad\quad$ $f - 1 = 3$

Explain your answer. ____

____

____

Try This

17. Solve $\frac{2z + 4}{5} = z - 1$

(*Hint:* Multiply both sides by 5.)

Solution ____________

Date Time

LESSON 9•5

Math Boxes

1. The area of the shaded part of the rectangle is 20 units^2.

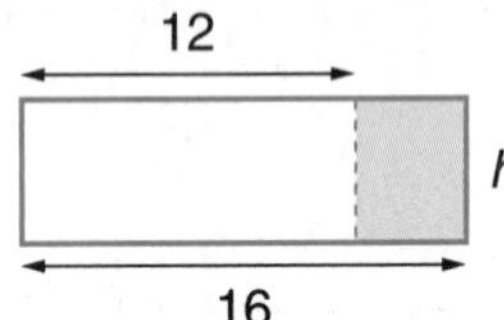

Write a number sentence to find the value of h.

Number sentence: ______________

Solve for h.

h = _____ units

2. Solve.

a. $\frac{1}{3}f - 6 = -8$

Solution ________

b. $-30 = b - 6 + 11b$

Solution ________

c. $4g + 4 = 2g + 36$

Solution ________

3. Triangles THG and TIN are similar.

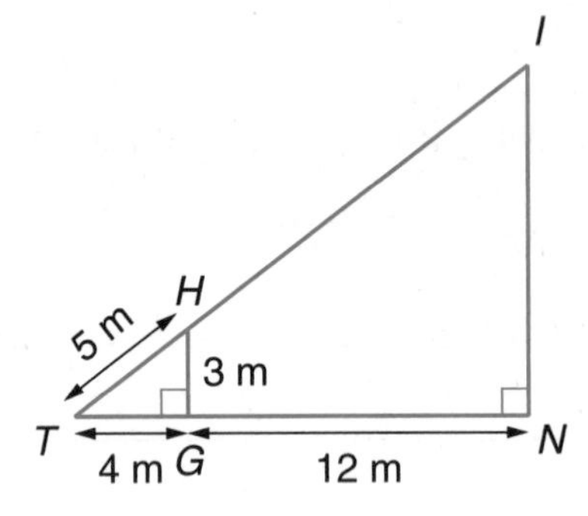

a. Length of $\overline{IN}$ = ________ m

b. Length of $\overline{HI}$ = ________ m

c. The size-change factor: $\frac{\text{triangle } THG}{\text{triangle } TIN}$ = ______:______

4. I am a quadrangle with 2 pairs of congruent adjacent sides. One of my diagonals is also my only line of symmetry. How many sides do I have?

____

Use your Geometry Template to draw this polygon in the space provided at the right.

5. The polygon below is a regular polygon. Find the measure of angle X without using a protractor.

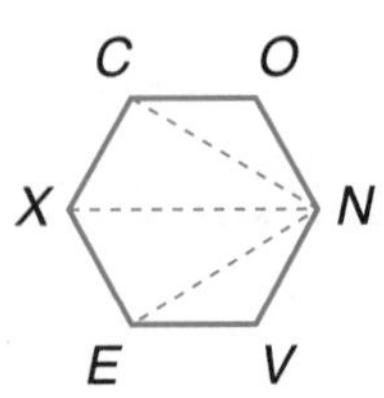

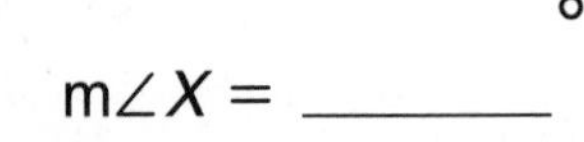

m∠X = ________°

Date Time

LESSON 9•6

Mobile Problems

The mobile shown in each problem is in balance.
The **fulcrum** of the mobile at the right is the center point of the rod.
A mobile will balance if $W * D = w * d$.

Write and solve an equation to answer each question.

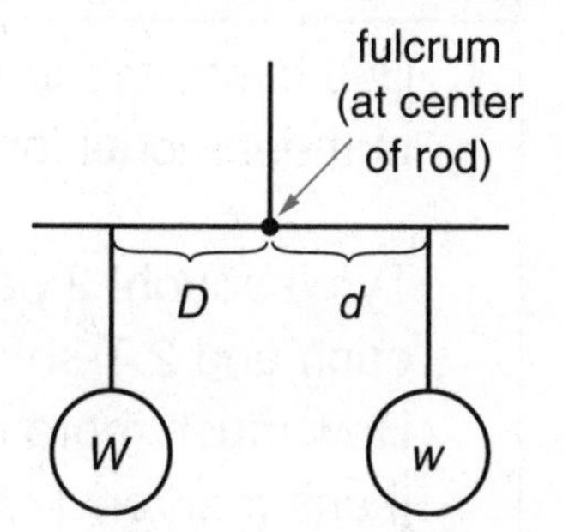

1. What is the distance from the fulcrum to the object on the right of the fulcrum?

   $W =$ ______ $D =$ ______ $w =$ ______ $d =$ ______

   Equation ____________________ Solution __________

   Distance ______ units

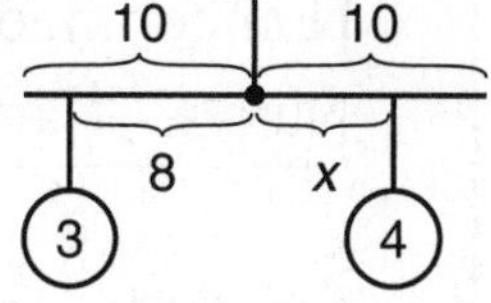

2. What is the weight of the object on the left of the fulcrum?

   $W =$ ______ $D =$ ______ $w =$ ______ $d =$ ______

   Equation ____________________ Solution __________

   Weight ______ units

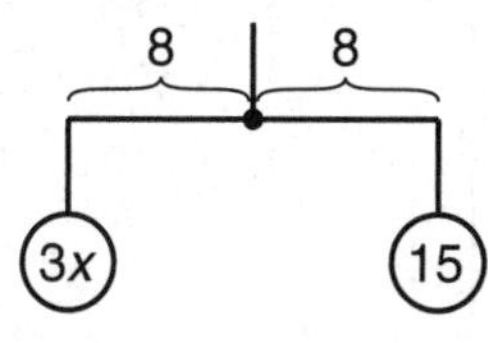

3. What is the distance from the fulcrum to each of the objects?

   $W =$ ______ $D =$ ______ $w =$ ______ $d =$ ______

   Equation ____________________ Solution __________

   Distance on the left of the fulcrum ______ units

   Distance on the right of the fulcrum ______ units

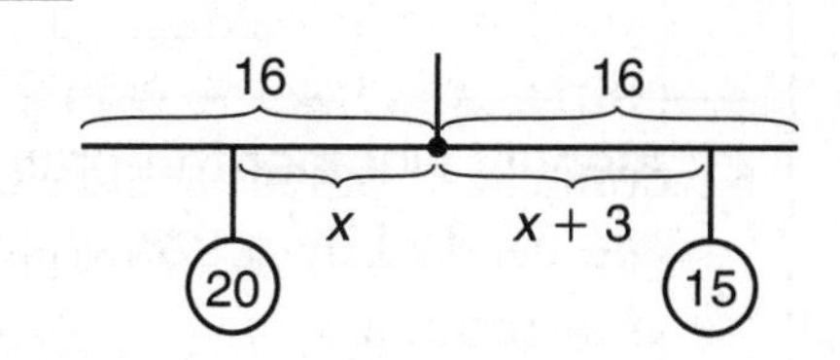

4. What is the weight of each object?

   $W =$ ______ $D =$ ______ $w =$ ______ $d =$ ______

   Equation ____________________ Solution __________

   Weight of the object on the left of the fulcrum ______ units

   Weight of the object on the right of the fulcrum ______ units

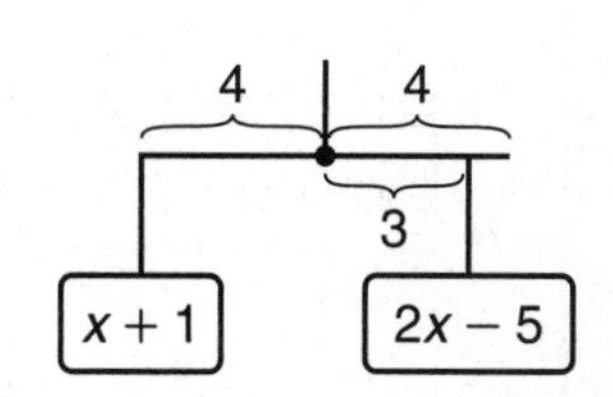

Date Time

LESSON 9•6

Math Boxes

1. Use the distributive property to write a number model for the problem. Then solve.

Tyrell bought 2 pairs of jeans for \$34.99 each and 2 T-shirts for \$19.99 each. How much more did he spend on jeans than on T-shirts?

Number model

____

Solution __________

2. Simplify each expression by combining like terms.

a. $9x + 12 - x$ ______________

b. $h - 14 - 2h + 8$ ______________

c. $8d - d - 5m$ ______________

d. $4w + 4t - 3w - 9t$ ______________

3. The ratio of facedown to faceup cards is 5:4. If there are 72 cards altogether, how many cards are faceup?

__________ cards

4. Draw the line(s) of symmetry for each figure below.

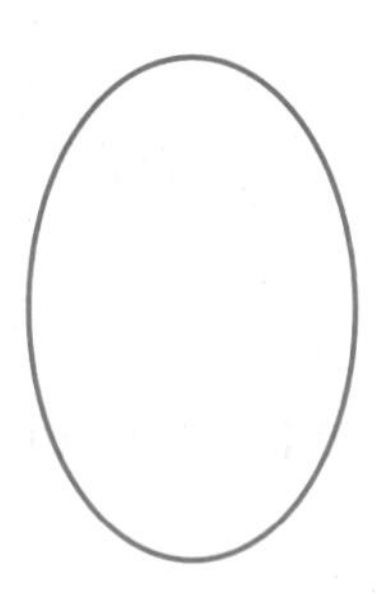

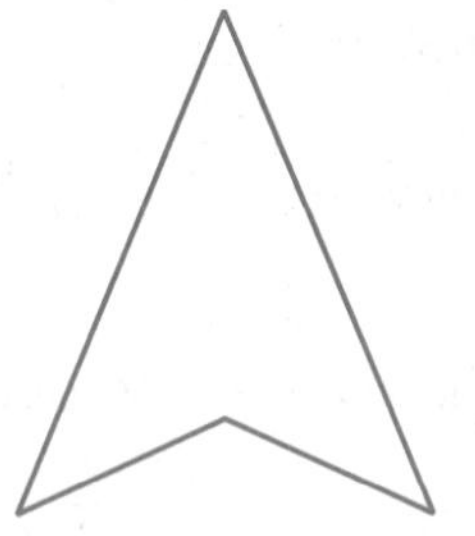

5. Identify whether the preimage (1) and image (2) are related by a translation, a reflection, or a rotation.

Write your answer on the line below.

____

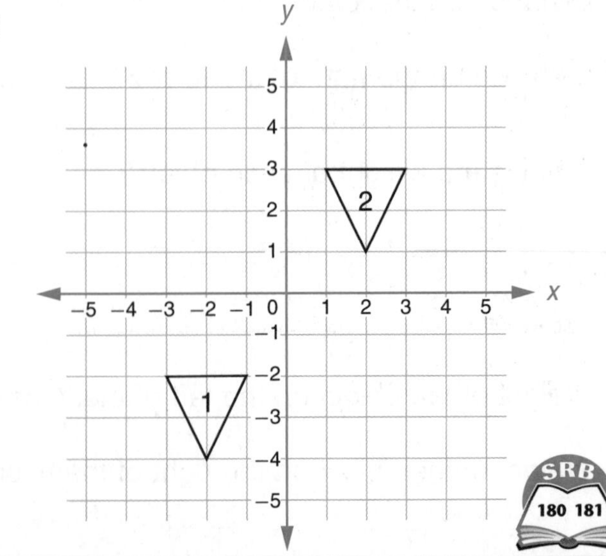

Date Time

LESSON 9·7 A Picnic Budget Spreadsheet

The following spreadsheet gives budget information for a class picnic.

Class Picnic ($)

	A	B	C	D
1		budget for class picnic		
2				
3	quantity	food items	unit price	cost
4	6	packages of hamburgers	2.79	16.74
5	5	packages of hamburger buns	1.29	6.45
6	3	bags of potato chips	3.12	9.36
7	3	quarts of macaroni salad	4.50	13.50
8	4	bottles of soft drinks	1.69	6.76
9			subtotal	52.81
10			8% tax	4.23
11			total	57.04

1. What information is shown in Row 8?

____

2. What information is shown in column A—labels, numbers, or formulas? _______________

3. Cell D6 holds the following formula: = A6 ∗ C6.

   a. What formula is stored in cell D4? _______________

   b. What formula is stored in cell D8? _______________

4. Circle the formula stored in cell D9.

   = C4 + C5 + C6 + C7 + C8 = D4 + D5 + D6 + D7 + D8

5. a. What does the formula stored in cell D10 calculate? _______________

   b. Circle the formula stored in cell D10.

   = 0.08 ∗ C9 = 0.08 ∗ D9 = 8 ∗ D9

6. a. What does the formula stored in cell D11 calculate? _______________

   b. Write the formula stored in cell D11. _______________

7. a. Which spreadsheet cells would change if you increased the number of bags of potato chips to 4? _______________

   b. Calculate the number that would be shown in each of these cells.

____

Date Time

LESSON 9·7

Math Boxes

1. Use the distributive property to remove the parentheses. Then combine like terms.

   a. $-5 + 5(x + 4)$ ____________

   b. $3m + 2(5m - 7)$ ____________

   c. $4(6t + 9) - 10t$ ____________

   d. $8k - 6(3 - 2k)$ ____________

2. Solve each equation.

   a. $8b - 10 = 2b + 14$

   $b =$ ____________

   b. $4 - 6w = 3(\frac{1}{2} + \frac{w}{2})$

   $w =$ ____________

3. Circle the equation that describes the relationship between the numbers in the table at the right.

   A. $(x - 9)5 = y$

   B. $\frac{x-9}{5} = y$

   C. $(y + 5)9 = x$

   D. $5(y + 5) = x$

x	y
10	$\frac{1}{5}$
14	1
19	2
49	8

4. Without using a protractor, find the measure of each numbered angle below. Write each measure on the drawing.

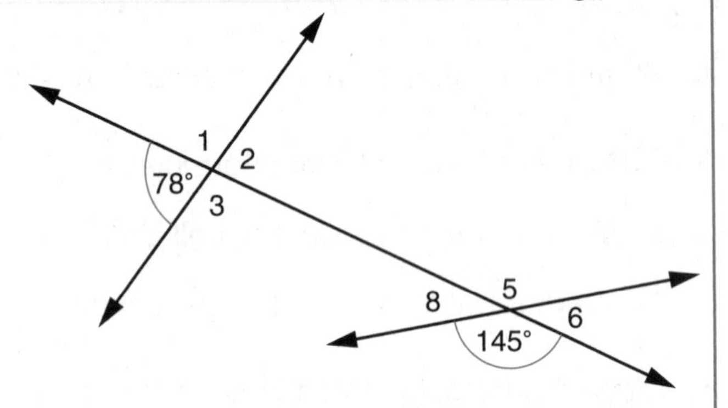

5. Using a compass, draw a circle that has a circumference of 15.7 cm. Label the diameter of your circle.

   Use 3.14 for π.

SRB 213

Date   Time

LESSON 9·8

Area Formulas

Calculate the area of each figure below. A summary of useful area formulas appears on page 377 of the *Student Reference Book.*

Measure dimensions to the nearest tenth of a centimeter. Record the dimensions next to each figure. You might need to draw and measure 1 or 2 line segments on a figure. Round your answers to the nearest square centimeter.

1.

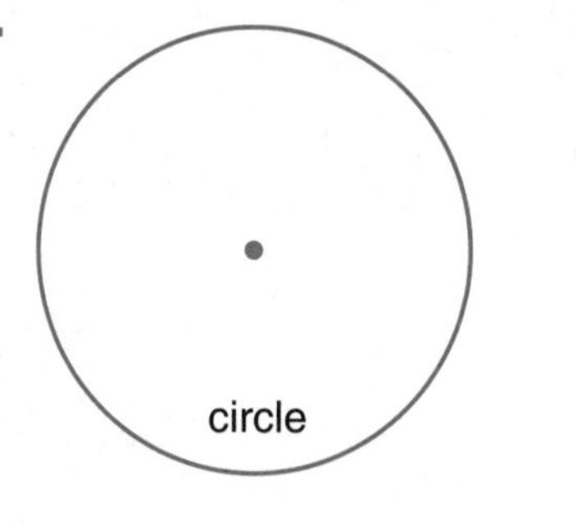

Area formula ____________________

Area ____________________ (unit)

2.

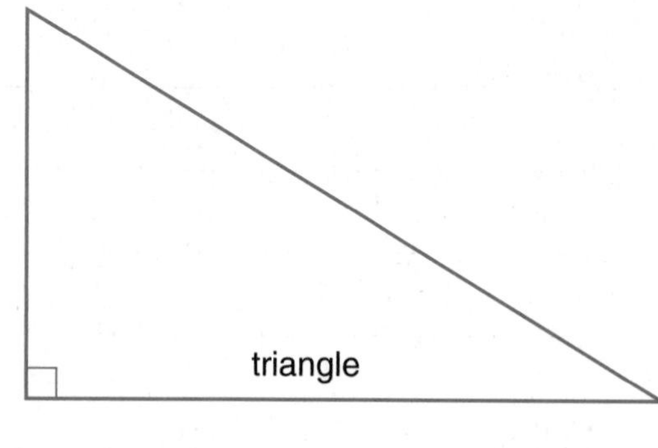

Area formula ____________________

Area ____________________ (unit)

3.

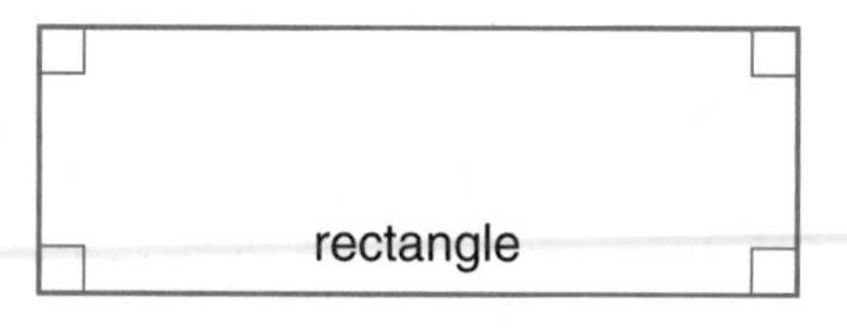

Area formula ____________________

Area ____________________ (unit)

4.

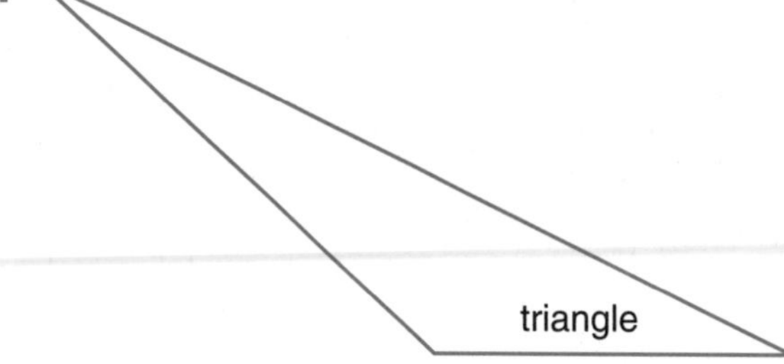

Area formula ____________________

Area ____________________ (unit)

5.

parallelogram

Area formula ____________________

Area ____________________ (unit)

Try This

6.

3 cm

h

trapezoid

b

Area formula ____________________

Area ____________________ (unit)

LESSON 9·8

Perimeter, Circumference, and Area

Solve each problem. Explain your answers.

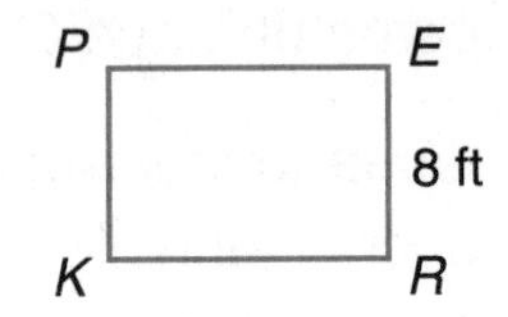

1. Rectangle *PERK* has a perimeter of 40 feet.

   Length of side *PE* ______ (unit)

   Area of rectangle *PERK* ______ (unit)

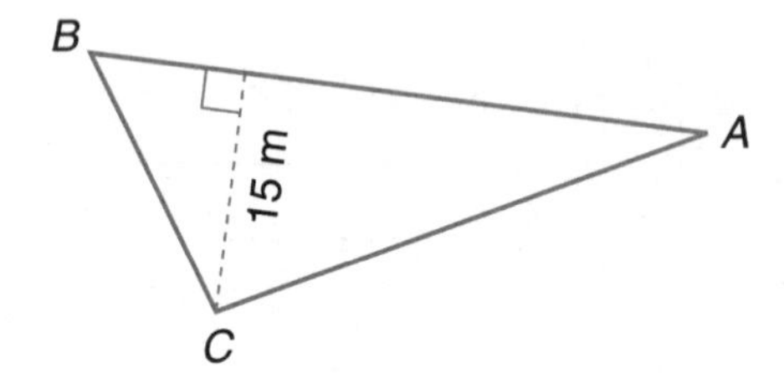

2. The area of triangle *BAC* is 300 meters^2.
   What is the length of side *AB?*

   Length of side *AB* ______ (unit)

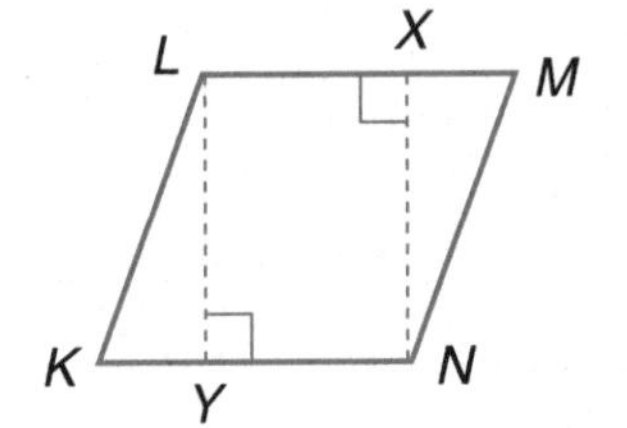

3. The area of parallelogram *LMNK* is 72 square inches.

   The length of side *LX* is 6 inches, and the length of side *KY* is 3 inches.

   What is the length of $\overline{LY}$?

   Length of $\overline{LY}$ ______ (unit)

LESSON 9·8

Perimeter, Circumference, and Area *cont.*

4. The area of triangle *ACE* is 42 square yards. What is the area of rectangle *BCDE?*

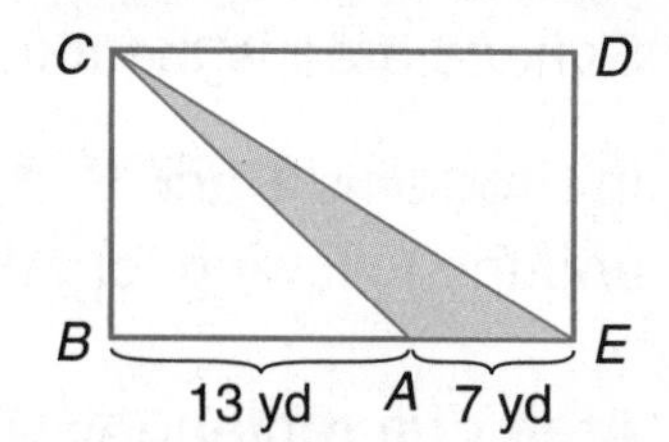

Area of rectangle *BCDE* ____________

For Problems 5 and 6, use 3.14 for π.

5. To the nearest percent, about what percent of the area of the square is covered by the area of the circle?

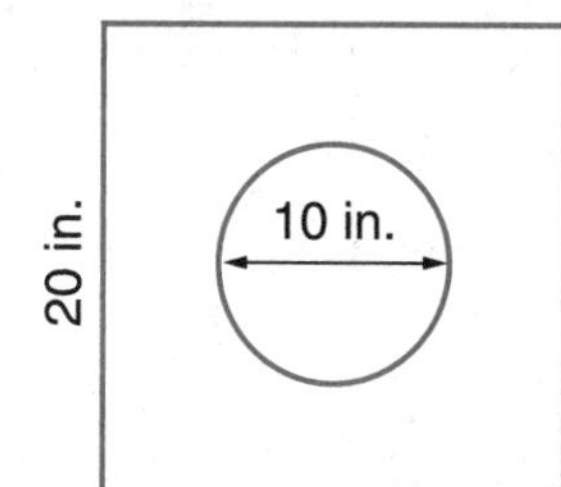

Answer ____________

6. Which path is longer: once around the figure 8—from *A* to *B* to *C* to *B* and back to *A*—or once around the large circle?

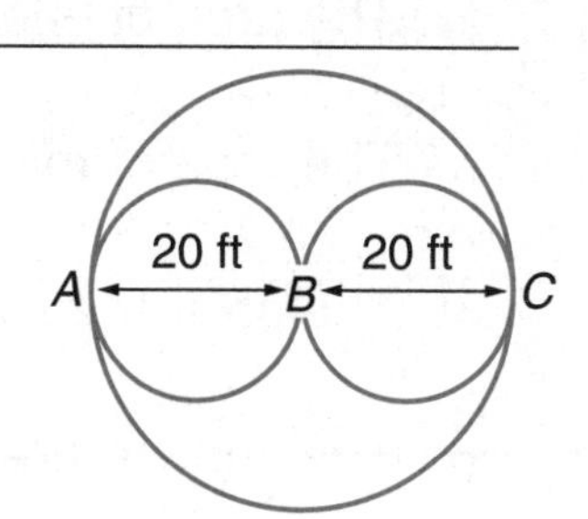

Date Time

LESSON 9•8 Surface Area

Surface area is the total area of all the surfaces of a 3-dimensional object.

In Problems 1 and 2, a 3-dimensional solid is represented by a net. Use the net and the following formulas to answer the questions.

Area of a rectangle: $A = b * h$

A is the area of the rectangle.
b is the length of its base.
h is the height of the rectangle.

Area of a triangle: $A = \frac{1}{2} * (b * h)$

A is the area of the triangle.
b is the length of its base.
h is the height of the triangle.

1. The net below represents a right triangular prism.

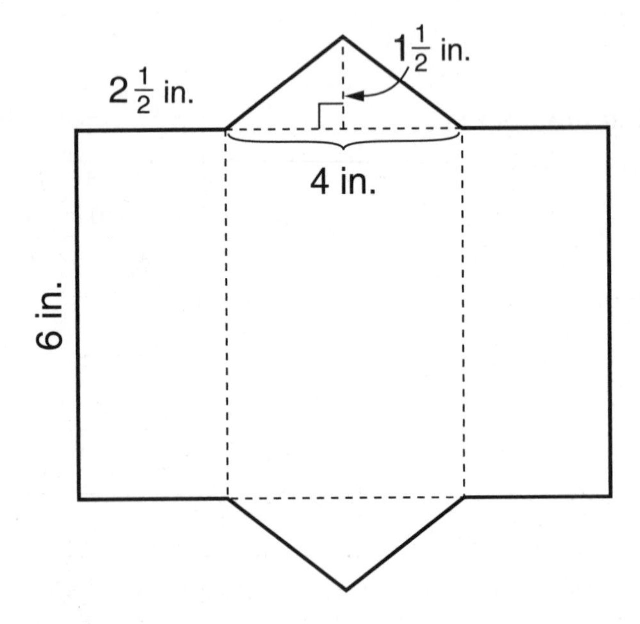

a. The area of each triangular face is ________ in.^{2}.

b. The area of the larger rectangular face is ________ in.^{2}.

c. The area of each of the smaller rectangular faces is ________ in.^{2}.

d. The surface area of the triangular prism is ________ in.^{2}.

LESSON 9·8

Surface Area *continued*

2. The net below represents a square pyramid. The triangular faces are all congruent.

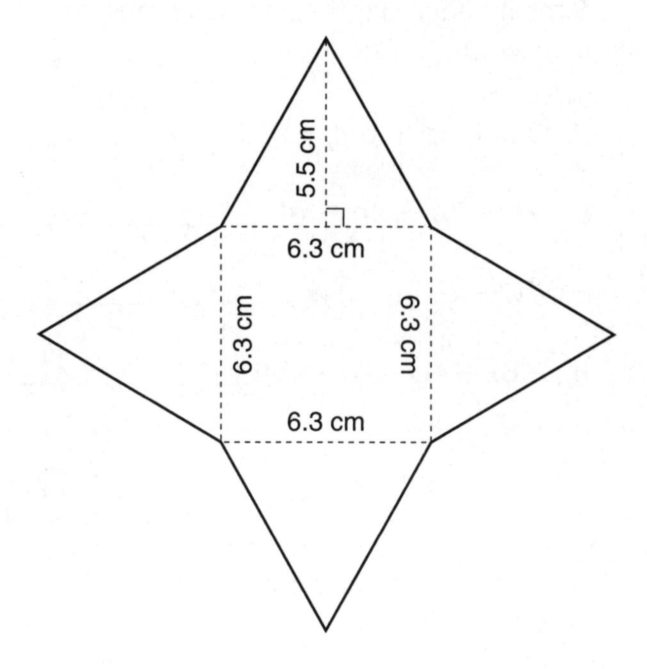

a. The area of each triangular face is ______________ cm^2.

b. The area of the square face is ______________ cm^2.

c. The surface area of the square pyramid is ______________ cm^2.

3. Roxanne cut out the shape below from a piece of cardboard. She will fold it to make a right rectangular prism. She plans to decorate it by completely covering the outside with paper. How much paper does Roxanne need to decorate the prism?

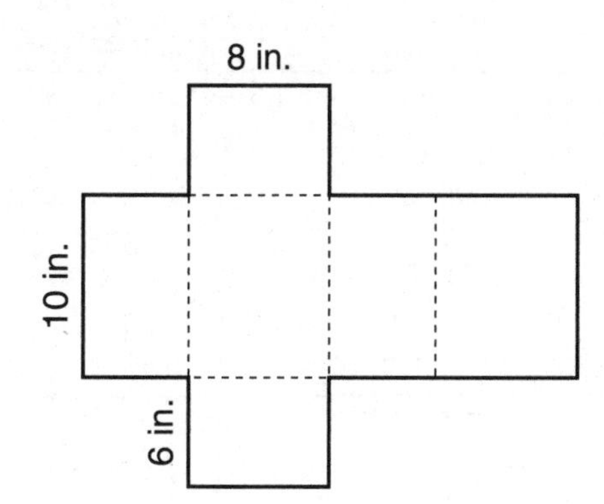

Roxanne needs __________ $in.^2$ of paper to decorate the prism.

Date Time

LESSON 9·8

Math Boxes

1. Use the distributive property to write a number model for the problem. Then solve.

Niesha bought 3 sets of screwdrivers at \$19.49 each and 3 boxes of screws at \$4.98 each. What was the total cost of Niesha's purchase?

Number model

____

Solution ________

2. Simplify each expression by combining like terms.

a. $3m + 4p + 6m + 9$ ________

b. $2k + 7n - 9n - 4k$ ________

c. $\frac{3}{2}w - (-\frac{4}{5}) + \frac{5}{2}w$ ________

d. $-6x - 6y - 6x - 7y$ ________

3. Seven out of 9 cards are faceup. If 56 cards are faceup, how many cards are there altogether?

________ cards

4. Draw the line(s) of symmetry for each figure below.

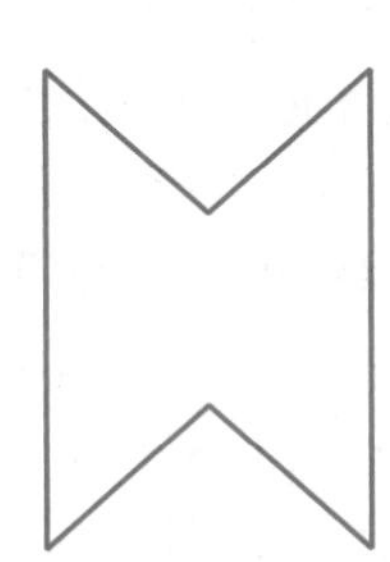

5. Identify whether the preimage (1) and image (2) are related by a translation, a reflection, or a rotation.

Write your answer on the line below.

____

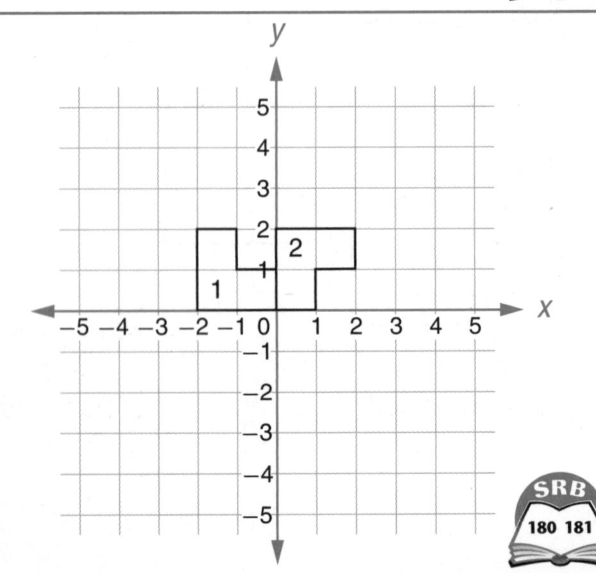

LESSON 9•9

Data Review

1. Below are the scores for a spelling test in Ms. Jenning's sixth-grade class:

72%	96%	88%	96%	80%	68%	44%
76%	96%	68%	56%	76%	96%	92%
80%	88%	68%	56%	100%	100%	88%
68%	96%	92%	96%	76%	80%	88%

a. Make a stem-and-leaf plot of the scores.

b. Find the following landmarks:

maximum ________ median ________

mode(s) ________ minimum ________

2. First Bank and Trust raised the interest rate on savings accounts 4 times in 1 year. To the right is a step graph of the interest rates for the year. Use the graph to answer the questions.

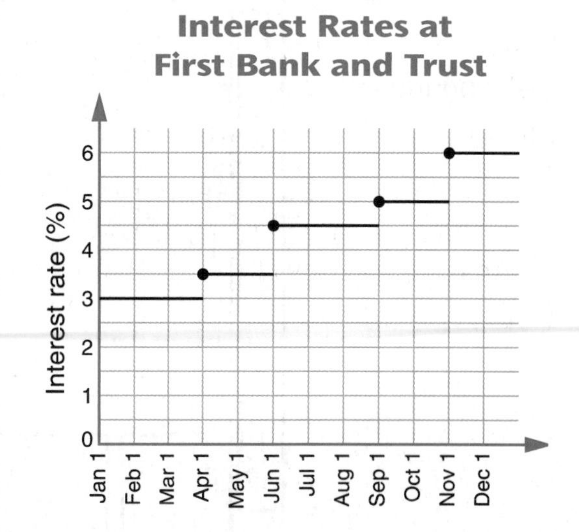

a. What was the interest rate in July? ________

b. For how many months did the interest rate stay at 4.5%? ________

c. By how much did the interest rate increase from February to October? ________

3. Which graph below most likely displays the number of cellular phone subscribers (in millions)? Graph ____

Graph A

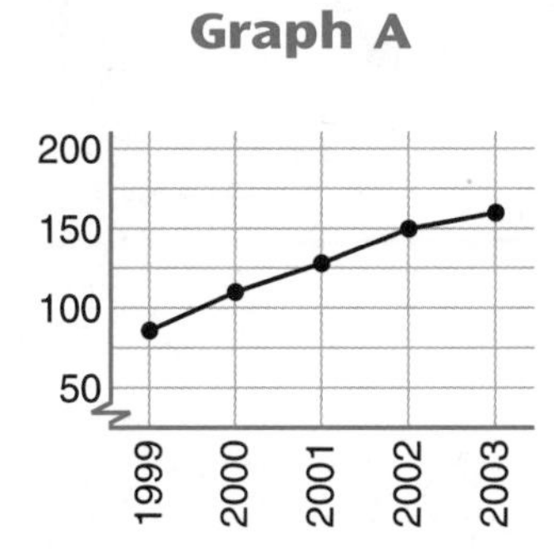

Graph B

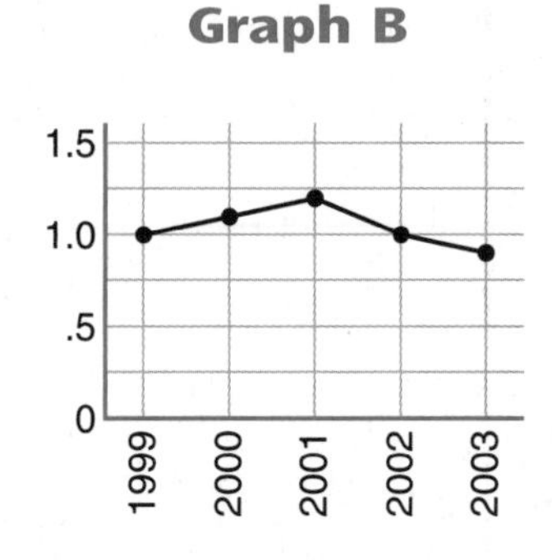

Graph C

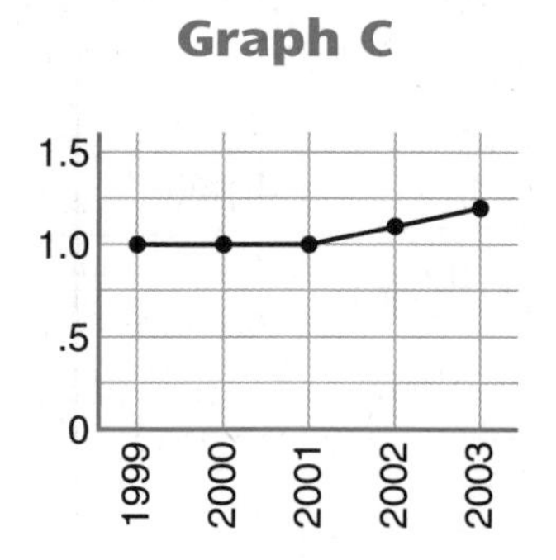

Date Time

LESSON 9•9

Calculating the Volume of the Human Body

An average adult human male is about 69 inches (175 centimeters) tall and weighs about 170 pounds (77 kilograms). The drawings below show how a man's body can be approximated by 7 cylinders, 1 rectangular prism, and 1 sphere.

The drawings use the scale 1 mm:1 cm. This means that every length of 1 millimeter in the drawing represents 1 centimeter of actual body length. The drawing below is 175 millimeters high. Therefore, it represents a male who is 175 centimeters tall.

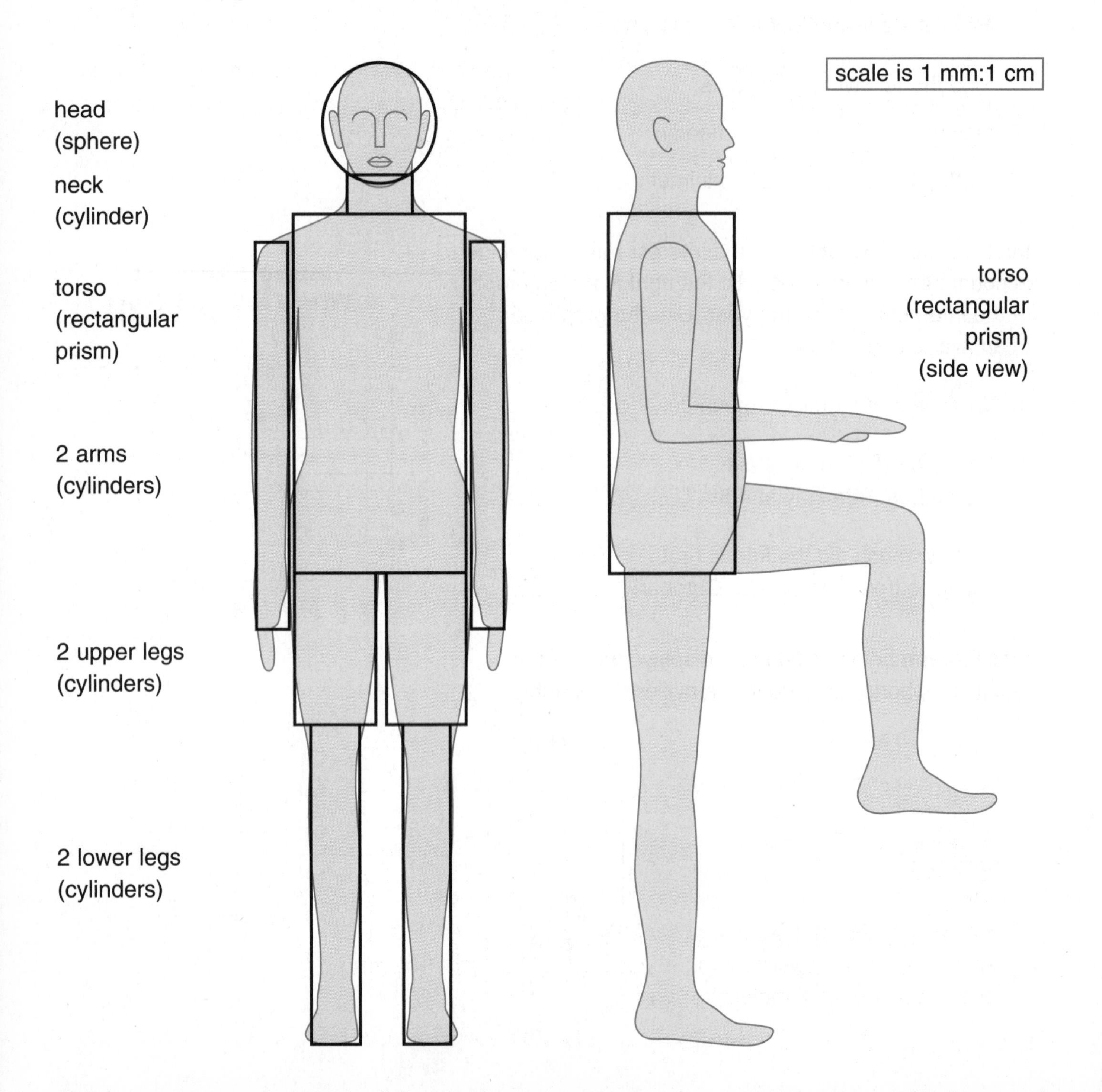

Date Time

LESSON 9·9

Calculating the Volume of the Human Body *cont.*

1. a. Use a centimeter ruler to estimate the measures of each geometric figure shown on page 350. Record your estimates on the drawings and in the table below. Be sure to record the actual body dimensions. For example, if you measure the length of an arm as 72 millimeters, record this as 72 centimeters because the scale of the drawing is 1 mm:1 cm.

   b. Calculate the volume of each body part and record it in the table. You will find a summary of useful volume formulas on page 378 in your *Student Reference Book.*

   For the arm, upper leg, and lower leg, multiply the volume by 2. Add to find the total volume of an average adult male's body. Your answer will be in cubic centimeters.

Body Part and Shape	Actual Body Dimensions (cm)	Volume (Round to the nearest 1,000 cm^3.)
Head (sphere)	radius:	$* 1 =$
Neck (cylinder)	radius: height:	$* 1 =$
Torso (rectangular prism)	length: width: height:	$* 1 =$
Arm (cylinder)	radius: height:	$* 2 =$
Upper leg (cylinder)	radius: height:	$* 2 =$
Lower leg (cylinder)	radius: height:	$* 2 =$
	Total Volume:	About

2. One liter is equal to 1,000 cubic centimeters. Use this fact to complete the following statement: I estimate that the total volume of an average adult male's body is about ________ liters.

3. A men's size 7 regulation basketball has a radius of about 12 cm.

   a. Find the volume of this type of basketball, rounded to the nearest 1,000 cm^3. ____________ (unit)

   b. Fill in the blank. The volume of a men's size 7 regulation basketball is about ________% the volume of an average adult male's head.

Date Time

LESSON 9•9

Math Boxes

1. Use the distributive property to remove parentheses. Then combine like terms.

   a. $3(t + 3) + 4t$ ____________

   b. $19 - 4(5y + 1) - 4y$ ____________

   c. $10 + 7m - 2(3m + 5)$ ____________

   d. $-7(2p - 1) + 3(8 - p)$ ____________

2. Solve each equation.

   a. $-2(g + 6) = g + 3 + 2g$

   $g =$ ____________

   b. $2(2x + \frac{1}{2}) = 3(x - \frac{2}{3})$

   $x =$ ____________

3. Circle the equation that describes the relationship between the numbers in the table at the right.

   A. $(x * 4) - 3 = y$

   B. $(4 * x) + 3 = y$

   C. $(y * 5) - 3 = x$

   D. $(4 * y) + 3 = x$

x	*y*
$\frac{1}{4}$	−2
$\frac{1}{2}$	−1
4	13
10	37

4. Without using a protractor, find the measure of each numbered angle below. Write each measure on the drawing. Lines *a* and *b* are parallel.

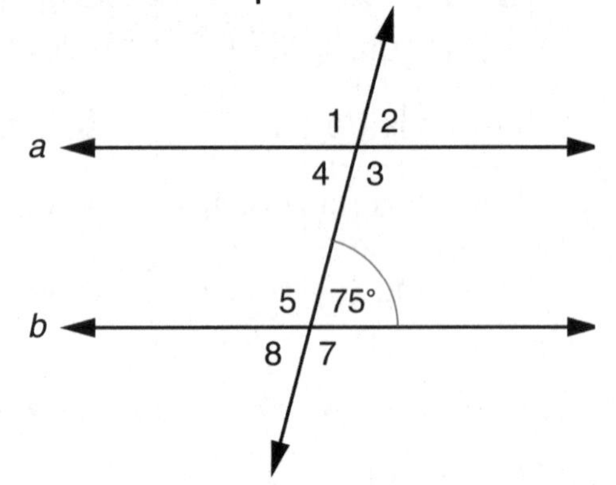

5. a. Using a compass, draw 2 concentric circles in the space at the right. The radius of the small circle is 1.5 cm. The radius of the large circle is 2 cm.

   b. What is the area of the ring between the 2 circles? Use 3.14 for π.

   About ____________ cm^2

Date Time

LESSON 9·10

Math Boxes

1. Calculate the area of the parallelogram.

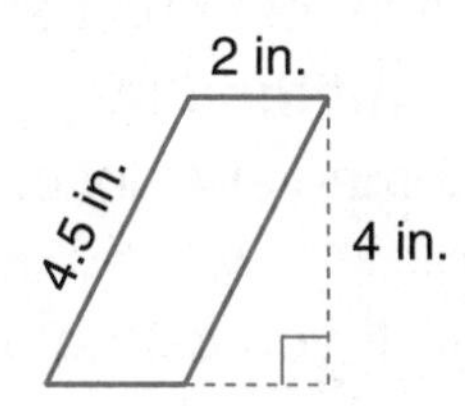

Area = ______ (unit)

2. Solve.

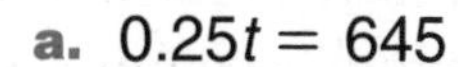

a. $0.25t = 645$ ______

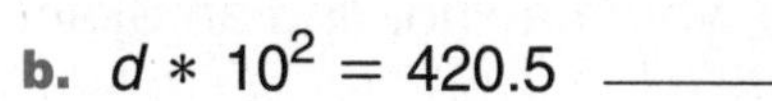

b. $d * 10^2 = 420.5$ ______

c. $\frac{1}{8}f = \frac{3}{16}$ ______

d. $\sqrt{h} = 20$ ______

3. The table at the right shows about how much a person weighing 100 pounds on Earth would weigh on each of the planets in our solar system.

a. On which planet would a person weigh about $\frac{1}{6}$ as much as on Mercury?

______

b. On which planet would a person weigh about $2\frac{1}{2}$ times what the person weighs on Earth?

______

Planet	Weight (lb)
Mercury	37
Venus	88
Earth	100
Mars	38
Jupiter	264
Saturn	115
Uranus	93
Neptune	122
Pluto	6

4. a. Draw and label an obtuse angle *CAT*. Measure it.

$\angle CAT$ measures ______°.

b. Draw and label a reflex angle *NOD*. Measure it.

$\angle NOD$ measures ______°.

Date Time

LESSON 9·10

Solving Equations by Trial and Error

If you substitute a number for the variable in an equation and the result is a true number sentence, then that number is a solution of the equation. One way to solve an equation is to try several **test numbers** until you find the solution. Each test number can help you close in on an exact solution. Using this **trial-and-error method** for solving equations, you may not find an exact solution, but you can come very close.

Example: Find a solution of the equation $\frac{1}{x} + x = 4$ by trial and error. If you cannot find an exact solution, try to find a number that is very close to an exact solution.

The table shows the results of substituting several test numbers for x.

x	$\frac{1}{x}$	$\frac{1}{x} + x$	Compare $(\frac{1}{x} + x)$ to 4.
1	1	2	Less than 4
2	0.5	2.5	Still less than 4, but closer
3	0.3	3.3	Less than 4, but even closer
4	0.25	4.25	Greater than 4

These results suggest that we try testing numbers for x that are between 3 and 4.

x	$\frac{1}{x}$	$\frac{1}{x} + x$	Compare $(\frac{1}{x} + x)$ to 4.
3.9	0.256...	4.156...	> 4
3.6	$0.2\overline{7}$	$3.8\overline{7}$	< 4

Now it's your turn. Try other test numbers. See how close you can get to 4 for the value of $\frac{1}{x} + x$.

x	$\frac{1}{x}$	$\frac{1}{x} + x$	Compare $(\frac{1}{x} + x)$ to 4.

My closest solution: ____________

Date Time

LESSON 9·10

Solving Equations by Trial and Error *cont.*

Find numbers that are closest to the solutions of the equations. Use the suggested test numbers to get started. Round approximate solutions to the nearest thousandth.

1. Equation: $\sqrt{y} + y = 10$

y	$\sqrt{y}$	$\sqrt{y} + y$	Compare $(\sqrt{y} + y)$ to 10.
0	0	0	< 10
5	2.236	7.236	
9	3		

My closest solution: ______

2. Equation: $x^2 - 3x = 8$

y	x^2	$3x$	$x^2 - 3x$	Compare $(x^2 - 3x)$ to 8.
4				
6				
5				

My closest solution: ______

Date Time

Using Formulas to Solve Problems

To solve a problem using a formula, you can substitute the known quantities for variables in the formula and solve the resulting equation.

Example: A formula for converting between Celsius and Fahrenheit temperatures is $F = 1.8C + 32$, where C represents the Celsius temperature and F represents the Fahrenheit temperature.

◆ Use the formula to convert 30°C to degrees Fahrenheit.

	$F = 1.8C + 32$
Substitute 30 for C in the formula.	$F = (1.8 * 30) + 32$
Solve the equation.	$F = 86$
Answer:	30°C = 86°F

◆ Use the formula to convert 50°F to degrees Celsius.

	$F = 1.8C + 32$
Substitute 50 for F in the formula.	$50 = (1.8 * C) + 32$
Solve the equation.	$10 = C$
Answer:	50°F = 10°C

1. The formula $W = 570a - 850$ expresses the relationship between the average number of words children know and their ages (for ages 2 to 8). The variable W represents the number of words known, and a represents age in years.

   a. About how many words might a $3\frac{1}{4}$-year-old child know? __________

   b. About how old might a child be who knows about 1,700 words? __________

2. A bowler whose average score is less than 200 is given a handicap. The **handicap** is a number of points added to a bowler's score for each game. A common handicap formula is $H = 0.8 * (200 - a)$, where H is the handicap and a is the average score.

   a. What is the handicap of a bowler whose average score is 160? __________

   b. What is the average score of a bowler whose handicap is 68? __________

3. An adult human female's height can be estimated from the length of her tibia (shinbone) by using the formula $H = 2.4 * t + 75$, where H is the height in centimeters and t is the length of the tibia in centimeters.

   a. Estimate the height of a female whose tibia is 31 centimeters long. __________

   b. Estimate the length of a female's tibia if she is 175 centimeters tall. __________

Date ______________________ Time ______________________

LESSON 9·11 Volume Problems

Solve each problem. You may need to look up formulas in your *Student Reference Book.* Use substitution to check your answers.

1. The volume of the desk drawer shown at the right is 1,365 $in.^3$
Find the depth (d) of the drawer.

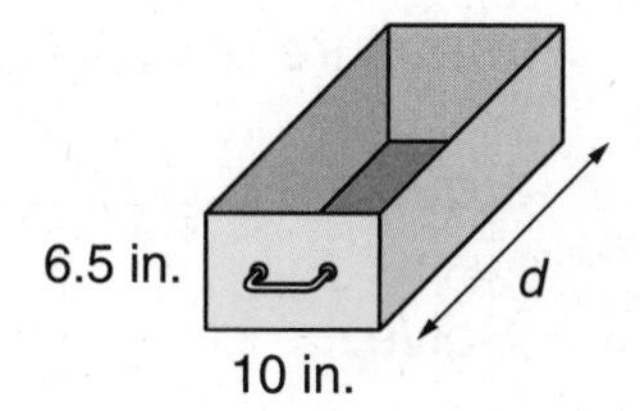

Formula ______________________

Substitute ______________________

Solve ______________________

____

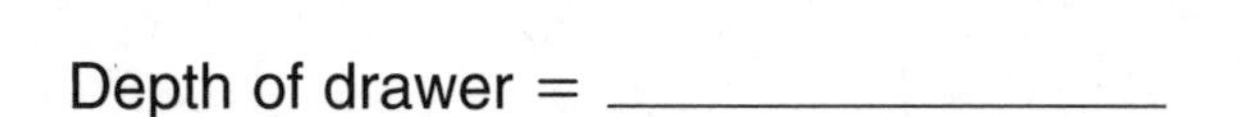

Depth of drawer = ______________

2. The cylindrical can holds about 4 liters (4 liters = 4,000 cm^3).
Find the height (h) of the can to the nearest centimeter. Use 3.14 for π.

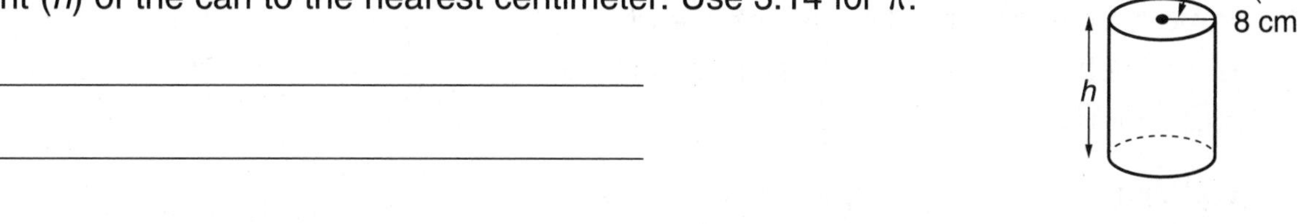

8 cm

h

Formula ______________________

Substitute ______________________

Solve ______________________

____

The can's height is about ______________.

3. A soccer ball has a 9-inch diameter.

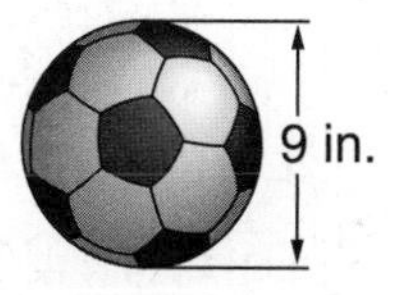

a. What is the shape of the smallest box that will hold the ball? ______________

b. What are the dimensions of this box? ______________________

c. Compare the volume of the box to the volume of the ball. Is the volume of the box more or less than twice the volume of the ball? ______________
(*Reminder:* A formula for finding the volume of a sphere is $V = \frac{4}{3} * \pi * r^3$.)

Explain your answer. ______________________

____

____

Date Time

LESSON 9·11 Angle, Perimeter, and Area Problems

Solve each problem. Use substitution to check your answers.

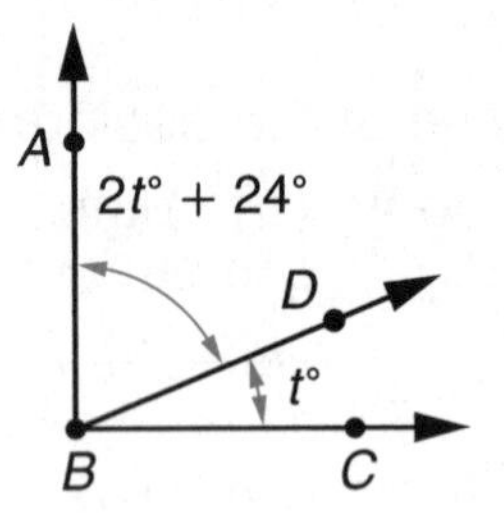

1. $\angle ABC$ is a right angle. What is the degree measure of $\angle CBD$? Of $\angle ABD$?

   Equation ______________________

   Solve.

   $m\angle CBD =$ ________ ° $m\angle ABD =$ ________ °

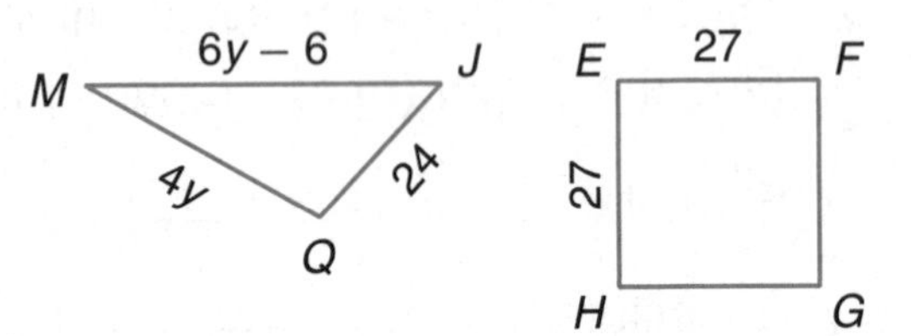

2. Triangle *MJQ* and square *EFGH* have the same perimeter. The dimensions are given in millimeters. What are the lengths of sides *MQ* and *MJ* in triangle *MJQ*?

   Equation ______________________

   Solve.

   Length of $\overline{MQ}$ = ________ (unit)   Length of $\overline{MJ}$ = ________ (unit)

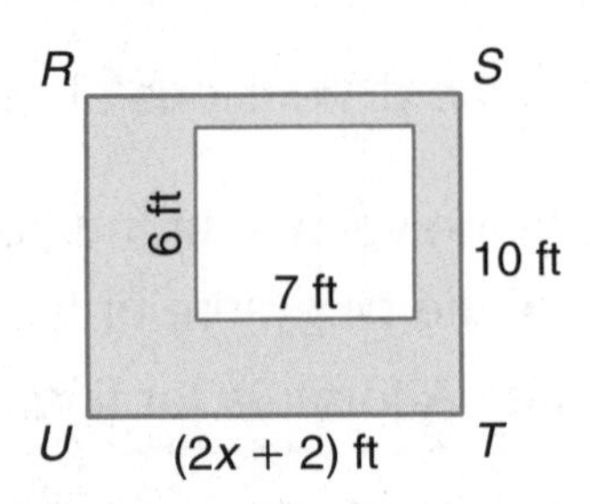

3. The area of the shaded part of rectangle *RSTU* is 78 ft^2. Find the length of side *TU*.

   Equation ______________________

   Solve.

   Length of side *TU* = ________ (unit)

Date Time

LESSON 9•11

Surface Area of Rectangular Prisms

The surface area (*SA*) of a rectangular prism is the sum of the areas of its faces.

Study the examples below.

Example 1:

Use a net to find the surface area (*SA*) of the rectangular prism.

- Draw and label a net.
- Find the area of each rectangle in the net.

$w = 4$ cm

3 cm * 4 cm = 12 cm^2

$l = 6$ cm

6 cm * 3 cm = 18 cm^2

6 cm * 4 cm = 24 cm^2

6 cm * 3 cm = 18 cm^2

6 cm * 4 cm = 24 cm^2

$h = 3$ cm

3 cm * 4 cm = 12 cm^2

- Find the sum of the areas.

18 + 24 + 18 + 24 + 12 + 12 = 108

Example 2:

Use a formula to find the surface area of the rectangular prism.

- Identify the length (l), width (w), and height (h) of the prism.
- Because there are 3 pairs of opposite faces and opposite faces have the same area, you can use the following formula.

$$\begin{aligned} SA &= 2lw + 2lh + 2wh \\ &= 2(24) + 2(18) + 2(12) \\ &= 48 + 36 + 24 \\ &= 108 \end{aligned}$$

The surface area of the rectangular prism is 108 cm^2.

Use the formula $SA = 2lw + 2lh + 2wh$ to find the surface area of each rectangular prism.

1.

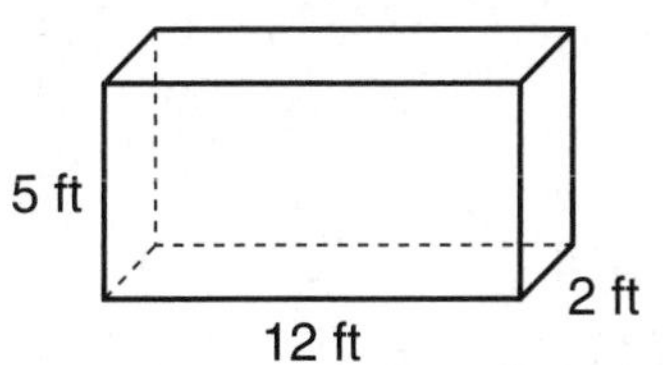

SA = __________ ft^2

2.

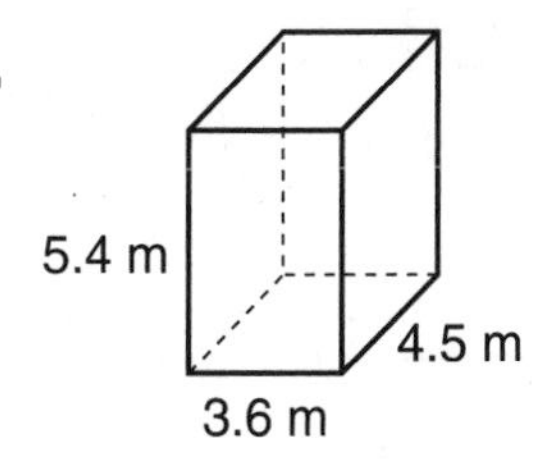

SA = __________ m^2

3. Which has the greater surface area, a cube with $s = 8$ m or a rectangular prism with dimensions $l = 6$ m, $w = 5$ m, and $h = 12$ m?

____

____

Date Time

LESSON 9•11

Surface Area of Cylinders

The surface area (*SA*) of a cylinder is the sum of the areas of its 2 circular bases and the area of its curved surface.

Example:

Find the surface area (*SA*) of the cylinder.

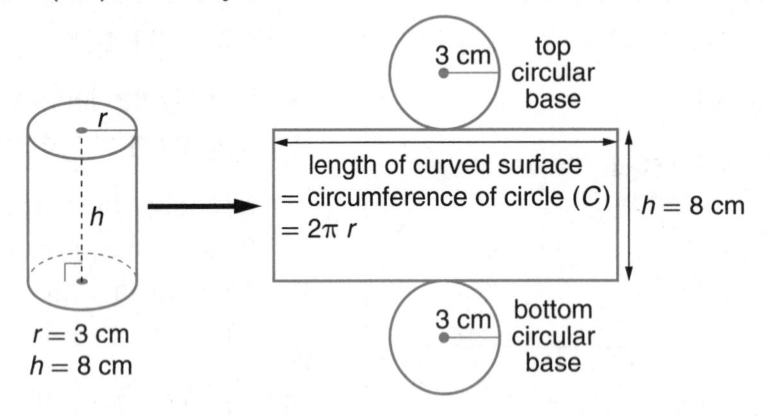

$$\begin{aligned} SA &= (2 * \text{area of circular base}) + (\text{area of curved surface}) \\ &= (2 * \text{area of circular base}) + (\text{circumference of base} * \text{cylinder height}) \\ &= (2 * \pi * r^2) + ((2 * \pi * r) * h) \\ &\approx (2 * 3.14 * (3\text{ cm})^2) + ((2 * 3.14 * 3\text{ cm}) * 8\text{ cm}) \\ &\approx (56.52\text{ cm}^2) + (150.72\text{ cm}^2) \\ &\approx 207.24\text{ cm}^2 \end{aligned}$$

The surface area of the cylinder is 207.24 cm^2.

Use the formula $SA = (2 * \pi * r^2) + ((2 * \pi * r) * h)$ and 3.14 for π to find the surface area of each cylinder.

1.

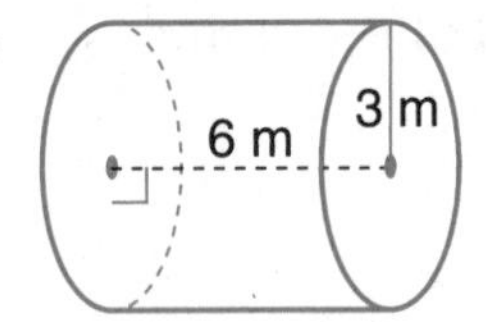

SA = ______________ m^2

2.

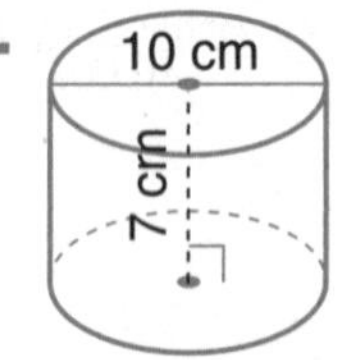

SA = ______________ cm^2

3. A can of tomato juice is a cylinder with a radius of about 7.5 cm and a height of about 20 cm. What is the area of a label that fits around the curved surface of the can with no overlap? Use 3.14 for π.

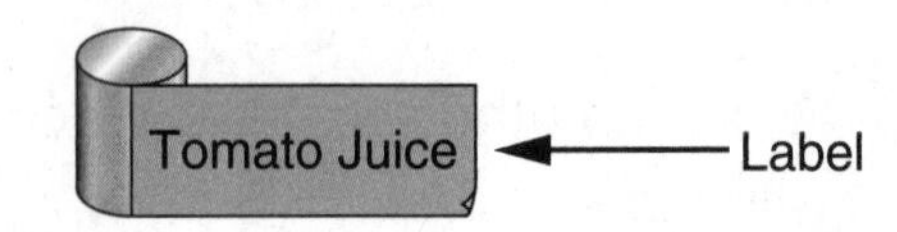

Area of label = ______________ cm^2

Date Time

LESSON 9•11 Math Boxes

1. Find the perimeter of the figure shown below. Use $\frac{22}{7}$ for π.

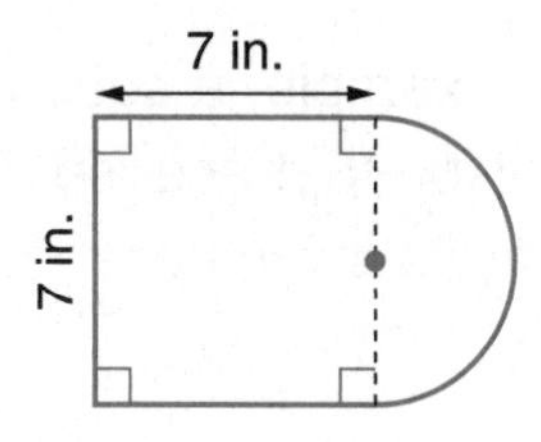

Perimeter = ________ (unit)

2. An aquarium with a rectangular base measuring 36 in. by 12 in. has a volume of 6,912 in.^{3} Find the height (h) of this aquarium.

h = ________ in.

3. Write an equation for each statement. Then solve.

   a. 20% of x is 24.

   Equation ________

   Solution ________

   b. 60% of n is 75.

   Equation ________

   Solution ________

4. Use the order of operations to evaluate each expression.

   a. $15 + 2^2 - 8 \div 4$ ________

   b. $9 * (6 + 2) - (-5)$ ________

   c. $52 - 8 \div 2$ ________

   d. $3[16 - (3 + 7) \div 5]$ ________

5. Fill in each shape to make a recognizable figure. See the example at the right.

a.

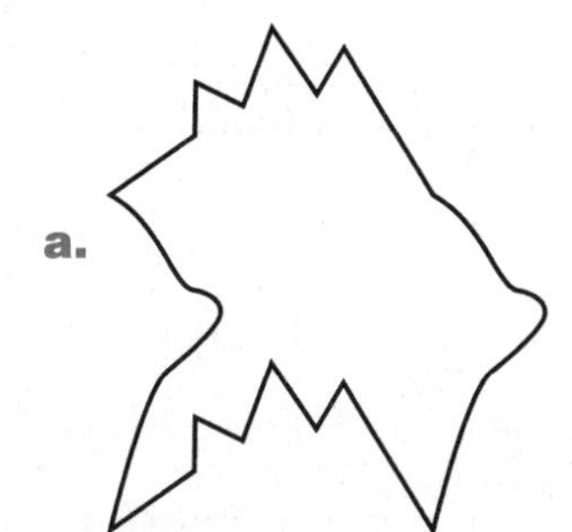

b.

LESSON 9·12

Squares and Square Roots of Numbers

Math Message

You know that the **square of a number** is equal to the number multiplied by itself. For example, $5^2 = 5 * 5 = 25$.

A **square root** of a number *n* is a number whose square is *n*. For example, a square root of 25 is 5, because $5^2 = 5 * 5 = 25$. A square root of 25 is also -5, because $(-5) * (-5) = (-5)^2 = 25$. Every positive number has 2 square roots, which are opposites of each other.

We use the symbol $\sqrt{}$ to write positive square roots. $\sqrt{25}$ is read as *the positive square root of 25.*

1. Write the positive square root of each number.

   a. $\sqrt{81}$ = ______ b. $\sqrt{100}$ = ______ c. $\sqrt{100^2}$ = ______

2. What is the square root of zero? ______

3. Explain why the square root of a negative number does not exist in the real number system.

   ____

   ____

To find the positive square root of a number with a calculator, use the [√] key. For example, to find the square root of 25, enter [√] 25 [)] [Enter]. The display will show 5.

4. Use a calculator. Round your answers for **a** and **d** to the nearest hundredth.

   a. $\sqrt{17}$ = ______ b. $\sqrt{17} * \sqrt{17}$ = ______ c. $\sqrt{\pi} * \sqrt{\pi}$ = ______

   d. $\sqrt{\pi}$ = ______ e. $(\sqrt{17})^2$ = ______ f. $\sqrt{\frac{1}{16}}$ = ______

5. The length of a side of a square is $\sqrt{6.25}$ centimeters. What is the area of the square? ______ (unit)

6. The area of a square is 21 square inches. What is the length of a side to the nearest tenth of an inch? ______ (unit)

7. The radius of a circle is $\sqrt{20}$ feet. What is its area to the nearest tenth of a foot? About ______ (unit)

Date ____________________ Time ____________________

LESSON 9•12 Verifying the Pythagorean Theorem

In a right triangle, the side opposite the right angle is called the **hypotenuse.** The other two sides are called the **legs of the triangle.**

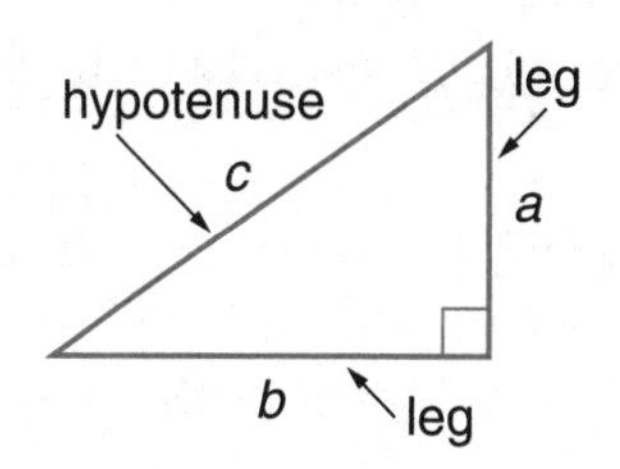

Think about the following statement:
If a and b are the lengths of the legs of a right triangle and c is the length of the hypotenuse, then $a^2 + b^2 = c^2$.

This statement is known as the **Pythagorean theorem.**

1. To verify that the Pythagorean theorem is true, use a blank sheet of paper that has square corners. Draw diagonal lines to form 4 right triangles, one at each corner. Then measure the lengths of the legs and the hypotenuse of each right triangle, to the nearest millimeter. Record the lengths in the table below. Then complete the table.

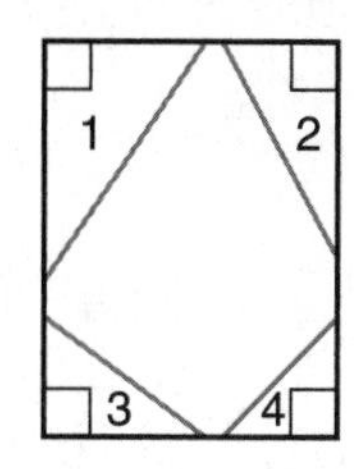

Triangle	Leg (a)	Leg (b)	Hypotenuse (c)	$a^2 + b^2$	c^2
1					
2					
3					
4					

2. Compare $(a^2 + b^2)$ to c^2 for each of the triangles you drew. Why might these two numbers be slightly different?

____

____

____

3. Use the Pythagorean theorem to find c^2 for the triangle at the right. Then find the length c.

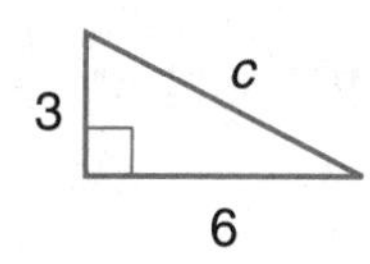

$c^2 =$ ____________ units^2 $\quad$ c is about ____________ units.

Date Time

LESSON 9•12

Using the Pythagorean Theorem

In Problems 1–6, use the Pythagorean theorem ($a^2 + b^2 = c^2$) to find each missing length. Round your answer to the nearest tenth.

1.

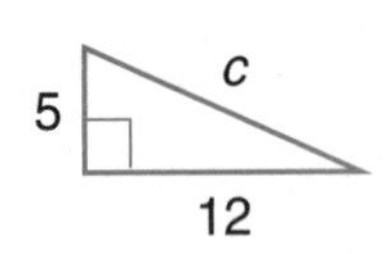

Equation $c^2 = 5^2 + 12^2$

$c^2 =$ ______ $c =$ ______

2.

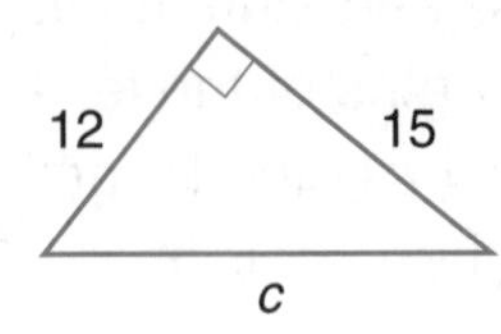

Equation ______________

$c^2 =$ ______ $c =$ ______

3.

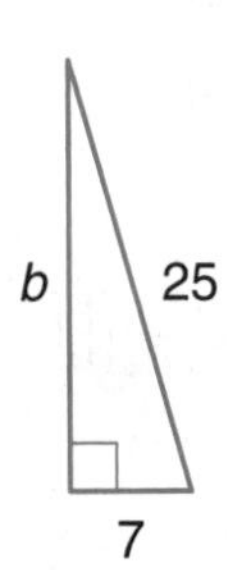

Equation ______________

$b^2 =$ ______ $b =$ ______

4.

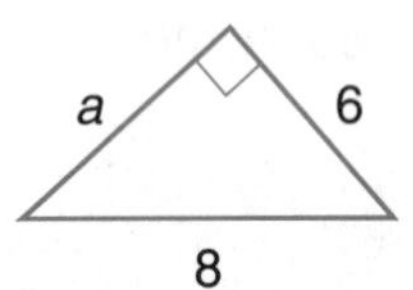

Equation ______________

$a^2 =$ ______ $a =$ ______

5.

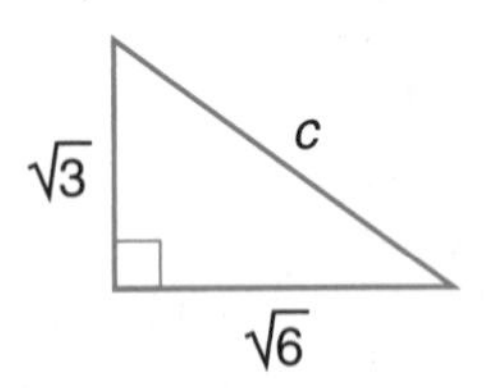

Equation ______________

$c^2 =$ ______ $c =$ ______

6.

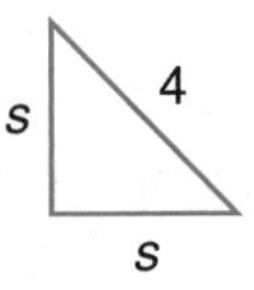

Equation ______________

$s^2 =$ ______ $s =$ ______

7. Is the triangle shown a right triangle? ______

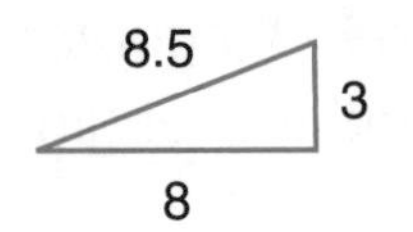

Explain. ______________

____

____

____

Date Time

LESSON 9•12 Math Boxes

1. What is the area of the shaded region in the figure shown below?

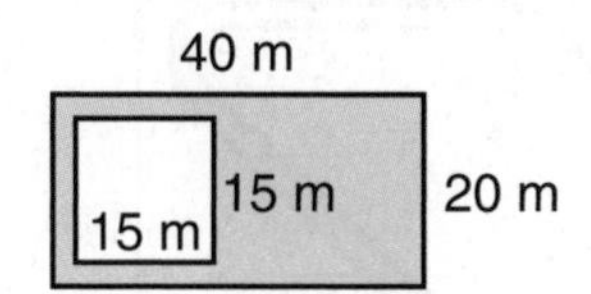

Area = ______ (unit)

SRB 215

2. Solve.

a. $w * 10^{-2} = 28.2$ ______

b. $420 * k = 140$ ______

c. $\frac{5}{2} - p = \frac{7}{4}$ ______

d. $(\sqrt{11})^2 = n$ ______

SRB 251 252

3. The table at the right shows the approximate number of calories a 150-pound person uses per hour while performing various activities.

For which activity does the person use about $\frac{2}{3}$ the number of calories used in jumping rope? Fill in the circle next to the best answer.

○ **A.** swimming

○ **B.** walking

○ **C.** volleyball

○ **D.** basketball

Activity	Calories Per Hour
Swimming (25 yd/min)	275
Walking (3 mph)	320
Volleyball	350
Basketball	500
Jumping rope	750
Running (7 mph)	920

SRB 257

4. Measure the angles.

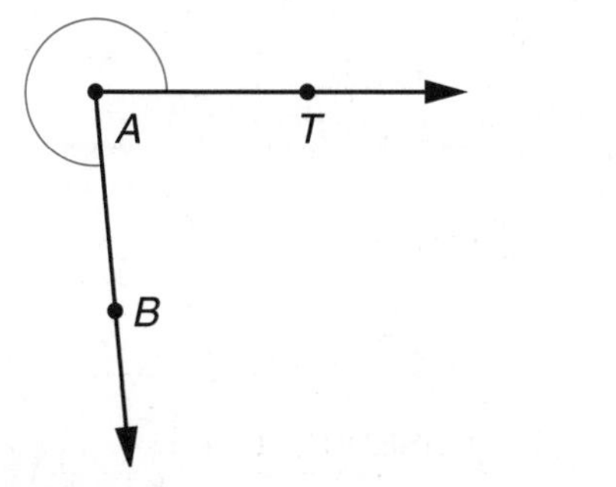

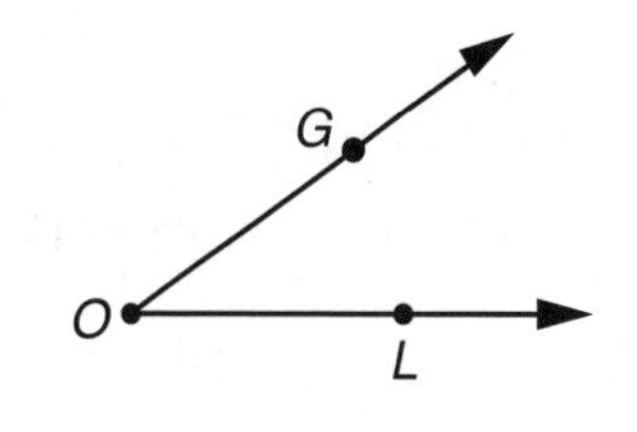

a. Reflex $\angle BAT$ measures ______°.

b. $m\angle LOG$ is about ______°.

SRB 160 230 231

Date Time

LESSON 9·13

Similar Figures and the Size-Change Factor

The 2 butterfly clamps shown below are similar because they each have the same shape. One clamp is an enlargement of the other. The size-change factor tells the amount of enlargement.

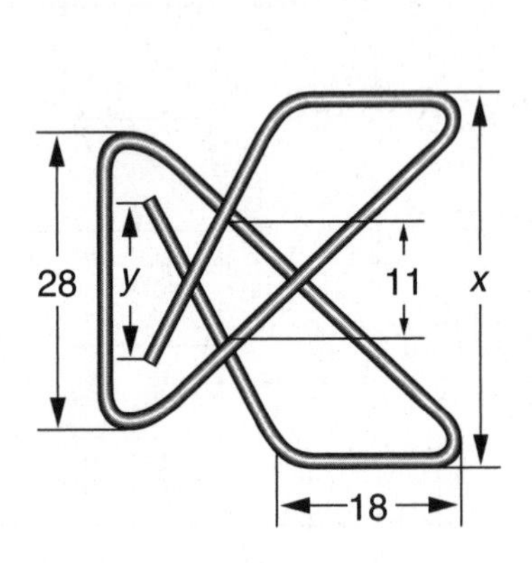

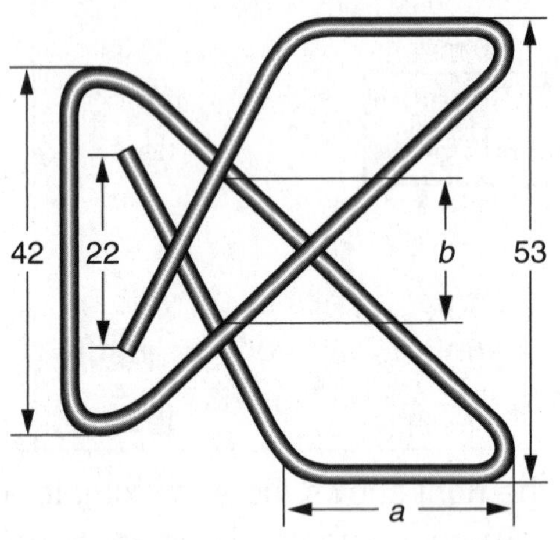

unit: millimeters (mm)

1. The size-change factor for the clamps shown above is ______________.

In Problems 2–5, use the size-change factor to find the missing lengths.

2. $a =$ ______________ mm = ______________ cm

3. $b =$ ______________ mm = ______________ cm

4. $x =$ ______________ mm = ______________ cm

5. $y =$ ______________ mm = ______________ cm

6. If a butterfly clamp is straightened, it forms a long, thin cylinder. When the small clamp is straightened, it is 21 cm long, and the thickness (diameter) of the clamp is 0.15 cm. Its radius is 0.075 cm. Calculate the volume of the small clamp. Use the formula $V = \pi r^2 h$.

   Volume of small clamp = ______________ cm^3 (to the nearest thousandth cm^3)

7. Find the length, thickness (diameter), and volume of the large clamp.

   Length = ______________ cm Diameter = ______________ cm

   Volume of large clamp = ______________ cm^3 (to the nearest thousandth cm^3)

Date Time

LESSON 9·13

Indirect Measurement Problems

In the following problems, you will use indirect measurement to determine the heights and lengths of objects that you cannot directly measure.

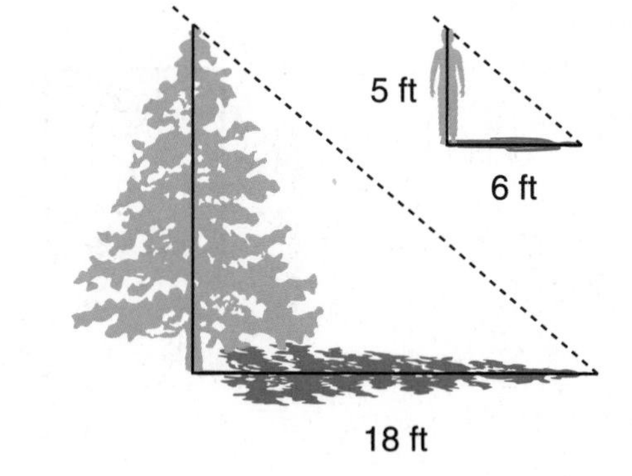

1. A tree is too tall to measure, but it casts a shadow that is 18 feet long. Ike is standing near the tree. He is 5 feet tall and casts a shadow that is 6 feet long.

The light rays, the tree, and its shadow form a triangle that is **similar** to the triangle formed by the light rays, Ike, and his shadow.

What is the size-change factor of the triangles? ____________

About how tall is the tree? ____________

2. Ike's dad is 6 feet tall. He is standing near the Washington Monument, which is 555 feet tall. Ike's dad casts a 7-foot shadow. About how long a shadow does the Washington Monument cast? (*Hint:* Draw sketches that include the above information.) ____________

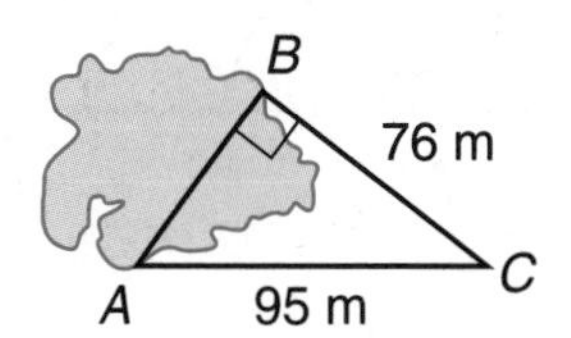

3. A surveyor wants to find the distance between points A and B on opposite ends of a lake. He sets a stake at point C so that angle ABC is a right angle. By measuring, he finds that $\overline{AC}$ is 95 meters long and $\overline{BC}$ is 76 meters long.

How far across the lake is it from point A to point B? ____________

Math Boxes

1. Find the area of the shaded region in the figure below. Use $\frac{22}{7}$ for π.

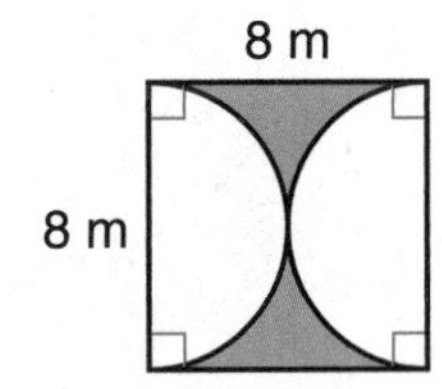

Area = ________ (unit)

2. Which expression represents the volume of the cylinder shown below? Circle the best answer.

A 800π

B 200π

C 80π

D 40π

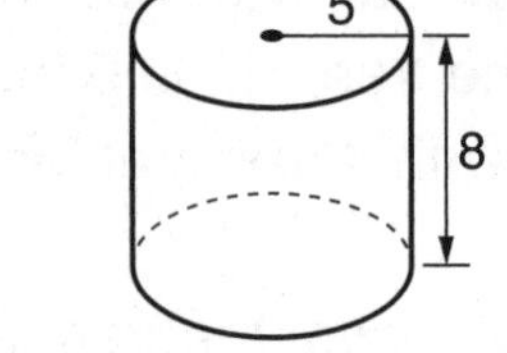

3. Write an equation for each statement. Then solve.

a. 125% of x is 625.

Equation ________

Solution ________

b. n% of 76 is 19.

Equation ________

Solution ________

4. Use the order of operations to evaluate each expression.

a. $(-3)^2 + 5 * 2^3$ ________

b. $(-1)^{50} * (-1)^{49}$ ________

c. $-6 + 2 * 3^3$ ________

d. $(-1)^5 * (2^4 - 13)^2$ ________

5. Fill in each shape to make a recognizable figure. See the example at the right.

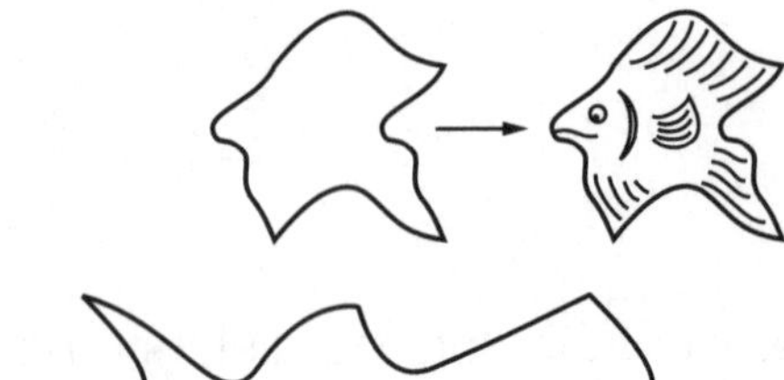

a.

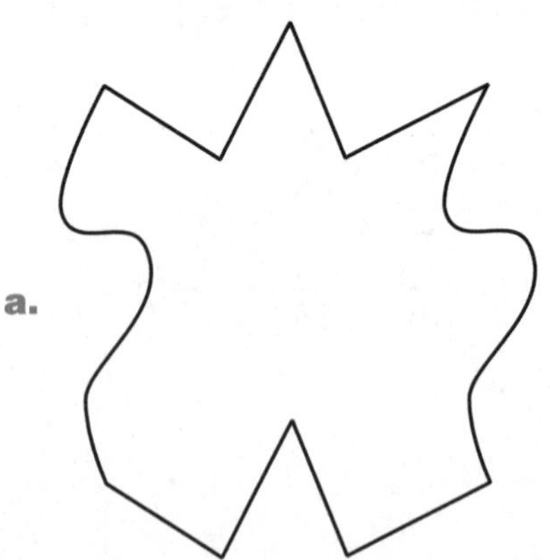

b.

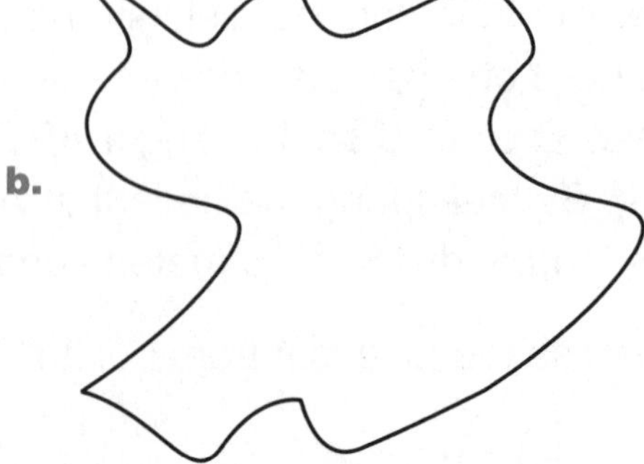

Date Time

LESSON 9•14

Math Boxes

1. Draw and label an obtuse angle *HIJ*. Then measure it.

m∠*HIJ* is about ________°.

2. Which of the hexagons shown below is a regular hexagon? Circle one.

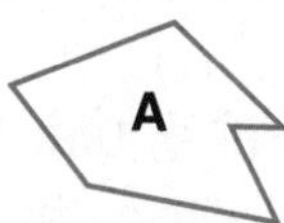

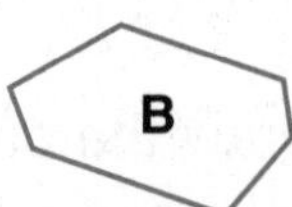

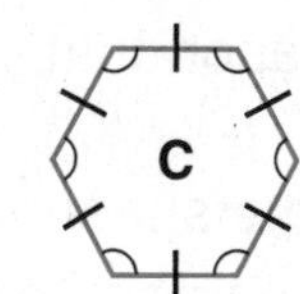

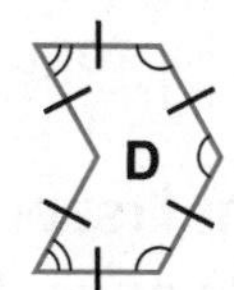

3. Use the diagonals to find the sum of the interior angle measures of the figure shown below.

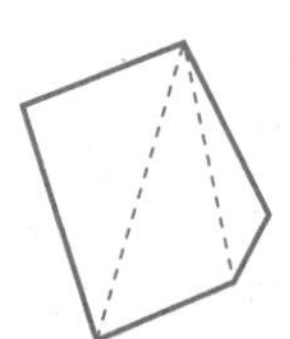

Sum of the interior angle measures = ________°.

4. Draw lines of symmetry for each figure shown below.

a.

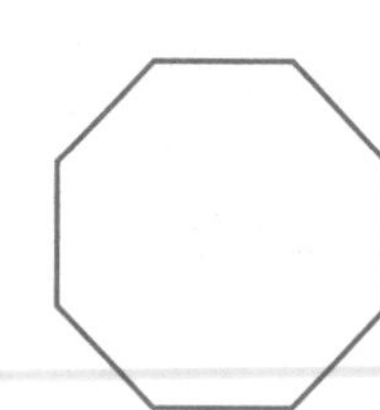

b.

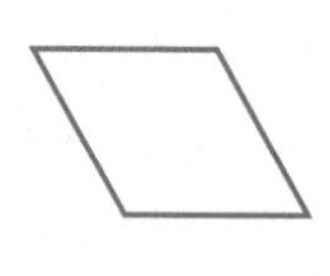

5. Identify whether the preimage (1) and image (2) are related by a translation, a reflection, or a rotation.

Write your answer on the line below.

____

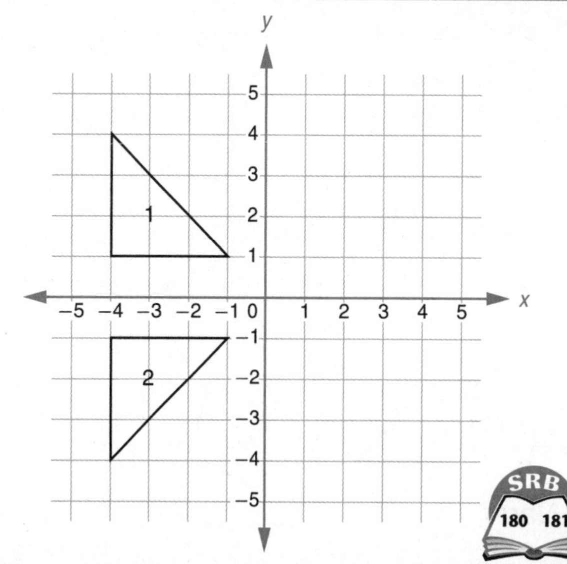

Date Time

Semiregular Tessellations

A **semiregular tessellation** is made up of two or more kinds of regular polygons. In a semiregular tessellation, the arrangement of angles around each **vertex point** looks the same. There are exactly 8 different semiregular tessellations. One is shown below.

Find and draw the other 7 semiregular tessellations. The only polygons that are possible in semiregular tessellations are equilateral triangles, squares, regular hexagons, regular octagons, and regular dodecagons. Use your Geometry Template and the template of a regular dodecagon that your teacher will provide.

Experiment first on a separate sheet of paper. Then draw the tessellations below and on the next page. Write the name of each tessellation.

1.

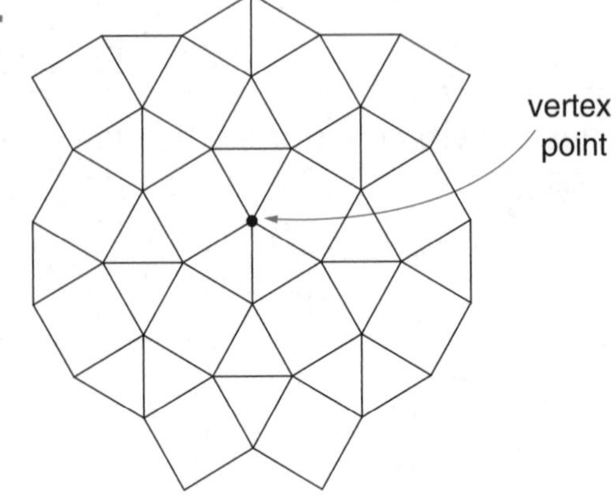

Name 3.3.4.3.4

2.

Name ____________

3.

Name ____________

4.

Name ____________

Date Time

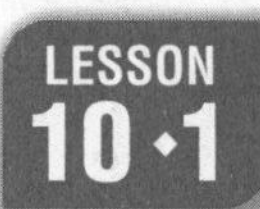

Semiregular Tessellations *continued*

5.

Name ____________________

6.

Name ____________________

7.

Name ____________________

8.

Name ____________________

Date Time

LESSON 10•1 Math Boxes

1. You can use the formula $s = 180 * (n - 2)$ to find the sum of the interior angle measures of a polygon having n sides.

For example, the sum of the interior angle measures (s) of a 7-sided polygon (n) is $180 * (7 - 2) = 180 * 5 = 900$, or 900°.

Suppose the sum of the interior angle measures of a polygon is 1,800°. How many sides does this polygon have?

______ sides

2. Without using a protractor, find the measure of each numbered angle.

E F 108° D 1 G 2 P 38° C H

m∠1 = ______ m∠2 = ______

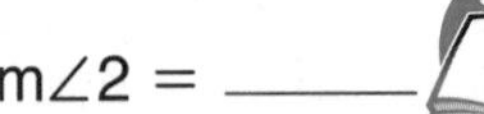

3. Multiply or divide. Write your answers in simplest form.

a. $3\frac{8}{9} * 4\frac{5}{6} =$ ______

b. ______ $= \frac{1}{5} * \frac{38}{3}$

c. ______ $= \frac{24}{15} \div \frac{1}{2}$

d. ______ $= \frac{3}{7} * \frac{22}{3}$

e. $\frac{24}{8} \div \frac{12}{7} =$ ______

4. Complete the table. Write a number sentence to describe the relationship between the numbers in the table.

x	y
12	−4
6	
	$-\frac{1}{3}$
−3	1
	6

Number sentence ______

5. Write an expression for the perimeter of the figure. Combine like terms.

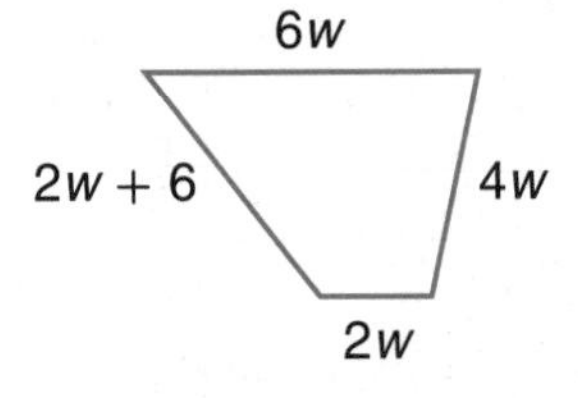

Perimeter = ______

6. Tell whether each of the following is true or false.

a. 1 cm > 1 in. ______

b. 1 m > 1 yd ______

c. 200 cm > 1 ft ______

d. 10 mm > 1 in. ______

e. 5 ft < 100 cm ______

LESSON 10•2

My Tessellation

On a separate sheet of paper, create an Escher-type translation tessellation using the procedure described on page 360 of the *Student Reference Book.* Experiment with several tessellations until you create one that you especially like.

Trace your final tessellation template in the space below.

In the space below, use your tessellation template to record what your tessellation looks like. Add details or color to your final design.

Date Time

LESSON 10•2 Translations, Reflections, Rotations

Use the grid to help you answer the questions below.

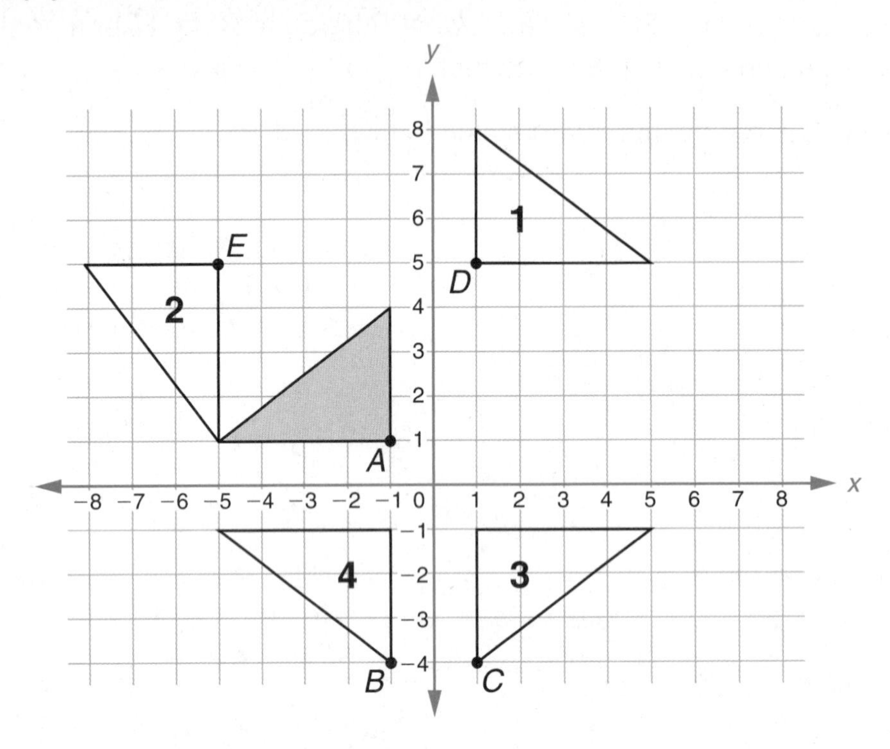

1. Identify which triangle results from performing the following transformations on the shaded right triangle.

   a. ______ A 90° counterclockwise rotation around point (–5,1)

   b. ______ A reflection over the *x*-axis

   c. ______ A reflection over the *y*-axis, followed by a reflection over the *x*-axis

   d. ______ A reflection over the *y*-axis, followed by a translation 4 units up

2. Use absolute value to find the distance between the following points. Write a number sentence to show how you found your answer.

   a. *A* and *B* ______________________________

   b. *B* and *C* ______________________________

   c. *C* and *D* ______________________________

   d. *D* and *E* ______________________________

Date Time

LESSON 10•2 Math Boxes

1. Apply the order of operations to evaluate each expression.

   a. $15 - 3.3 * 4 =$ ________

   b. $\frac{20}{4} * 5 + (-8) * 2 =$ ________

   c. $0.01 + 0.01 * 10 + 0.01 =$ ________

   d. $8 * (2 + -5) - 4 =$ ________

   e. $7 * 3^2 - \frac{10}{2} =$ ________

2. Estimate each quotient or product.

   a. $5.25 \div 2.003$ About ________

   b. $4.29 * 67.1$ About ________

   c. $80.25 \div 18.93$ About ________

   d. $52.31 * 19.9$ About ________

3. Convert between number-and-word and standard notations.

   a. 2,500,000 ________________

   b. 0.3 thousand ________________

   c. 7,400,000,000,000

   ____

   d. 1,234.5 million

   ____

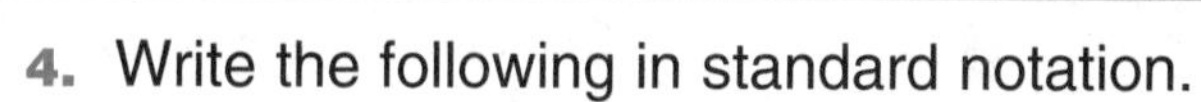

4. Write the following in standard notation.

   a. $5.38 * 10^7 =$ ________________

   b. $6.91 * 10^{-3} =$ ________________

   c. $3.04 * 10^0 =$ ________________

   d. ________________ $= 9.9011 * 10^5$

   e. $7.2 * 10^{-6} =$ ________________

5. Write an algebraic expression for the following situation.

   Rafael is twice as old as his brother Jorge was 3 years ago. Jorge is *j* years old now. How old is Rafael?

   Expression ________________

6. Fill in the blanks to complete each number sentence.

   a. 8(40 + 5) = (____ ∗ 40) + (8 ∗ ____)

   b. (10 − 3)6 = (10 ∗ ____) − (3 ∗ ____)

   c. (9 ∗ 50) + (9 ∗ ____) = 9(____ + 4)

   d. (7 ∗ 20) − (7 ∗ 3) = ____(____ − ____)

SRB 248 249

LESSON 10•3

Rotation Symmetry

Cut out the figures on Activity Sheet 7. Cut along the dashed lines only. Using the procedure demonstrated by your teacher, determine the number of different ways in which you can rotate (but not flip) each figure so the image exactly matches the preimage. Record the order of rotation symmetry for each figure.

1.

Order of rotation symmetry ________

2.

Order of rotation symmetry ________

3.

Order of rotation symmetry ________

4.

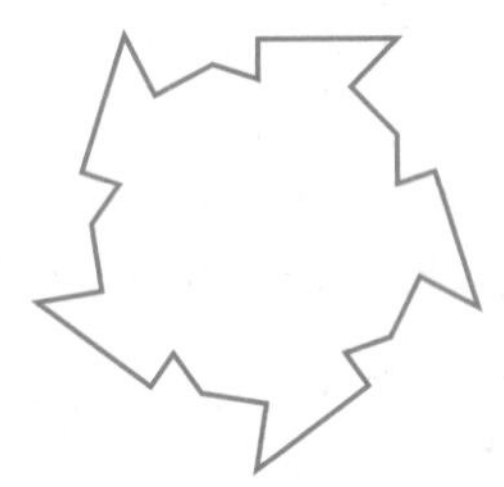

Order of rotation symmetry ________

Try This

5. This 10 of hearts has point symmetry. When the card is rotated 180°, it looks the same as it did in the original position.

original position

180° rotation

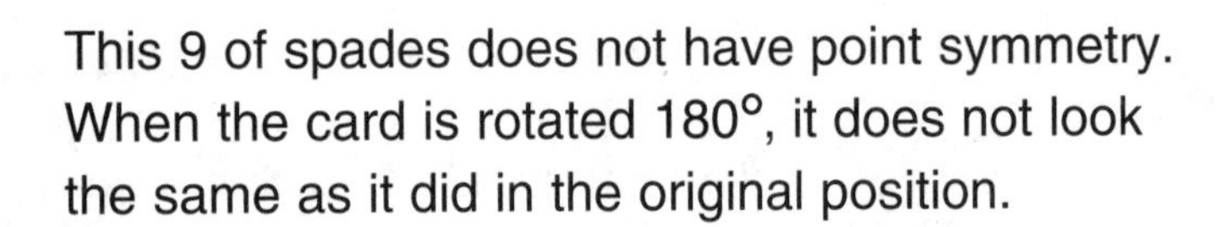

This 9 of spades does not have point symmetry. When the card is rotated 180°, it does not look the same as it did in the original position.

original position

180° rotation

Which of the cards in an ordinary deck of playing cards (not including face cards) have point symmetry?

____

____

To learn a magic trick that uses point symmetry with playing cards, see page 355 in the *Student Reference Book*.

Date Time

LESSON 10·3 Math Boxes

1. You can use the formula $s = 180 * (n - 2)$ to find the sum of the interior angle measures of a polygon having n sides.

Find the value of x in the figure below.

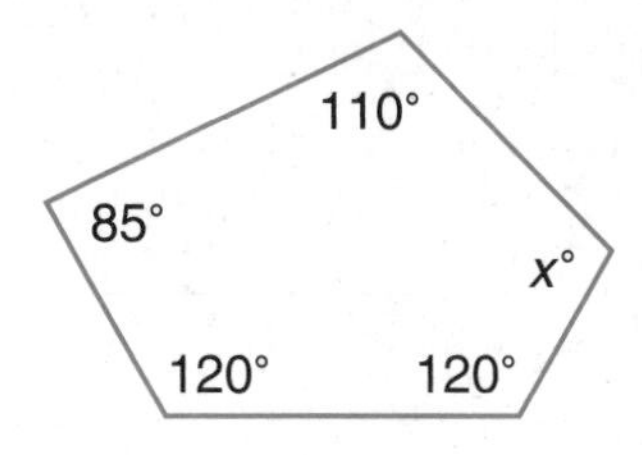

$x =$ ______ °

2. Write an equation that describes the angle relationships shown. Then solve it.

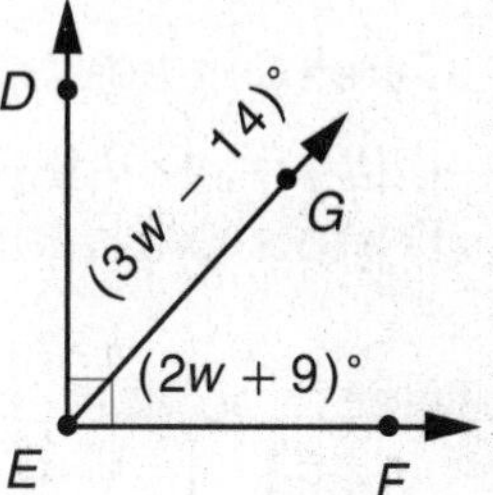

Equation ______

$w =$ ______

m∠$DEG =$ ______ °

m∠$GEF =$ ______ °

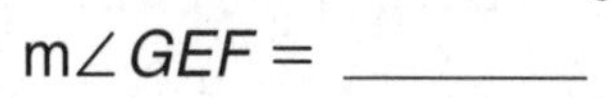

3. Multiply or divide. Write your answers in simplest form.

a. $1\frac{3}{7} * 2\frac{1}{5} =$ ______

b. ______ $= 3\frac{6}{8} * \frac{28}{6}$

c. ______ $= 5\frac{1}{10} \div 2\frac{5}{4}$

d. ______ $= \frac{46}{3} \div 20$

e. $5\frac{3}{8} * \frac{1}{8} =$ ______

4. Complete the table. Write a number sentence to describe the relationship between the numbers in the table.

x	y
100	
30	$\frac{3}{2}$
10	
–2	$-\frac{1}{10}$
	$-\frac{3}{4}$

Number sentence ______

5. The area of the parallelogram shown below is 63 cm^2. Which equation can you use to find its height? Circle the best answer.

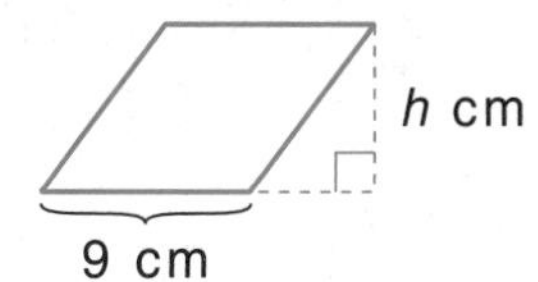

A. $\frac{9h}{2} = 63$ **B.** $2h + 18 = 63$

C. $9h = 63$ **D.** $\frac{9}{63} = h$

6. Tell whether each of the following is true or false.

a. 1 L > 1 pt ______

b. 1 mL < 1 fl oz ______

c. 1 L > 1 gal ______

d. 1 kg > 1 lb ______

e. 1 g < 1 oz ______

Date Time

LESSON 10·4

Shrinking Quarter

Math Message

1. Carefully trace the dime-size circle below onto the center of an $8\frac{1}{2}$ in. by 11 in. sheet of paper. Cut out the circle.
   Try to slip a quarter through the hole in the page without tearing the paper.

2. Were you able to do this trick? If you were, explain how you did it. If not, explain why you could not.

____

____

____

____

____

Date Time

LESSON 10•4

Math Boxes

1. Apply the order of operations to evaluate each expression.

   a. $4 * \frac{7}{2} + 7 =$ ________

   b. $8 + (-15) * 6 =$ ________

   c. $\frac{6^2}{9} + 3 * 4 =$ ________

   d. $8 + 7 - (-2) * 5 =$ ________

   e. $12 / 6 + 9 * 3 =$ ________

2. Estimate each quotient or product.

   a. $44.2 * 37$ About ________

   b. $708 \div 0.52$ About ________

   c. $625.7 \div 8.3$ About ________

   d. $99.4 * 3.7$ About ________

3. Convert between number-and-word and standard notations.

   a. 6,500,000 ________________

   b. 0.75 billion ________________

   c. 12,500,000,000,000

   ____

   d. 57.25 million

   ____

4. Write the following in standard notation.

   a. $2.73 * 10^5 =$ ________________

   b. $1.03 * 10^{-4} =$ ________________

   c. $9.855 * 10^0 =$ ________________

   d. 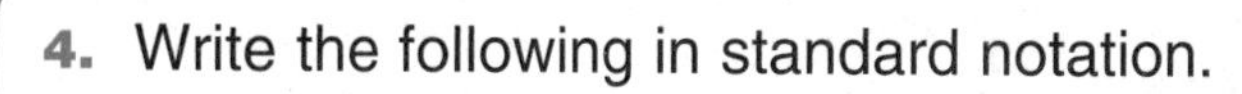 $= 4.226 * 10^6$

   e. $5.435 * 10^{-2} =$ ________________

5. Chen ran 2 miles more than two-thirds as far as Kayla ran. If Kayla ran k miles, how many miles did Chen run?

   Which algebraic expression can you use to answer the question? Fill in the circle next to the best answer.

   Ⓐ $\frac{2}{3}k + 2$ Ⓑ $k + \frac{2}{3}$

   Ⓒ  $2k + \frac{2}{3}$ Ⓓ $2k \div 3$

6. Fill in the blanks to complete each number sentence.

   a. $9(20 + 7) = (\_\_\_\_ * 20) + (9 * \_\_\_\_)$

   b. $(50 - 5)3 = (50 * \_\_\_\_) - (5 * \_\_\_\_)$

   c. $(8 * 70) + (8 * \_\_\_\_) = 8(\_\_\_\_ + 7)$

   d. $(6 * 50) - (6 * 9) = \_\_\_\_(\_\_\_\_ - \_\_\_\_)$

Date Time

LESSON 10•4 Rubber-Sheet Geometry

You and your partner will need the following materials: 3 latex gloves, a straightedge, a pair of scissors, and a permanent marker.

Step 1 Cut the fingers and thumb off of each glove.

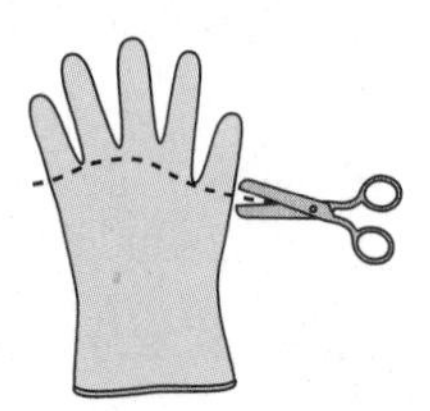

Step 2 Make a vertical cut through each of the three cylinders that remain. The cutting creates 3 rubber sheets.

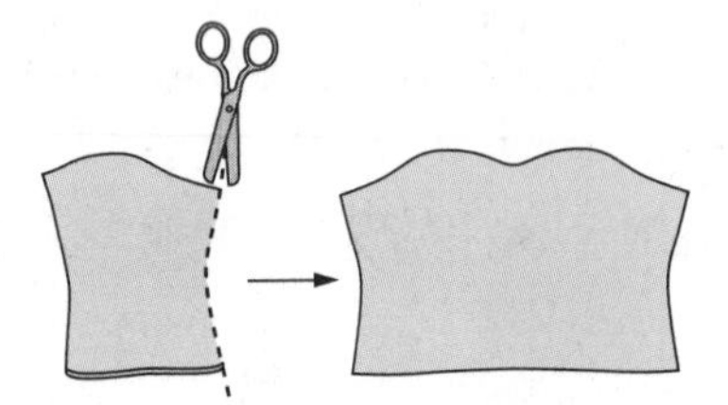

Step 3 Use a permanent marker and a straightedge to draw the following figures on the rubber sheets. Draw the figures large enough to fill most of the sheet.

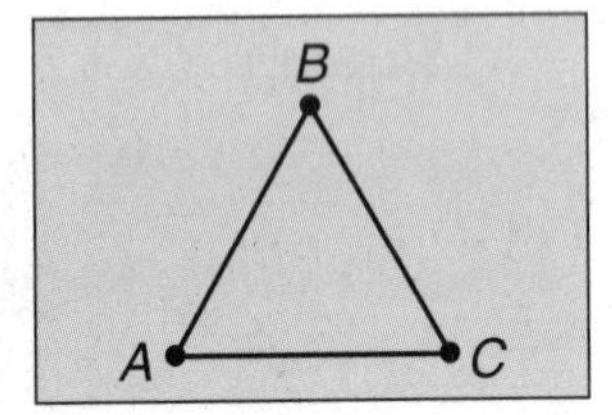

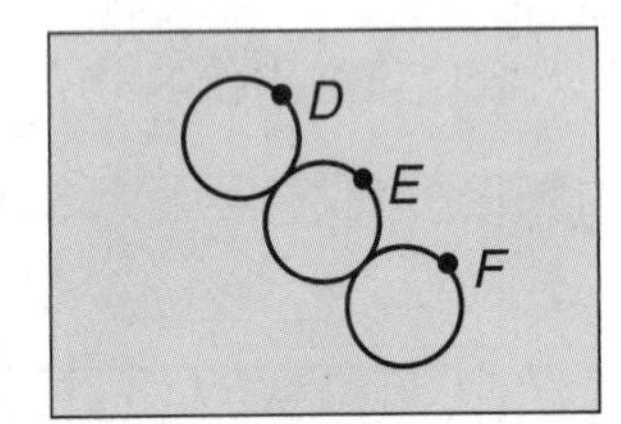

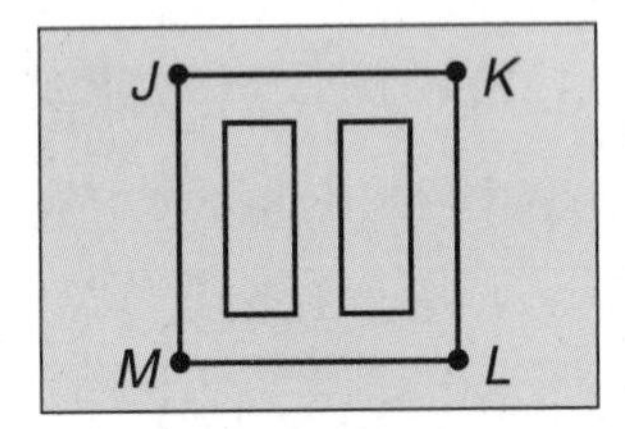

Step 4 Work with your partner to stretch the rubber sheets to see what other figures you can make.

Step 5 Complete journal page 381.

Date Time

LESSON 10•4

Rubber-Sheet Geometry *continued*

1. Experiment with the figures on your rubber sheets. Circle any of the transformed figures in the right column that are topologically equivalent to the corresponding original figure in the left column.

Original Figure	Transformed Figures
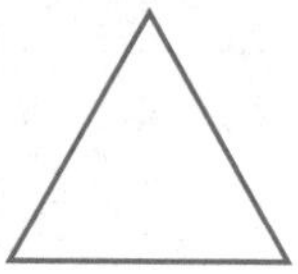	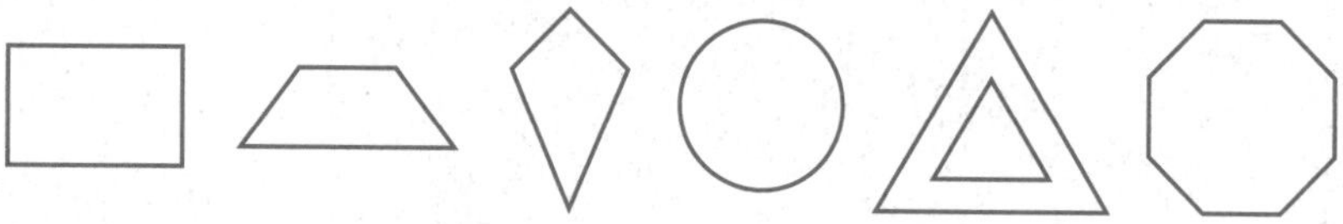
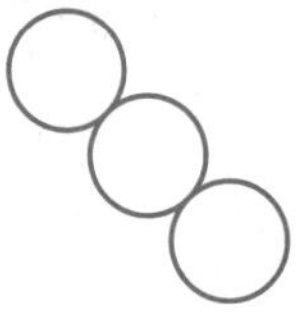	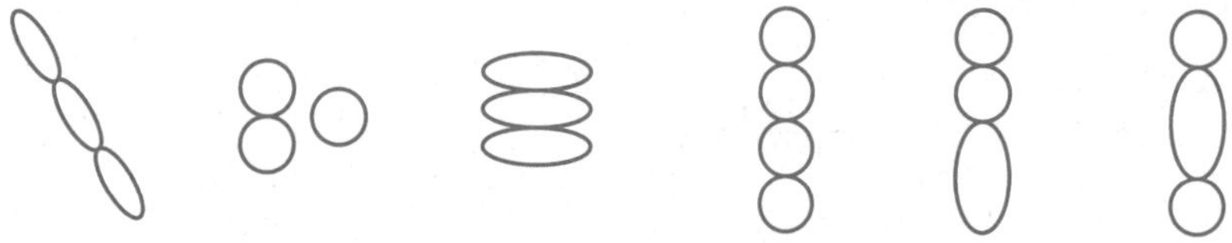
	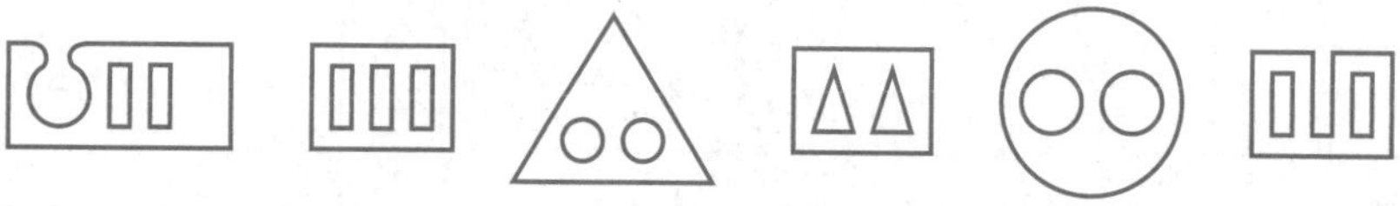

2. Choose one of the above figures that you did not circle and explain why it is not topologically equivalent to the original figure.

____

____

____

Date Time

LESSON 10•5

Math Boxes

1. Draw the line(s) of reflection symmetry for the figure below. Then determine its order of rotation symmetry.

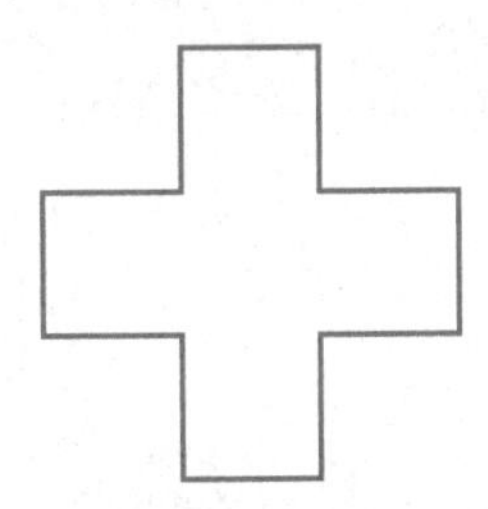

Order of rotation symmetry ______

2. Lines p and q are parallel. Write an equation you can use to find the value of x.

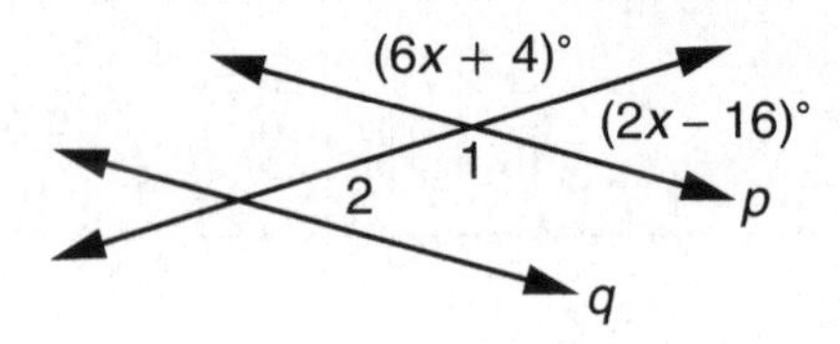

Equation ______

$x =$ ______°

$m\angle 1 =$ ______°

$m\angle 2 =$ ______°

3. Multiply or divide. Write your answers in simplest form.

a. $1\frac{3}{5} * (-2\frac{1}{2}) =$ ______

b. ______ $= (-\frac{7}{12})(-\frac{3}{84})$

c. ______ $= 5\frac{1}{10} \div 2\frac{5}{4}$

d. ______ $= -1\frac{1}{3} \div (-\frac{5}{9})$

e. $-3\frac{2}{3} \div (-2\frac{4}{9}) =$ ______

4. Complete the table. Write a number sentence to describe the relationship between the numbers in the table.

x	y
10	23
4	
$\frac{1}{2}$	4
	3
–3	–3

Number sentence ______

5. Tennis balls with a diameter of 2.5 in. are packaged 3 to a can. The can is a cylinder. Find the volume of the space in the can that is *not* occupied by tennis balls. Assume that the balls touch the sides, top, and bottom of the can.

Use the formula $V = \frac{4}{3}\pi r^3$ and 3.14 for π. Round your answer to the nearest hundredth.

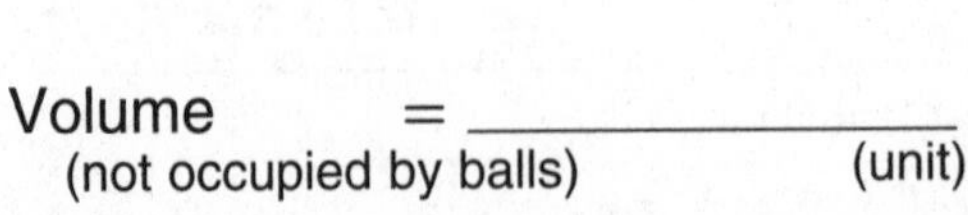

Volume (not occupied by balls) = ______ (unit)

6. Complete.

a. $1\ ft^2 =$ ______ $in.^2$

b. $1\ m^2 =$ ______ cm^2

c. $1\ yd^2 =$ ______ ft^2

d. $1\ ft^3 =$ ______ $in.^3$

e. $1\ yd^3 =$ ______ ft^3

Date Time

LESSON 10•5

Making a Möbius Strip

Follow the steps below to make a Möbius strip.

Materials

- ☐ a sheet of newspaper or adding machine tape
- ☐ scissors
- ☐ tape
- ☐ a bright color crayon, marker, or pencil

Step 1 Cut a strip of newspaper about $1\frac{1}{2}$ inches wide and as long as possible, or cut a strip of adding machine tape about 2 feet long.

1.5 in. 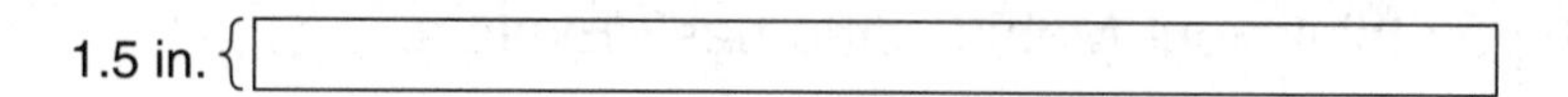

Step 2 Put the ends of the strip together as though you were making a simple loop.

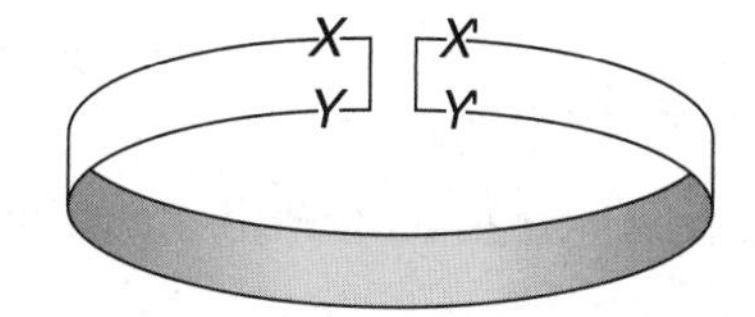

Step 3 Give one end of the strip a half-twist and tape the two ends together.

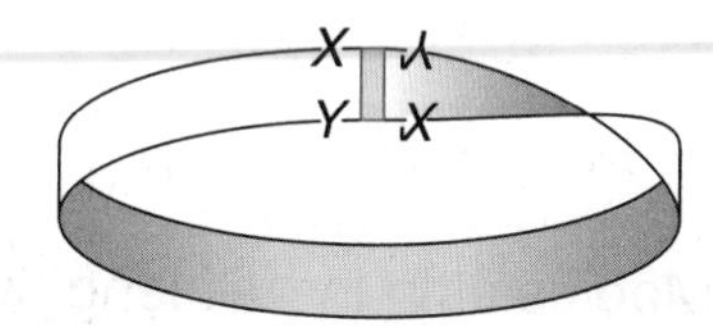

You have just made what mathematicians call a **Möbius strip.** How is it different from the simple loop of paper used in the Math Message? Do you notice anything special about it?

____

____

____

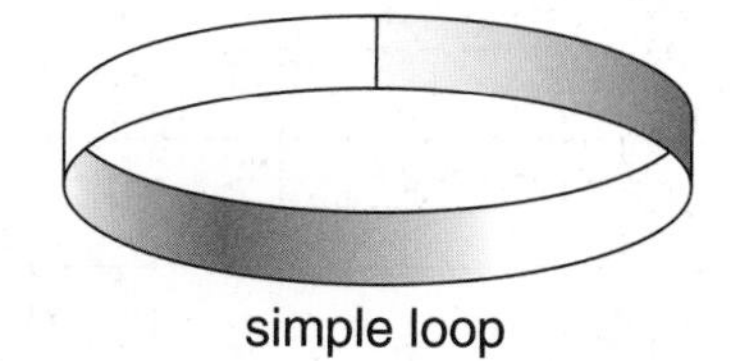

simple loop

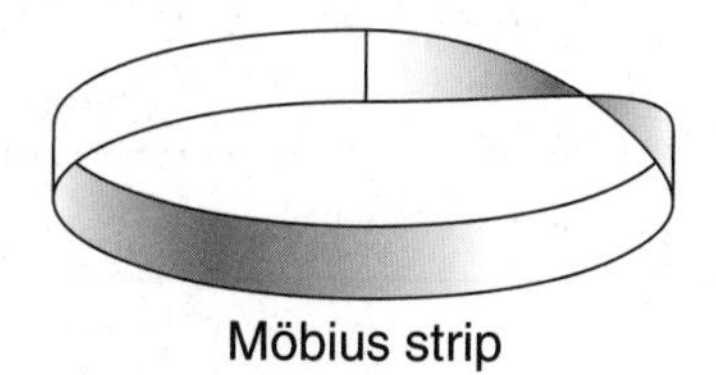

Möbius strip

Date Time

LESSON 10•5

Experimenting with Möbius Strips

1. How many sides do you think your Möbius strip has? ________ side(s)

2. Use a marker to shade one side of your Möbius strip.

3. Now how many sides do you think your Möbius strip has? Explain.

____

4. How many edges do you think your Möbius strip has? ________ edge(s)

5. Use your marker to color one edge of your Möbius strip.

6. Now how many edges do you think your Möbius strip has? Explain.

____

____

Cutting Möbius strips leads to some surprising results.

7. Predict what will happen if you cut your Möbius strip in half lengthwise.

____

____

8. Now cut your Möbius strip in half lengthwise. How many strips did you get? ________ strip(s)

   Compare the lengths and widths of the new strip and the original strip.

   Describe your observations. ____

   ____

   How many half-twists does your new strip have? ________ half-twist(s)

9. Make another Möbius strip and cut it one-third of the way from the edge. You may find it helpful to draw lines on the strip before cutting.

   What happened? ____

   ____

   ____

Date Time

LESSON 10•5

Experimenting with Möbius Strips *continued*

10. Make another Möbius strip and a simple loop. Then tape the loop and the Möbius strip together at right angles.

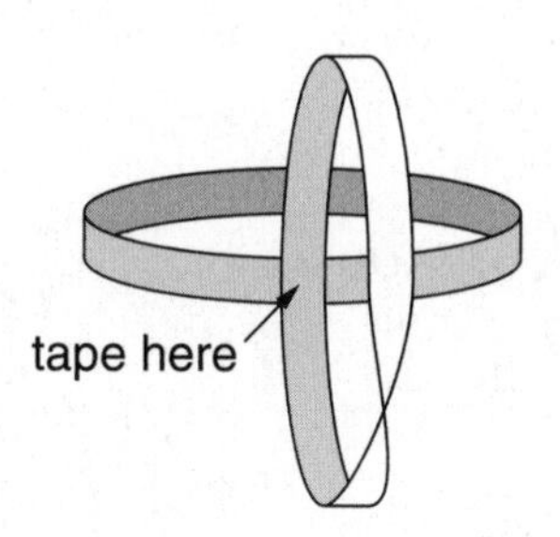

Cut the Möbius strip and the loop in half lengthwise. What happened?

____

____

11. Experiment with cutting Möbius strips in half and in thirds lengthwise. Try putting two or more half-twists in the band before you tape it. Describe what you did, as well as your results.

____

____

____

____

12. Shortly after the first Earth Day on April 22, 1970, a producer of recycled paperboard sponsored a contest. Contestants presented designs that symbolized the company's recycling process. Gary Anderson, a student at the University of Southern California, won the top prize. Anderson's symbol was a three-chasing-arrows Möbius strip.

Explain what you think Anderson was trying to say about recycling with his symbol.

____

____

Date Time

LESSON 10•5

Reflection and Rotation Symmetry

For each figure, draw the line(s) of reflection symmetry, if any. Then determine the order of rotation symmetry for the figure.

1.

Order of rotation symmetry ______

2.

Order of rotation symmetry ______

3.

Order of rotation symmetry ______

4.

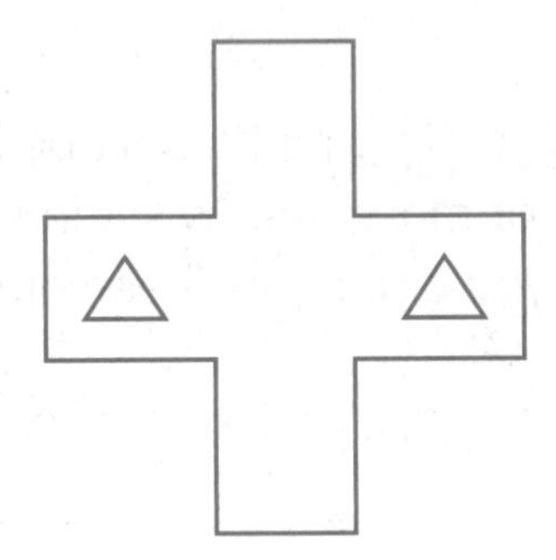

Order of rotation symmetry ______

5.

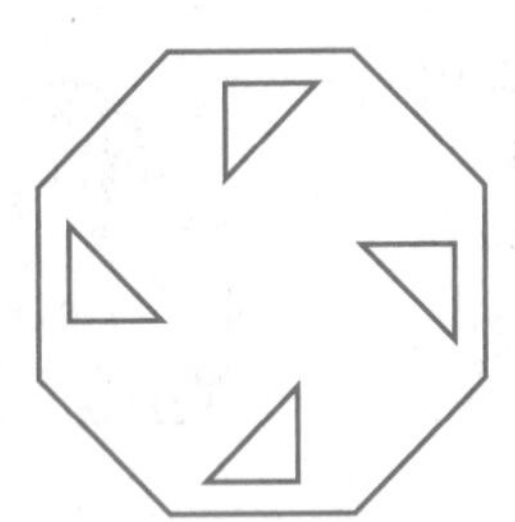

Order of rotation symmetry ______

6.

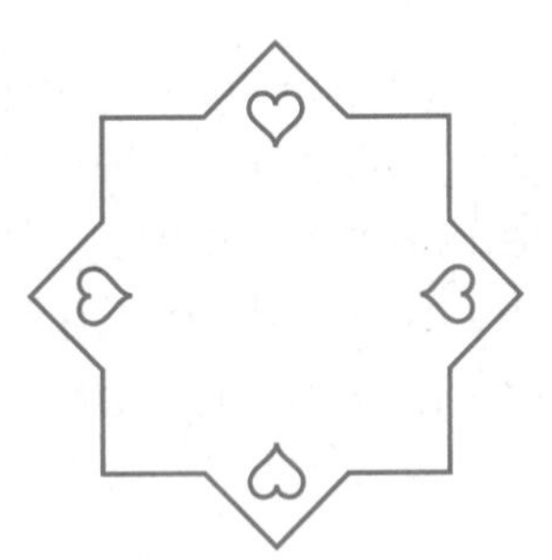

Order of rotation symmetry ______

U.S. Traditional Addition: Decimals 1

Algorithm Project 1

Use any strategy to solve the problem.

1. Avani spent \$3.79 on a sandwich. She spent \$1.47 on a glass of juice. How much money did Avani spend in all?

   ____

Estimate each sum. Then use U.S. traditional addition to solve each problem.

2. 6.515 + 2.443 Estimate: ______________

   6.515 + 2.443 = ______________

3. 89.26 + 4.9 Estimate: ______________

   ______________ = 89.26 + 4.9

4. 17.6 + 605.62 Estimate: ______________

   17.6 + 605.62 = ______________

5. \$72.88 + \$7.25 Estimate: ______________

   \$72.88 + \$7.25 = ______________

6. 44.97 + 23.075 Estimate: ______________

   44.97 + 23.075 = ______________

7. 3.066 + 8.38 Estimate: ______________

   ______________ = 3.066 + 8.38

Date Time

U.S. Traditional Addition: Decimals 2

Algorithm Project 1

Estimate each sum. Then use U.S. traditional addition to solve each problem.

1. Rafael had $26.75. Later he found 68¢ in one of his pockets. How much money does Rafael have now?

Estimate: ______________ Solution: ______________

2. 74.26 + 7.88 Estimate: ______________

74.26 + 7.88 = ______________

3. 5.201 + 2.49 Estimate: ______________

5.201 + 2.49 = ______________

4. 66.59 + 9.177 Estimate: ______________

______________ = 66.59 + 9.177

5. 80.556 + 6.767 Estimate: ______________

80.556 + 6.767 = ______________

6. $389.99 + $7.78 Estimate: ______________

______________ = $389.99 + $7.78

7. 75.695 + 64.906 Estimate: ______________

75.695 + 64.906 = ______________

 Go to www.everydaymathonline.com for additional practice pages.

Date Time

U.S. Traditional Addition: Decimals 3

Algorithm Project 1

Estimate the sum. Then use U.S. traditional addition to solve the problem.

1. Kwan bought a flashlight for $9.75 and batteries for $6.98. How much did Kwan spend altogether?

   Estimate: ______________ Solution: ______________

2. Write a number story for $45.48 + $9.66. Solve your number story.

   ____

   ____

   ____

   ____

Fill in the missing digits in the addition problems.

3.
```
   1  1  1   1
   5  8  ☐ . 9  4
+     6  7 . 2  7
-----------------
   ☐  5  2 . ☐  1
```

4.
```
   1  ☐     1  1
   3  2  7 . 8  ☐  9
+     9  5 . 9  9  ☐
--------------------
   ☐  2  ☐ . 8  5  9
```

5.

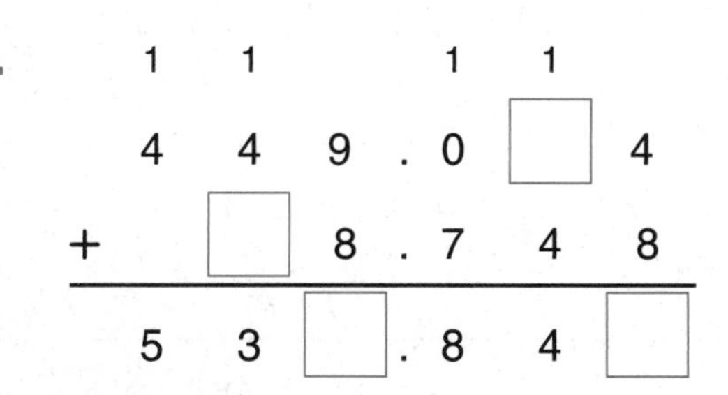

6.
```
                  1
   1  2  3 . 7  8  9
+  5  ☐  ☐ . ☐  0  ☐
--------------------
   ☐  6  9 . 8  9  0
```

U.S. Traditional Addition: Decimals 4

Algorithm Project 1

Estimate the sum. Then use U.S. traditional addition to solve the problem.

1. Jessie's father keeps a very accurate record of Jessie's height. Jessie was 96.45 cm tall when she was 4 years old. Since then, she has grown 54.08 cm. How tall is Jessie now?

   Estimate: ______________ Solution: ______________

2. Write a number story for 5.73 + 209.79. Solve your number story.

   ____

   ____

   ____

   ____

Fill in the missing digits in the addition problems.

3.
```
   1  1  1   1
   2  6  [ ] . 5  6
+     9  9   . 4  6
-------------------
   3  [ ] 3  . [ ] 2
```

4.
```
[ ]  1       1
     7  6 . [ ]  9
+ 5  3 [ ] . 0   1
------------------
  6 [ ] 4  . 5  [ ]
```

5.

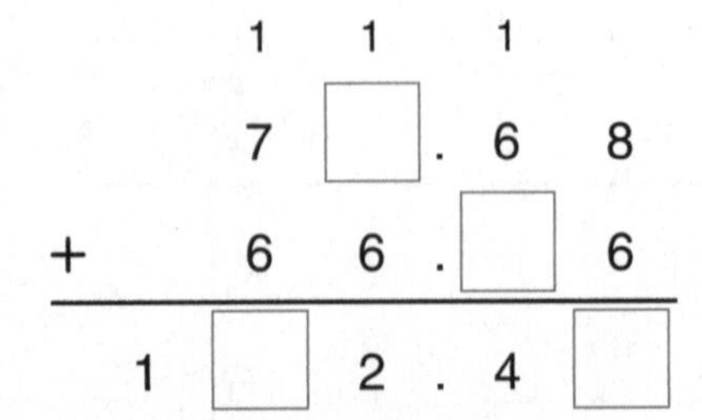

```
     1  1    1
     7  [ ] . 6  8
+    6  6   . [ ] 6
-------------------
  1 [ ] 2   . 4  [ ]
```

6.
```
   1  1  1   1
   3  8  9 . 9  8   3
+  3  4  7 . 7 [ ] [ ]
---------------------
  [ ] 3 [ ] . [ ] 3  7
```

Date Time

PROJECT 2

U.S. Traditional Subtraction: Decimals 1

Algorithm Project 2

Use any strategy to solve the problem.

1. Eleni has \$8.52 in her pocket. Paul has \$4.87 in his pocket. How much more does Eleni have?

____

Estimate each difference. Then use U.S. traditional subtraction to solve each problem.

2. 8.635 – 2.431 Estimate: _______

8.635 – 2.431 = _______

3. 3.07 – 1.48 Estimate _______

3.07 – 1.48 = _______

4. 7.1 – 2.855 Estimate: _______

_______ = 7.1 – 2.855

5. \$96.25 – \$68.73 Estimate _______

\$96.25 – \$68.73 = _______

6. 72.004 – 59.965 Estimate: _______

_______ = 72.004 – 59.965

7. 6.484 – 3.49 Estimate _______

6.484 – 3.49 = _______

Date Time

PROJECT 2

U.S. Traditional Subtraction: Decimals 2

Algorithm Project 2

Estimate each difference. Then use U.S. traditional subtraction to solve each problem.

1. Tisha bought a pair of jeans. The jeans originally cost $42.95. They were on sale for $28.76. How much money did Tisha save?

Estimate: ______ Solution: ______

2. 5.03 – 1.44 Estimate: ______

5.03 – 1.44 = ______

3. 8.772 – 4.501 Estimate: ______

8.772 – 4.501 = ______

4. 6.433 – 2.83 Estimate: ______

______ = 6.433 – 2.83

5. 663.7 – 9.85 Estimate: ______

663.7 – 9.85 = ______

6. $4.25 – $1.79 Estimate: ______

______ = $4.25 – $1.79

7. 90.031 – 65.674 Estimate: ______

90.031 – 65.674 = ______

 Go to www.everydaymathonline.com for additional practice pages.

Date Time

U.S. Traditional Subtraction: Decimals 3

Algorithm Project 2

Estimate the difference. Then use U.S. traditional subtraction to solve the problem.

1. Shawn bought a notebook and a pen. The total cost (before tax) was $5.46. The pen cost $1.89. How much did the notebook cost?

Estimate: ________________ Solution: ________________

2. Write a number story for $65.33 – $48.95. Solve your number story.

____

____

____

____

Fill in the missing numbers in the subtraction problems.

3.

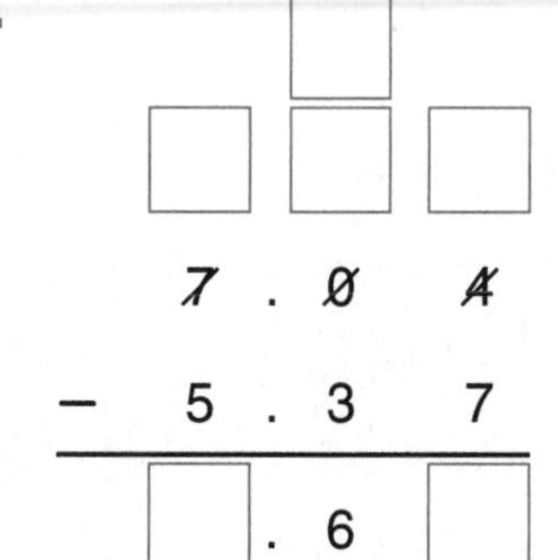

4.

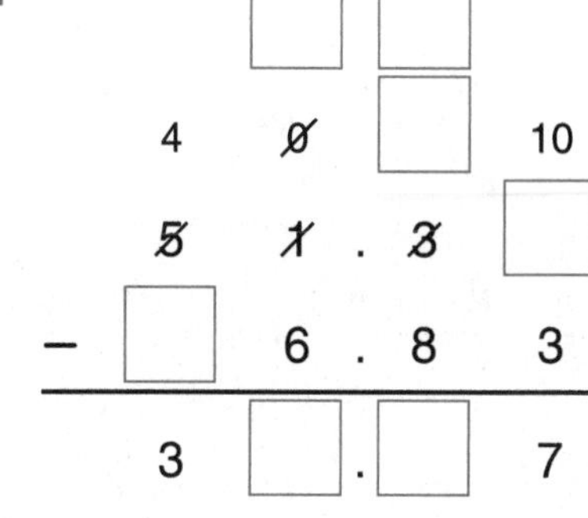

5.

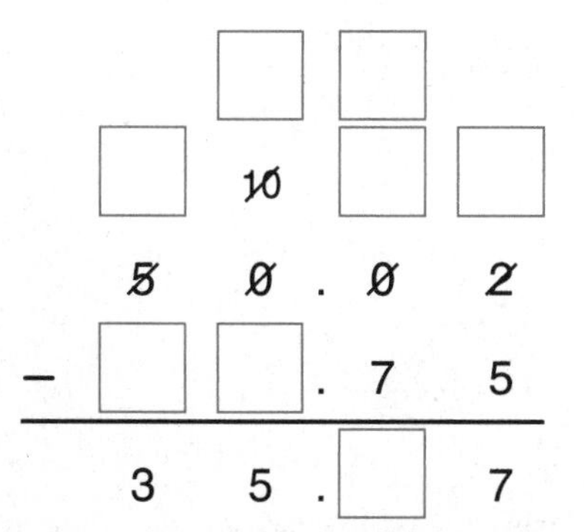

6.

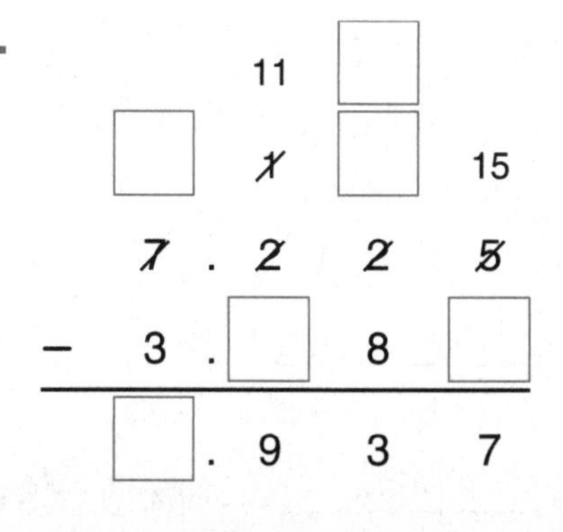

Date Time

U.S. Traditional Subtraction: Decimals 4

Algorithm Project 2

Estimate the difference. Then use U.S. traditional subtraction to solve the problem.

1. Isabel weighs 36.84 kg. Her father weighs 72.42 kg. How much less does Isabel weigh?

Estimate: ______________________ Solution: ______________

2. Write a number story for 7.12 − 3.45. Solve your number story.

____

____

____

____

Fill in the missing numbers in the subtraction problems.

3.

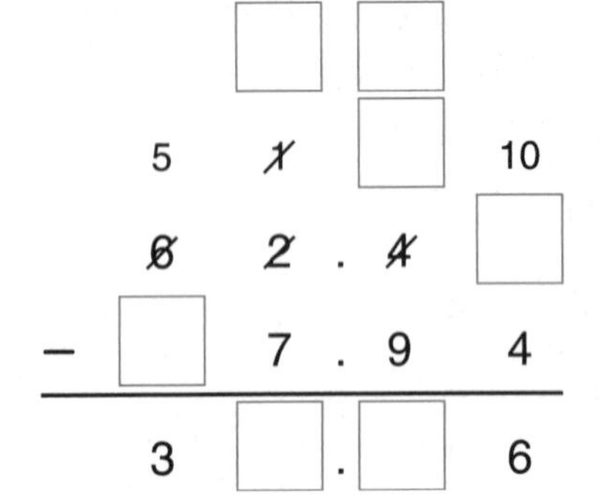

4.

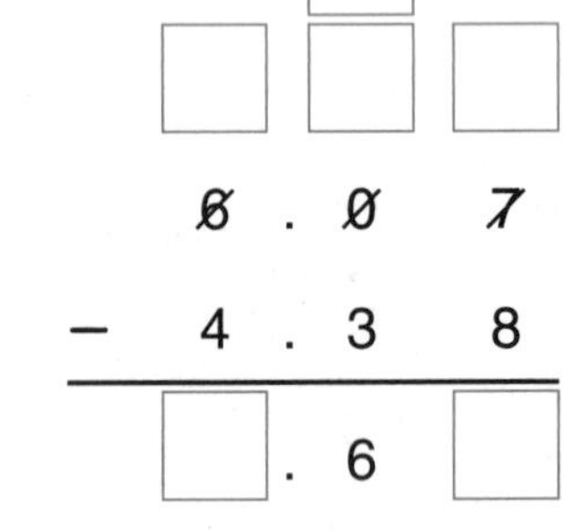

5.

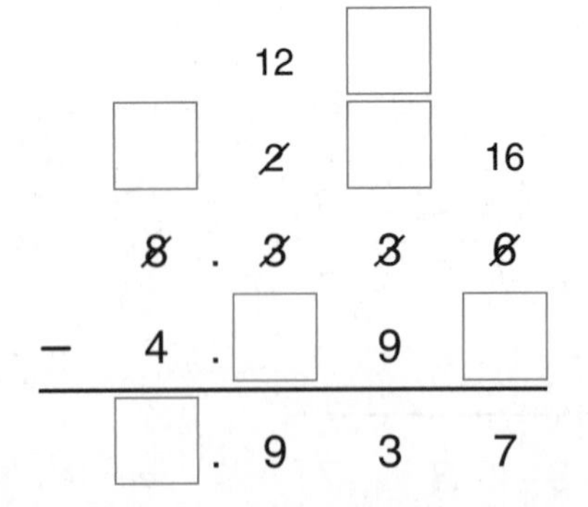

6.

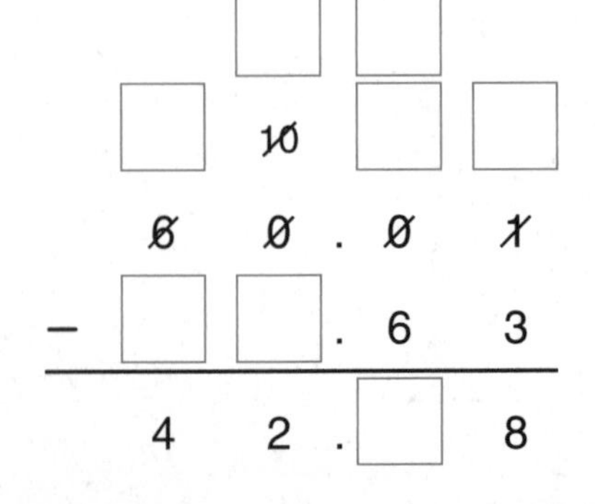

 Go to www.everydaymathonline.com for additional practice pages.

Date Time

U.S. Traditional Multiplication: Decimals 1

Algorithm Project 3

Use any strategy to solve the problem.

1. At Jerry's Fruit Market, cherries cost $3.49 per pound. How much would 3 pounds of cherries cost?

____

Use U.S. traditional multiplication to solve each problem. Use estimation or count decimal places to place the decimal point in your answers.

2. 90.07 * 96 = _______________

3. 2.9 * 6,067 = _______________

4. _______________ = 0.7 * 0.09

5. 0.08 * 0.09 = _______________

6. _______________ = 14.964 * 5.5

7. 3.955 * 0.8 = _______________

Date Time

PROJECT 3

U.S. Traditional Multiplication: Decimals 2

Algorithm Project 3

Use U.S. traditional multiplication to solve each problem. Use estimation or count decimal places to place the decimal point in your answers.

1. Find the area of the rectangle.

____

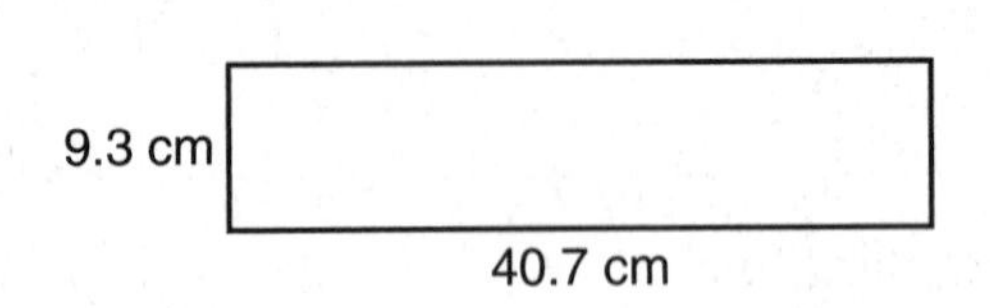

2. 9.19 * 0.9 = _______________

3. 0.26 * 0.43 = _______________

4. _______________ = 3.7 * 240

5. 0.95 * 0.238 = _______________

6. _______________ = 70.1 * 48

7. 32.337 * 3 = _______________

 Go to www.everydaymathonline.com for additional practice pages.

Date Time

U.S. Traditional Multiplication: Decimals 3

Algorithm Project 3

Use U.S. traditional multiplication to solve each problem. Use estimation or count decimal places to place the decimal point in your answers.

1. In a scale drawing of a house, the front door is 4.1 cm tall. The actual door is 48 times taller. What is the height of the actual door?

____

2. Write a number story for 60.03 * 96. Solve your number story.

____

____

____

____

Fill in the missing digits in the multiplication problems.

3.

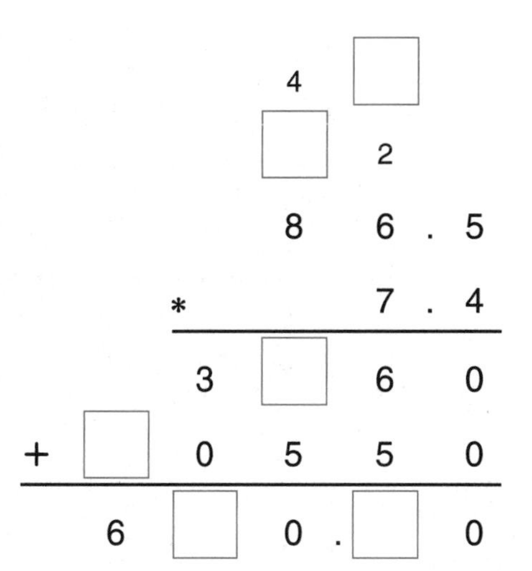

4.

```
       4
    6 . 0 5
*     0 . □
-----------
□ . 4 □ 5
```

5.

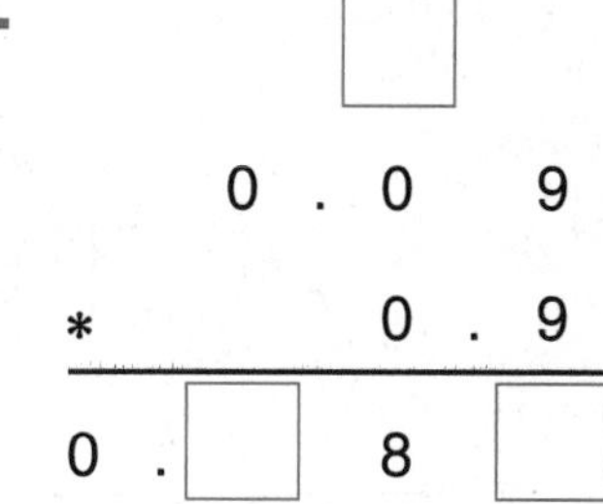

PROJECT 3

U.S. Traditional Multiplication: Decimals 4

Algorithm Project 3

Use U.S. traditional multiplication to solve each problem. Use estimation or count decimal places to place the decimal point in your answers.

1. Mandisa counted 206 cans of her favorite brand of almonds at the grocery store. One can of almonds weighs 9.5 ounces. What is the total weight of all 206 cans?

____

2. Write a number story for 4.13 * 21. Solve your number story.

____

____

____

____

Fill in the missing digits in the multiplication problems.

3.

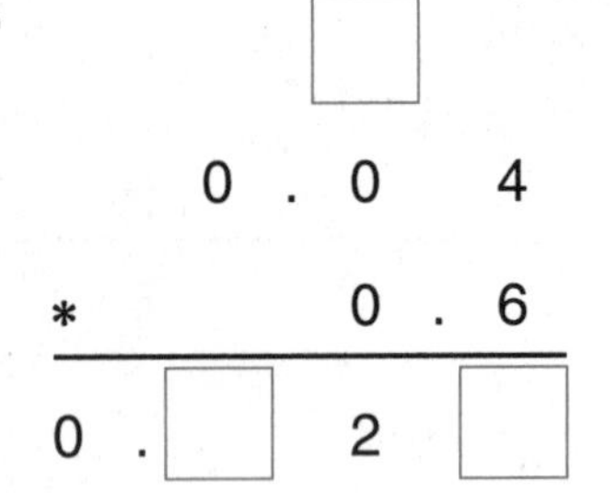

4.

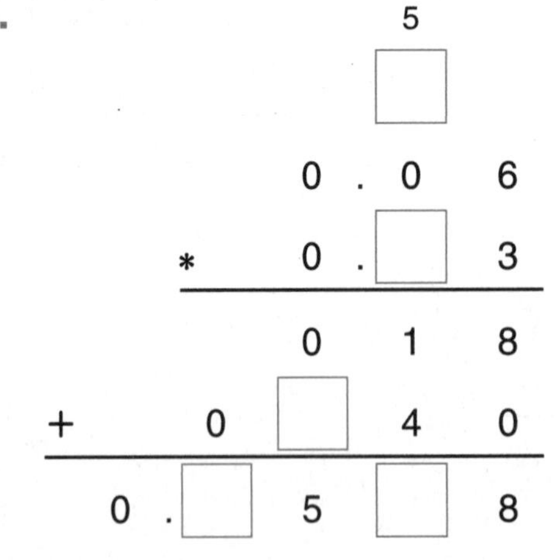

5.

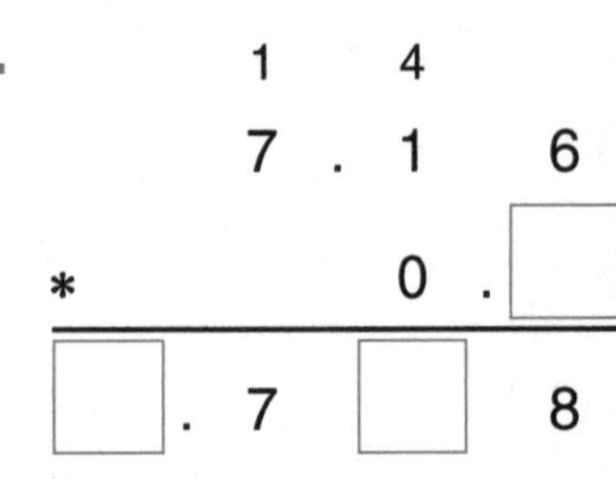

Date Time

U.S. Traditional Long Division

Algorithm Project 4

Use any strategy to solve the problem.

1. The four sixth-grade classes at Linda Vista Elementary School held a book sale to raise money for their classroom libraries. The sale raised \$464. How much should each class get?

____

Use U.S. traditional long division to solve each problem.

2. \$395 / 5 = _______________

3. 908 / 22 = _______________

4. 837 / 3 = _______________

5. 975 / 75 = _______________

Date Time

PROJECT 4

Long Division with Decimal Dividends

Algorithm Project 4

Use any strategy to solve the problem.

1. Three friends bought some supplies for a school project for \$14.07. How much should each friend pay?

______

Use U.S. traditional long division to solve each problem.

2. \$25.86 / 6 = ______

3. 1.071 / 7 = ______

4. 6.952 / 11 = ______

5. Rename $\frac{3}{8}$ as a decimal.

______

 Go to www.everydaymathonline.com for additional practice pages.

Date Time

Long Division with Decimal Divisors

Algorithm Project 4

Use any strategy to solve the problem.

1. Donuts cost $0.89 each at the Farmers' Market. How many donuts can be bought with $11.50?

______

For Problems 2–5, find equivalent problems with no decimals in the divisors. Then solve the equivalent problems.

2. 24 / 0.8 =

______
equivalent problem

3. 27.096 / 0.06 =

______
equivalent problem

4. 28.8 / 1.8 =

______
equivalent problem

5. 0.0084 / 0.3 =

______
equivalent problem

Date Time

PROJECT 4

Short Division

Algorithm Project 4

Short division is a fast way to divide with paper and pencil. It's like long division, but all the multiplying and subtracting is done mentally. Short division works best with single-digit divisors.

Study the examples below. Then use short division to solve Problems 1–6.

Example 1:

Long Division

		3	5	8
5	1	7	9	2
—	1	5		
		2	9	
	—	2	5	
			4	2
		—	4	0
				2

Short Division

		3	5	8	R2
5	1	7	$^{2}9$	$^{4}2$	

Example 2:

Long Division

	3	0	7
3	9	2	1
—	9		
	0	2	
	—	0	
		2	1
	—	2	1
			0

Short Division

	3	0	7
3	9	2	$^{2}1$

1.

5	3	8	9	2

2.

7	8	5	7	9

3.

3	6	8	9	0

4.

6	5	6	8	1

5.

4	5	7	0	2

6.

9	3	8	0	1

 Go to www.everydaymathonline.com for additional practice pages.

Date Time

Reference

Metric System

Units of Length

1 kilometer (km)	= 1,000 meters (m)
1 meter	= 10 decimeters (dm)
	= 100 centimeters (cm)
	= 1,000 millimeters (mm)
1 decimeter	= 10 centimeters
1 centimeter	= 10 millimeters

Units of Area

1 square meter (m^2)	= 100 square decimeters (dm^2)
	= 10,000 square centimeters (cm^2)
1 square decimeter	= 100 square centimeters
1 are (a)	= 100 square meters
1 hectare (ha)	= 100 ares
1 square kilometer (km^2)	= 100 hectares

Units of Volume

1 cubic meter (m^3)	= 1,000 cubic decimeters (dm^3)
	= 1,000,000 cubic centimeters (cm^3)
1 cubic decimeter	= 1,000 cubic centimeters

Units of Capacity

1 kiloliter (kL)	= 1,000 liters (L)
1 liter	= 1,000 milliliters (mL)

Units of Mass

1 metric ton (t)	= 1,000 kilograms (kg)
1 kilogram	= 1,000 grams (g)
1 gram	= 1,000 milligrams (mg)

Units of Time

1 century	= 100 years
1 decade	= 10 years
1 year (yr)	= 12 months
	= 52 weeks (plus one or two days)
	= 365 days (366 days in a leap year)
1 month (mo)	= 28, 29, 30, or 31 days
1 week (wk)	= 7 days
1 day (d)	= 24 hours
1 hour (hr)	= 60 minutes
1 minute (min)	= 60 seconds (sec)

U.S. Customary System

Units of Length

1 mile (mi)	= 1,760 yards (yd)
	= 5,280 feet (ft)
1 yard	= 3 feet
	= 36 inches (in.)
1 foot	= 12 inches

Units of Area

1 square yard (yd^2)	= 9 square feet (ft^2)
	= 1,296 square inches ($in.^2$)
1 square foot	= 144 square inches
1 acre	= 43,560 square feet
1 square mile (mi^2)	= 640 acres

Units of Volume

1 cubic yard (yd^3)	= 27 cubic feet (ft^3)
1 cubic foot	= 1,728 cubic inches ($in.^3$)

Units of Capacity

1 gallon (gal)	= 4 quarts (qt)
1 quart	= 2 pints (pt)
1 pint	= 2 cups (c)
1 cup	= 8 fluid ounces (fl oz)
1 fluid ounce	= 2 tablespoons (tbs)
1 tablespoon	= 3 teaspoons (tsp)

Units of Weight

1 ton (T)	= 2,000 pounds (lb)
1 pound	= 16 ounces (oz)

System Equivalents

1 inch is about 2.5 cm (2.54).

1 kilometer is about 0.6 mile (0.621).

1 mile is about 1.6 kilometers (1.609).

1 meter is about 39 inches (39.37).

1 liter is about 1.1 quarts (1.057).

1 ounce is about 28 grams (28.350).

1 kilogram is about 2.2 pounds (2.205).

1 hectare is about 2.5 acres (2.47).

Rules for Order of Operations

1. Do operations within parentheses or other grouping symbols before doing anything else.
2. Calculate all exponents.
3. Multiply or divide in order from left to right.
4. Add or subtract in order from left to right.

Date Time

Reference

Symbols

+	plus or positive
−	minus or negative
∗, ×	multiplied by
÷, /	divided by
=	is equal to
≠	is not equal to
<	is less than
>	is greater than
≤	is less than or equal to
≥	is greater than or equal to
x^n	*n*th power of *x*
$\sqrt{x}$	square root of *x*
%	percent
$a:b$, a/b, $\frac{a}{b}$	ratio of *a* to *b* or *a* divided by *b* or the fraction $\frac{a}{b}$
°	degree
(*a*,*b*)	ordered pair
$\overleftrightarrow{AS}$	line *AS*
$\overline{AS}$	line segment *AS*
$\overrightarrow{AS}$	ray *AS*
⊾	right angle
⊥	is perpendicular to
∥	is parallel to
△*ABC*	triangle *ABC*
∠*ABC*	angle *ABC*
∠*B*	angle *B*

Place-Value Chart

trillions	100B	10B	billions	100M	10M	millions	hundred-thousands	ten-thousands	thousands	hundreds	tens	ones	.	tenths	hundredths	thousandths
1,000 billions			1,000 millions			1,000,000s	100,000s	10,000s	1,000s	100s	10s	1s	.	0.1s	0.01s	0.001s
10^{12}	10^{11}	10^{10}	10^{9}	10^{8}	10^{7}	10^{6}	10^{5}	10^{4}	10^{3}	10^{2}	10^{1}	10^{0}	.	10^{-1}	10^{-2}	10^{-3}

Probability Meter

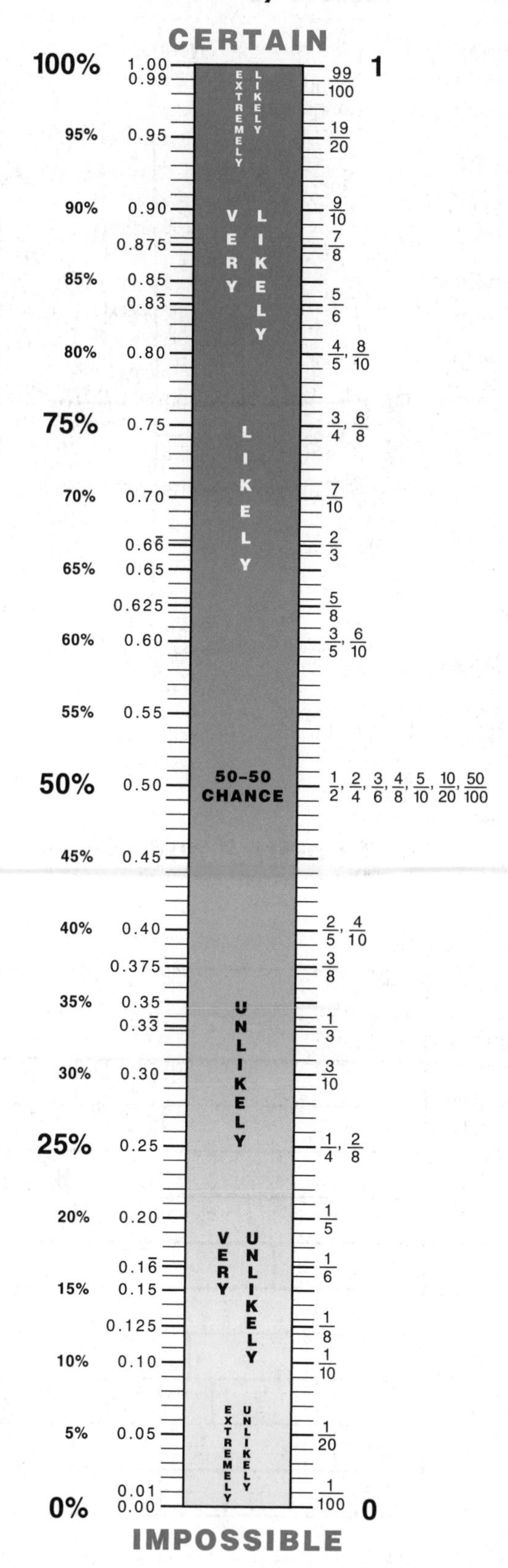

Date Time

Reference

Latitude and Longitude

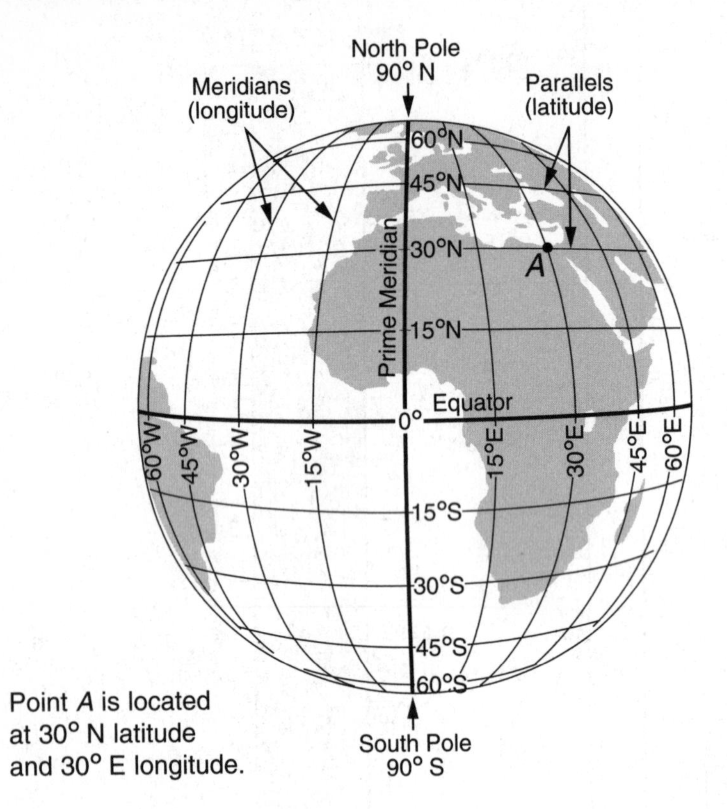

Point *A* is located at 30° N latitude and 30° E longitude.

Rational Numbers

Rule	Example
$\frac{a}{b} = \frac{n * a}{n * b}$	$\frac{2}{3} = \frac{4 * 2}{4 * 3} = \frac{8}{12}$
$\frac{a}{b} = \frac{a / n}{b / n}$	$\frac{8}{12} = \frac{8 / 4}{12 / 4} = \frac{2}{3}$
$\frac{a}{a} = a * \frac{1}{a} = 1$	$\frac{4}{4} = 4 * \frac{1}{4} = 1$
$\frac{a}{b} + \frac{c}{b} = \frac{a + c}{b}$	$\frac{3}{5} + \frac{1}{5} = \frac{3 + 1}{5} = \frac{4}{5}$
$\frac{a}{b} - \frac{c}{b} = \frac{a - c}{b}$	$\frac{3}{5} - \frac{1}{5} = \frac{3 - 1}{5} = \frac{2}{5}$
$\frac{a}{b} * \frac{c}{d} = \frac{a * c}{b * d}$	$\frac{1}{4} * \frac{2}{3} = \frac{1 * 2}{4 * 3} = \frac{2}{12}$

To compare, add, or subtract fractions:

1. Find a common denominator.
2. Rewrite fractions as equivalent fractions with the common denominator.
3. Compare, add, or subtract these fractions.

Fraction-Stick and Decimal Number-Line Chart

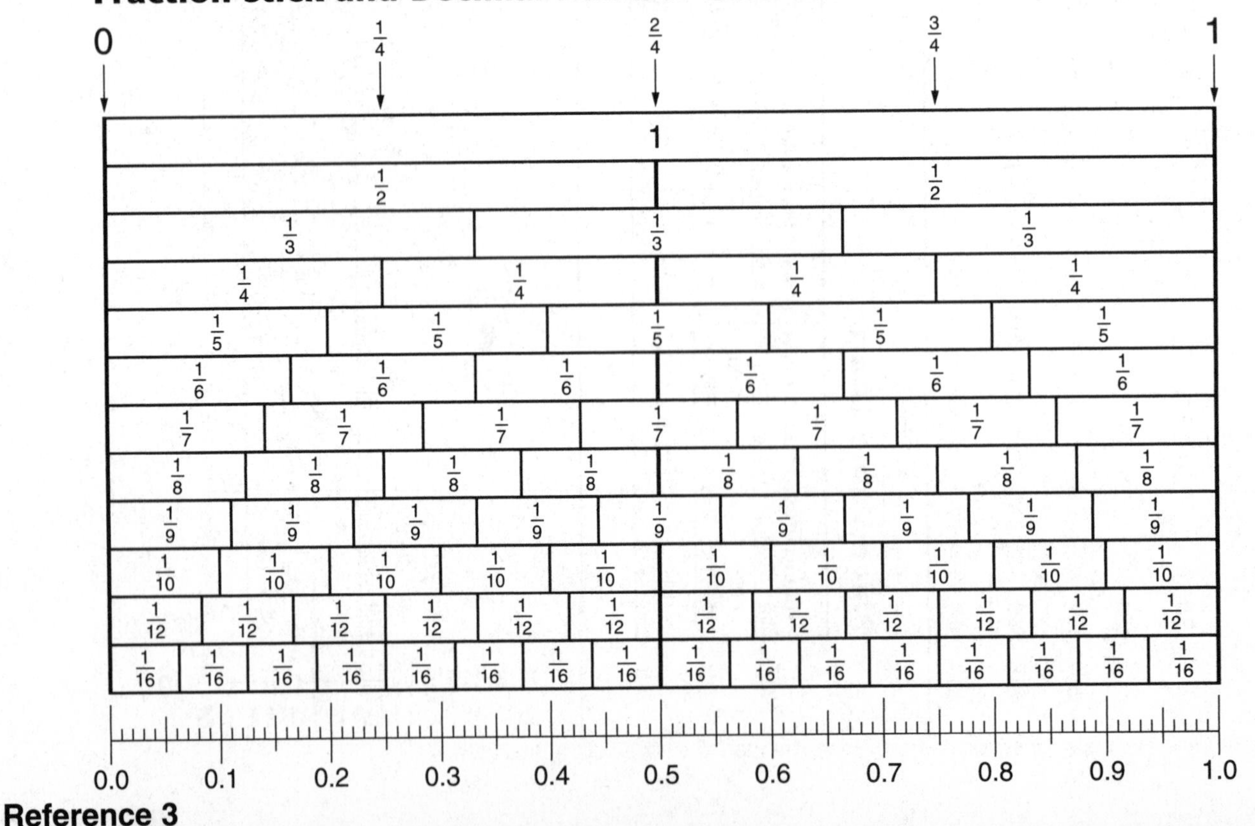

Reference

Equivalent Fractions, Decimals, and Percents

$\frac{1}{2}$	$\frac{2}{4}$	$\frac{3}{6}$	$\frac{4}{8}$	$\frac{5}{10}$	$\frac{6}{12}$	$\frac{7}{14}$	$\frac{8}{16}$	$\frac{9}{18}$	$\frac{10}{20}$	$\frac{11}{22}$	$\frac{12}{24}$	$\frac{13}{26}$	$\frac{14}{28}$	$\frac{15}{30}$	0.5	50%
$\frac{1}{3}$	$\frac{2}{6}$	$\frac{3}{9}$	$\frac{4}{12}$	$\frac{5}{15}$	$\frac{6}{18}$	$\frac{7}{21}$	$\frac{8}{24}$	$\frac{9}{27}$	$\frac{10}{30}$	$\frac{11}{33}$	$\frac{12}{36}$	$\frac{13}{39}$	$\frac{14}{42}$	$\frac{15}{45}$	$0.\overline{3}$	$33\frac{1}{3}\%$
$\frac{2}{3}$	$\frac{4}{6}$	$\frac{6}{9}$	$\frac{8}{12}$	$\frac{10}{15}$	$\frac{12}{18}$	$\frac{14}{21}$	$\frac{16}{24}$	$\frac{18}{27}$	$\frac{20}{30}$	$\frac{22}{33}$	$\frac{24}{36}$	$\frac{26}{39}$	$\frac{28}{42}$	$\frac{30}{45}$	$0.\overline{6}$	$66\frac{2}{3}\%$
$\frac{1}{4}$	$\frac{2}{8}$	$\frac{3}{12}$	$\frac{4}{16}$	$\frac{5}{20}$	$\frac{6}{24}$	$\frac{7}{28}$	$\frac{8}{32}$	$\frac{9}{36}$	$\frac{10}{40}$	$\frac{11}{44}$	$\frac{12}{48}$	$\frac{13}{52}$	$\frac{14}{56}$	$\frac{15}{60}$	0.25	25%
$\frac{3}{4}$	$\frac{6}{8}$	$\frac{9}{12}$	$\frac{12}{16}$	$\frac{15}{20}$	$\frac{18}{24}$	$\frac{21}{28}$	$\frac{24}{32}$	$\frac{27}{36}$	$\frac{30}{40}$	$\frac{33}{44}$	$\frac{36}{48}$	$\frac{39}{52}$	$\frac{42}{56}$	$\frac{45}{60}$	0.75	75%
$\frac{1}{5}$	$\frac{2}{10}$	$\frac{3}{15}$	$\frac{4}{20}$	$\frac{5}{25}$	$\frac{6}{30}$	$\frac{7}{35}$	$\frac{8}{40}$	$\frac{9}{45}$	$\frac{10}{50}$	$\frac{11}{55}$	$\frac{12}{60}$	$\frac{13}{65}$	$\frac{14}{70}$	$\frac{15}{75}$	0.2	20%
$\frac{2}{5}$	$\frac{4}{10}$	$\frac{6}{15}$	$\frac{8}{20}$	$\frac{10}{25}$	$\frac{12}{30}$	$\frac{14}{35}$	$\frac{16}{40}$	$\frac{18}{45}$	$\frac{20}{50}$	$\frac{22}{55}$	$\frac{24}{60}$	$\frac{26}{65}$	$\frac{28}{70}$	$\frac{30}{75}$	0.4	40%
$\frac{3}{5}$	$\frac{6}{10}$	$\frac{9}{15}$	$\frac{12}{20}$	$\frac{15}{25}$	$\frac{18}{30}$	$\frac{21}{35}$	$\frac{24}{40}$	$\frac{27}{45}$	$\frac{30}{50}$	$\frac{33}{55}$	$\frac{36}{60}$	$\frac{39}{65}$	$\frac{42}{70}$	$\frac{45}{75}$	0.6	60%
$\frac{4}{5}$	$\frac{8}{10}$	$\frac{12}{15}$	$\frac{16}{20}$	$\frac{20}{25}$	$\frac{24}{30}$	$\frac{28}{35}$	$\frac{32}{40}$	$\frac{36}{45}$	$\frac{40}{50}$	$\frac{44}{55}$	$\frac{48}{60}$	$\frac{52}{65}$	$\frac{56}{70}$	$\frac{60}{75}$	0.8	80%
$\frac{1}{6}$	$\frac{2}{12}$	$\frac{3}{18}$	$\frac{4}{24}$	$\frac{5}{30}$	$\frac{6}{36}$	$\frac{7}{42}$	$\frac{8}{48}$	$\frac{9}{54}$	$\frac{10}{60}$	$\frac{11}{66}$	$\frac{12}{72}$	$\frac{13}{78}$	$\frac{14}{84}$	$\frac{15}{90}$	$0.1\overline{6}$	$16\frac{2}{3}\%$
$\frac{5}{6}$	$\frac{10}{12}$	$\frac{15}{18}$	$\frac{20}{24}$	$\frac{25}{30}$	$\frac{30}{36}$	$\frac{35}{42}$	$\frac{40}{48}$	$\frac{45}{54}$	$\frac{50}{60}$	$\frac{55}{66}$	$\frac{60}{72}$	$\frac{65}{78}$	$\frac{70}{84}$	$\frac{75}{90}$	$0.8\overline{3}$	$83\frac{1}{3}\%$
$\frac{1}{7}$	$\frac{2}{14}$	$\frac{3}{21}$	$\frac{4}{28}$	$\frac{5}{35}$	$\frac{6}{42}$	$\frac{7}{49}$	$\frac{8}{56}$	$\frac{9}{63}$	$\frac{10}{70}$	$\frac{11}{77}$	$\frac{12}{84}$	$\frac{13}{91}$	$\frac{14}{98}$	$\frac{15}{105}$	0.143	14.3%
$\frac{2}{7}$	$\frac{4}{14}$	$\frac{6}{21}$	$\frac{8}{28}$	$\frac{10}{35}$	$\frac{12}{42}$	$\frac{14}{49}$	$\frac{16}{56}$	$\frac{18}{63}$	$\frac{20}{70}$	$\frac{22}{77}$	$\frac{24}{84}$	$\frac{26}{91}$	$\frac{28}{98}$	$\frac{30}{105}$	0.286	28.6%
$\frac{3}{7}$	$\frac{6}{14}$	$\frac{9}{21}$	$\frac{12}{28}$	$\frac{15}{35}$	$\frac{18}{42}$	$\frac{21}{49}$	$\frac{24}{56}$	$\frac{27}{63}$	$\frac{30}{70}$	$\frac{33}{77}$	$\frac{36}{84}$	$\frac{39}{91}$	$\frac{42}{98}$	$\frac{45}{105}$	0.429	42.9%
$\frac{4}{7}$	$\frac{8}{14}$	$\frac{12}{21}$	$\frac{16}{28}$	$\frac{20}{35}$	$\frac{24}{42}$	$\frac{28}{49}$	$\frac{32}{56}$	$\frac{36}{63}$	$\frac{40}{70}$	$\frac{44}{77}$	$\frac{48}{84}$	$\frac{52}{91}$	$\frac{56}{98}$	$\frac{60}{105}$	0.571	57.1%
$\frac{5}{7}$	$\frac{10}{14}$	$\frac{15}{21}$	$\frac{20}{28}$	$\frac{25}{35}$	$\frac{30}{42}$	$\frac{35}{49}$	$\frac{40}{56}$	$\frac{45}{63}$	$\frac{50}{70}$	$\frac{55}{77}$	$\frac{60}{84}$	$\frac{65}{91}$	$\frac{70}{98}$	$\frac{75}{105}$	0.714	71.4%
$\frac{6}{7}$	$\frac{12}{14}$	$\frac{18}{21}$	$\frac{24}{28}$	$\frac{30}{35}$	$\frac{36}{42}$	$\frac{42}{49}$	$\frac{48}{56}$	$\frac{54}{63}$	$\frac{60}{70}$	$\frac{66}{77}$	$\frac{72}{84}$	$\frac{78}{91}$	$\frac{84}{98}$	$\frac{90}{105}$	0.857	85.7%
$\frac{1}{8}$	$\frac{2}{16}$	$\frac{3}{24}$	$\frac{4}{32}$	$\frac{5}{40}$	$\frac{6}{48}$	$\frac{7}{56}$	$\frac{8}{64}$	$\frac{9}{72}$	$\frac{10}{80}$	$\frac{11}{88}$	$\frac{12}{96}$	$\frac{13}{104}$	$\frac{14}{112}$	$\frac{15}{120}$	0.125	$12\frac{1}{2}\%$
$\frac{3}{8}$	$\frac{6}{16}$	$\frac{9}{24}$	$\frac{12}{32}$	$\frac{15}{40}$	$\frac{18}{48}$	$\frac{21}{56}$	$\frac{24}{64}$	$\frac{27}{72}$	$\frac{30}{80}$	$\frac{33}{88}$	$\frac{36}{96}$	$\frac{39}{104}$	$\frac{42}{112}$	$\frac{45}{120}$	0.375	$37\frac{1}{2}\%$
$\frac{5}{8}$	$\frac{10}{16}$	$\frac{15}{24}$	$\frac{20}{32}$	$\frac{25}{40}$	$\frac{30}{48}$	$\frac{35}{56}$	$\frac{40}{64}$	$\frac{45}{72}$	$\frac{50}{80}$	$\frac{55}{88}$	$\frac{60}{96}$	$\frac{65}{104}$	$\frac{70}{112}$	$\frac{75}{120}$	0.625	$62\frac{1}{2}\%$
$\frac{7}{8}$	$\frac{14}{16}$	$\frac{21}{24}$	$\frac{28}{32}$	$\frac{35}{40}$	$\frac{42}{48}$	$\frac{49}{56}$	$\frac{56}{64}$	$\frac{63}{72}$	$\frac{70}{80}$	$\frac{77}{88}$	$\frac{84}{96}$	$\frac{91}{104}$	$\frac{98}{112}$	$\frac{105}{120}$	0.875	$87\frac{1}{2}\%$
$\frac{1}{9}$	$\frac{2}{18}$	$\frac{3}{27}$	$\frac{4}{36}$	$\frac{5}{45}$	$\frac{6}{54}$	$\frac{7}{63}$	$\frac{8}{72}$	$\frac{9}{81}$	$\frac{10}{90}$	$\frac{11}{99}$	$\frac{12}{108}$	$\frac{13}{117}$	$\frac{14}{126}$	$\frac{15}{135}$	$0.\overline{1}$	$11\frac{1}{9}\%$
$\frac{2}{9}$	$\frac{4}{18}$	$\frac{6}{27}$	$\frac{8}{36}$	$\frac{10}{45}$	$\frac{12}{54}$	$\frac{14}{63}$	$\frac{16}{72}$	$\frac{18}{81}$	$\frac{20}{90}$	$\frac{22}{99}$	$\frac{24}{108}$	$\frac{26}{117}$	$\frac{28}{126}$	$\frac{30}{135}$	$0.\overline{2}$	$22\frac{2}{9}\%$
$\frac{4}{9}$	$\frac{8}{18}$	$\frac{12}{27}$	$\frac{16}{36}$	$\frac{20}{45}$	$\frac{24}{54}$	$\frac{28}{63}$	$\frac{32}{72}$	$\frac{36}{81}$	$\frac{40}{90}$	$\frac{44}{99}$	$\frac{48}{108}$	$\frac{52}{117}$	$\frac{56}{126}$	$\frac{60}{135}$	$0.\overline{4}$	$44\frac{4}{9}\%$
$\frac{5}{9}$	$\frac{10}{18}$	$\frac{15}{27}$	$\frac{20}{36}$	$\frac{25}{45}$	$\frac{30}{54}$	$\frac{35}{63}$	$\frac{40}{72}$	$\frac{45}{81}$	$\frac{50}{90}$	$\frac{55}{99}$	$\frac{60}{108}$	$\frac{65}{117}$	$\frac{70}{126}$	$\frac{75}{135}$	$0.\overline{5}$	$55\frac{5}{9}\%$
$\frac{7}{9}$	$\frac{14}{18}$	$\frac{21}{27}$	$\frac{28}{36}$	$\frac{35}{45}$	$\frac{42}{54}$	$\frac{49}{63}$	$\frac{56}{72}$	$\frac{63}{81}$	$\frac{70}{90}$	$\frac{77}{99}$	$\frac{84}{108}$	$\frac{91}{117}$	$\frac{98}{126}$	$\frac{105}{135}$	$0.\overline{7}$	$77\frac{7}{9}\%$
$\frac{8}{9}$	$\frac{16}{18}$	$\frac{24}{27}$	$\frac{32}{36}$	$\frac{40}{45}$	$\frac{48}{54}$	$\frac{56}{63}$	$\frac{64}{72}$	$\frac{72}{81}$	$\frac{80}{90}$	$\frac{88}{99}$	$\frac{96}{108}$	$\frac{104}{117}$	$\frac{112}{126}$	$\frac{120}{135}$	$0.\overline{8}$	$88\frac{8}{9}\%$

Note: The decimals for sevenths have been rounded to the nearest thousandth.

Date Time

First to 100 Problem Cards

How many inches are in x feet? How many centimeters are in x meters? 1	How many quarts are in x gallons? 2	What is the smallest number of x's you can add to get a sum greater than 100? 3	Is $50 * x$ greater than 1,000? Is $\frac{x}{10}$ less than 1? 4
$\frac{1}{2}$ of $x = ?$ $\frac{1}{10}$ of $x = ?$ 5	$1 - x = ?$ $x + 998 = ?$ 6	If x people share 1,000 stamps equally, how many stamps will each person get? 7	What time will it be x minutes from now? What time was it x minutes ago? 8
It is 102 miles to your destination. You have gone x miles. How many miles are left? 9	What whole or mixed number equals x divided by 2? 10	Is x a prime or a composite number? Is x divisible by 2? 11	The time is 11:05 A.M. The train left x minutes ago. What time did the train leave? 12
Bill was born in 1939. Freddy was born the same day but x years later. In what year was Freddy born? 13	Which is larger: $2 * x$ or $x + 50$? 14	There are x rows of seats. There are 9 seats in each row. How many seats are there in all? 15	Sargon spent x cents on apples. If she paid with a $5 bill, how much change should she get? 16

Date Time

First to 100 Problem Cards *continued*

The temperature was 25°F. It dropped x degrees. What is the new temperature? 17	Each story in a building is 10 feet high. If the building has x stories, how tall is it? 18	Which is larger: $2 * x$ or $\frac{100}{x}$? 19	$20 * x = ?$ 20
Name all the whole-number factors of x. 21	Is x an even or an odd number? Is x divisible by 9? 22	Shalanda was born on a Tuesday. Linda was born x days later. On what day of the week was Linda born? 23	Will had a quarter plus x cents. How much money did he have in all? 24
Find the perimeter and area of this square. x cm, x cm 25	What is the median of these weights? 5 pounds, 21 pounds, x pounds. What is the range? 26	$x°$, ?° 27	$x^2 = ?$ 50% of $x^2 = ?$ 28
$(3x + 4) - 8 = ?$ 29	x out of 100 students voted for Ruby. Is this more than 25%, less than 25%, or exactly 25% of the students? 30	There are 200 students at Wilson School. x% speak Spanish. How many students speak Spanish? 31	People answered a survey question either Yes or No. x% answered Yes. What percent answered No? 32

Date Time

Rotation Symmetry

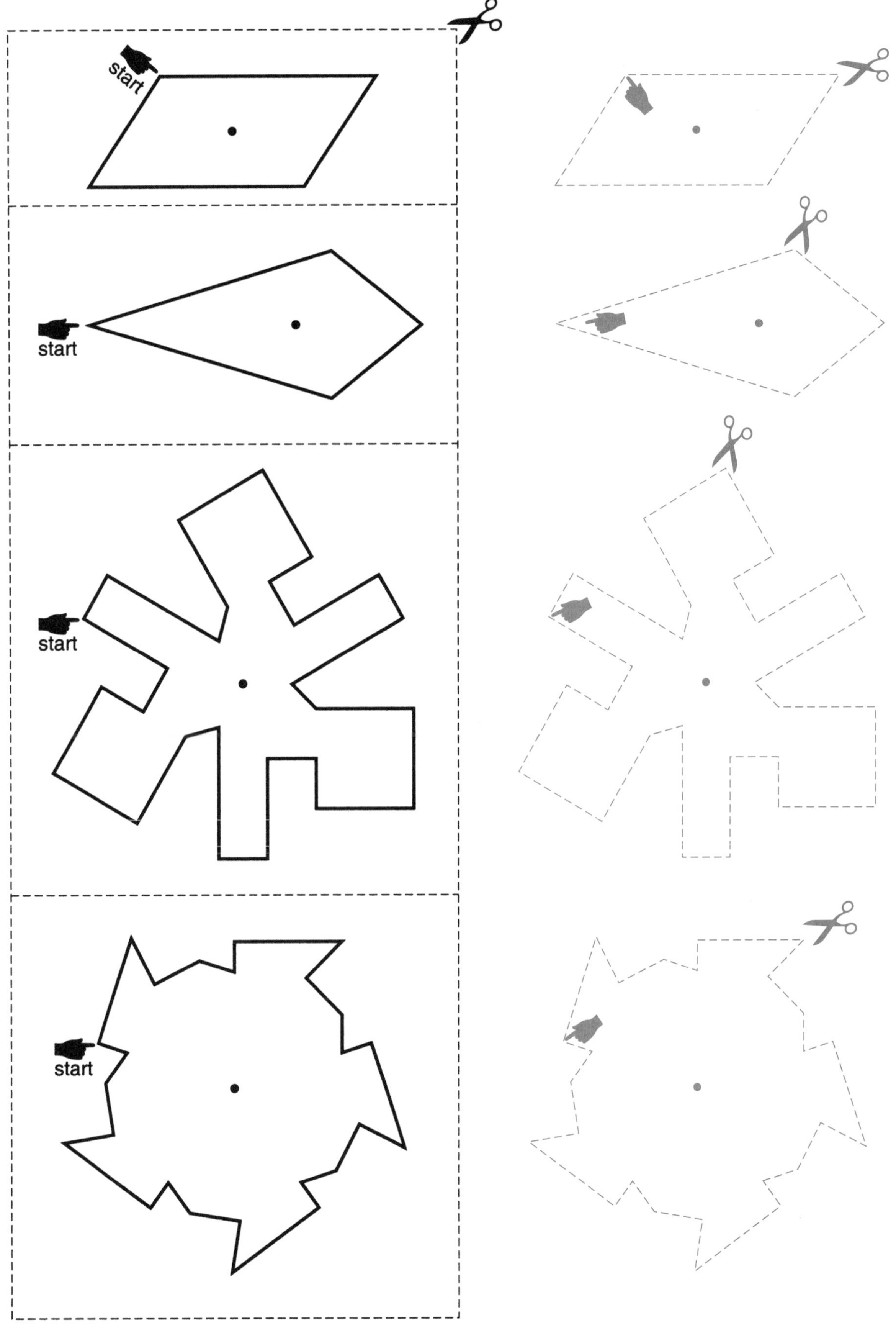

Centre for Educational Research and Innovation (CERI)

BECOMING ADULT IN A CHANGING SOCIETY

by
James S. Coleman and Torsten Husén

ORGANISATION FOR ECONOMIC CO-OPERATION AND DEVELOPMENT

Pursuant to article 1 of the Convention signed in Paris on 14th December, 1960, and which came into force on 30th September, 1961, the Organisation for Economic Co-operation and Development (OECD) shall promote policies designed:

- to achieve the highest sustainable economic growth and employment and a rising standard of living in Member countries, while maintaining financial stability, and thus to contribute to the development of the world economy;
- to contribute to sound economic expansion in Member as well as non-member countries in the process of economic development; and
- to contribute to the expansion of world trade on a multilateral, non-discriminatory basis in accordance with international obligations.

The Signatories of the Convention on the OECD are Austria, Belgium, Canada, Denmark, France, the Federal Republic of Germany, Greece, Iceland, Ireland, Italy, Luxembourg, the Netherlands, Norway, Portugal, Spain, Sweden, Switzerland, Turkey, the United Kingdom and the United States. The following countries acceded subsequently to this Convention (the dates are those on which the instruments of accession were deposited): Japan (28th April, 1964), Finland (28th January, 1969), Australia (7th June, 1971) and New Zealand (29th May, 1973).

The Socialist Federal Republic of Yugoslavia takes part in certain work of the OECD (agreement of 28th October, 1961).

Publié en français sous le titre:

DEVENIR ADULTE
DANS UNE SOCIÉTÉ EN MUTATION

The CERI project on the Transition to Adulthood has been developed along two complementary approaches. One has focused on a conceptual and research-oriented analysis of the broader issues related to the transition to adulthood. The other has concentrated on a more empirical investigation of a wide range of institutional innovations in Member countries with a bearing on the transition to adulthood and covering both education and training as well as broader community-based institutions.

This publication embodies the results of the conceptual and research-oriented part of the project. It contains three principal parts: the first part presents a broad and inter-disciplinary descriptive overview of changing concepts and viewpoints relating to the transition to adulthood. The second part is a more analytical attempt to explain current relationships and changes among the three major institutions of work, family and school, as they affect the transition to adulthood. The third and final part identifies a number of possible policy responses, within and outside education, to the problems analysed in the earlier two parts. It also points to areas where further research is needed.

The present report has been written by Professor James Coleman, University of Chicago, United States, and Professor Torsten Husén, University of Stockholm, Sweden, in close consultation with the Secretariat. It has been reviewed by the CERI Governing Board and has been derestricted under the responsibility of the Secretary-General.

Also available

YOUTH EMPLOYMENT IN FRANCE. Recent Strategies (November 1984)
(81 84 09 1) ISBN 92-64-12629-5 112 pages £4.50 US$9.00 F45.00

OECD EMPLOYMENT OUTLOOK (September 1984)
(81 84 08 1) ISBN 92-64-12621-X 118 pages £6.50 US$13.00 F65.00

THE NATURE OF YOUTH UNEMPLOYMENT: AN ANALYSIS FOR POLICY-MAKERS (July 1984)
(81 84 07 1) ISBN 92-64-12573-6 224 pages £9.50 US$19.00 F95.00

EDUCATION AND WORK. The Views of the Young (July 1983)
(96 83 02 1) ISBN 92-64-12464-0 122 pages £4.80 US$9.75 F48.00

YOUTH WITHOUT WORK. Three Countries Approach the Problem. Report by Shirley Williams and other experts (September 1981)
(81 81 02 1) ISBN 92-64-12240-0 256 pages £6.80 US$15.00 F68.00

THE WELFARE STATE IN CRISIS (September 1981)
(81 81 01 1) ISBN 92-64-12192-7 274 pages £7.00 US$17.50 F70.00

Prices charged at the OECD Publications Office.

THE OECD CATALOGUE OF PUBLICATIONS and supplements will be sent free of charge on request addressed either to OECD Publications Office, 2, rue André-Pascal, 75775 PARIS CEDEX 16, or to the OECD Sales Agent in your country.

TABLE OF CONTENTS

INTRODUCTION

Since the late 1960s youth unemployment has been steadily rising in OECD countries. More generally today we are faced with a *new* situation with regard to helping young people to find their adult roles. There is a growing suspicion that the troubled transition to adulthood depends upon the failure of existing institutions, particularly the regular school, to cope adequately with their socialisation task, either singularly or in their mutual relationships. If this is so, assigning a widened role to existing institutions, unmodified, must seriously be brought into question. In the same context the question ought to be raised as to what kind of conditions in present-day society tend to make some young people seem, in economic terms, "superfluous".

It appears to us that too much attention has been focused too exclusively on the employment problem. This is understandable both because elaborate and detailed labour market statistics lend themselves to economic analyses and because employment problems translate directly into income problems. The malfunctioning of the economy for youth, as the malfunctioning of other institutions such as the school and family, have been more in the background.

Our task in preparing a research-oriented report on problems of transition to adulthood is to focus mainly on problems of youth socialisation. In doing so the report cannot be confined to problems pertaining to adolescent and youth problems proper without considering the studies of "primary socialisation" that occurs in the family as well as in pre-school and primary school. This makes it imperative to consider major changes that have occurred in the ecology of education over the last few decades, for instance changes in the family constellation.

We are aware of at least three limitations that beset our report. The first one relates to our respective fields of interest and specialisation – sociology and psychology – and we have by no means been able to cover all important and relevant scholarly inquiries dealing with the transition problems. Secondly, given the fact that our report aims at providing background material for policy deliberations, we have been forced to narrow our focus with regard both to content and analytic treatment. Thirdly, we are keenly aware that the generalisations we have arrived at apply to the different OECD countries to a highly varying degree.

The decades since World War II have seen a number of social changes in the highly industrialised countries with extensive consequences for youth. Although these changes are interrelated, seven aspects of change are particularly important for them:

i) The postwar baby boom created cohorts of youth in the late 1960s and on into the 1970s which were much larger than their predecessors;

ii) There has been extensive growth in upper secondary and postsecondary education in all OECD countries, as well as democratisation. This has not encompassed all youth, however, leaving what might be called a new underclass, a minority of youth with school failures and low levels of education;

iii) The period of the late 1960s and early 1970s was marked by a turbulence involving youth which has left its mark on many social institutions, particularly the educational ones;

iv) There have been extensive changes in family formation and family organisation. Birth rates have fallen sharply with many couples choosing to have no children at all. There has been a decline in the proportion of households that have been regarded as "standard": two parents, together with their children. The proportion of single-person households has grown greatly, as has the proportion of single-parent households resulting from divorce or from childbirth outside marriage. The proportion of couples living together without marriage, and other forms of unconventional households have also grown greatly. All these changes signify family organisation that is greatly different from that of the recent past;

v) Youth unemployment has become a matter of serious concern and it appears to be endemic, leaving a large number of young people in a stage of "redundancy";

vi) The number of women with children of school age who have gone out to work outside the home has over a short period grown manifold in some countries;

vii) Finally, there has been a growth of deviant behaviour among youth, including drug use and crime. While this differs in different OECD countries, the pattern is sufficiently widespread that it can be regarded as a general change.

There are two different views about the long-term importance of these changes for youth, and for social policies affecting youth. One is that most of these changes are, like the baby boom of the 1950s and the youth revolt of the late 1960s, temporary in character, and within a relatively short period, institutions, social norms, and patterns of behaviour will return to what they were before this era of change. If this view is correct, the implication for social policy is that policies need only be remedial and ameliorative, designed to reduce any harmful impacts of current conditions until such time as there is a return to the institutions, norms, and behaviour that existed before the changes took place.

The second view is diametrically opposed to this. According to this view, certain of the institutional and normative changes are fundamental, and will not be replaced by a reversion to earlier conditions. This view holds that the fundamental changes are sufficiently important and far-reaching that policies must regard them as given, and as likely to be even further extended and dispersed in the future. The implication is that policies should be designed to be compatible with a new social structure that has emerged, and to aid the process of transition from childhood to adulthood within this new social structure. This means that the policies must assume that changes in the social and psychological environments within which youth grow to adulthood create new and different problems for youth than existed in the old social structure.

There is, we think, reason to believe that the second of these views is the more valid. In particular, changes in the family and changes in the structure of the economy appear to us so profound, and so far-reaching in their consequences for children and youth, that they will generate increasingly serious dislocations and problems for youth. This report is therefore based on the second view, and constitutes an attempt to describe and analyse the character of these changes, as well as to provide an overview of the various directions of research and theory about youth and the transition to adulthood that can help in understanding these changes.

Accepting the view that the changes have indeed created a new social structure within which youth come into adulthood, we feel that the innovations in institutions designed to facilitate that transition must be equally fundamental.

1. CURRENT PROBLEMS IN THE TRANSITION

First, youth unemployment is a matter of serious concern, particularly since it appears to be endemic: youth has increasingly become a stage of "redundancy". The need for youth in the labour force has sharply declined, and in recent years even the ability of the economy to absorb new cohorts reaching adulthood has declined. Of special concern is the emergence of what we call a "new underclass", a minority of students who leave school at an early stage without having acquired the knowledge and skills necessary to cope with the complexities of modern society, and therefore have small chances of becoming meaningfully employed. But even for youth in general and, not least among those with advanced education, there is a troublesome discrepancy between aspirations inculcated by prolonged schooling and what they are likely to achieve in terms of status and economic remuneration.

In the life pattern of the individual, family and work represent two arenas which provide anchoring and identity. With the weakening of the role of the family and the mounting difficulties of transition to working life, the school serves as an intermediate institution with a heavy burden of increased functions. There is often a lack of articulation between what the family and the school do in educating a growing person. Even more, a clear articulation is lacking between the school and the economy. Establishing connections between school and workplace is a major problem facing educators today.

Thus in our view there is a strong need to rethink the functions of the family, the school and the workplace in socialising young people into adult society, and to find appropriate links between them.

Young people today from later teens to early 20s in important respects live in a no-man's land. The attachment to family has decreased. School in the later teens no longer acts *in loco parentis.* For many, employment is temporary and sporadic. Even the question of who should be financially responsible for young people in this transition period is unresolved.

Issues such as these call for a reappraisal of the role of the institutions in society of today. The very institutions of "work" and "schooling" may need to be reappraised and re-defined.

In his analytical report (OECD, 1981) from the OECD conference on "Social Policies in the 1980s" Professor A.H. Halsey concludes that the OECD countries are going through "a renegotiation of the division of labour between institutions and individuals which adds up to a new phase of transition for industrial society" (p. 21). The conventional "triangle", the economy, the family, and the State, where the economy produces, the family consumes, and the State redistributes, is not adequate any longer. The traditional meanings have changed. The economy educates as well as produces, the family produces (with most women in the labour force), and the State is drawn into the productive system as a major provider of services. This also means that the relationships between the three have changed profoundly. Altogether, it constitutes a new division of labour between institutions in the welfare society.

2. SCOPE OF THE EXAMINATION

We have, in preparing this report, been faced with two closely connected problems of comprehensiveness, one related to countries under consideration and the other to the mode of analysis.

Youth problems and the transition from youth to adulthood differ from one country to another, even within the group of OECD Member countries, most of which are on an advanced level of industrialisation with market economies.

There are striking differences between countries, for instance, in terms of youth unemployment. Correspondingly, there are variables among OECD countries which could serve to explain differences in youth unemployment. The structure of the economy varies considerably between the countries under consideration here, as does the role of the labour movement. The role of the school in social stratification also varies, even though there is high commonality among countries. In the report, we attempt to cover the range of variation to be found in OECD countries, but not that found in less developed countries nor in non-market economies.

We have avoided focusing on a single issue – for instance youth unemployment – but have tried to give a more comprehensive perspective on youth problems as they appear in the highly technological society of today. Nor do we see these problems as arising from a single source which would allow one overriding explanation to account for the "problems of youth". On the contrary, we see the problems emerging from changed patterns of interaction between youth, the institutions in charge of socialising them, and society at large.

The functional connection that exists in the preindustrial rural society between education and work, and the connection between the family within which one is born and the family one forms as an adult, have been broken in our highly industrialised society. This has created a problem of transition between poorly connected institutions in a "client-oriented society", a problem made serious by the poor connections. A major question for the future is how the loss of these functional connections may be repaired or compensated for in a way that facilitates the transition of youth to adulthood.

As part of the task of this report, we have drawn upon an extensive body of scholarly knowledge which has been developed by youth research (see, for example, Rosenmayr, 1976 for a review). This work began with detailed ethnographic studies, as can be found in such as *Middletown, Elmstown,* and *Youth Plainsville,* continued with investigations of youth culture, such as the study in American high schools by Coleman (1961), and with extensive socialisation research conducted with various methodological approaches, such as the study in the Federal Republic of Germany (Hurrelmann, 1978). We will not attempt a review of the major studies in the field, but will draw on this research where it appears useful.

3. YOUTH AS A STAGE AND A STYLE OF LIFE

In 1904, the American psychologist G. Stanley Hall announced what he regarded as a new stage of life, "adolescence", falling between childhood and adulthood. Adolescence began with puberty and ended in the middle or later teens, which was when school ended for most persons.

In the post World War II period in OECD countries, there has emerged another stage falling between adolescence and adulthood, a stage of "youth". Attention to what appeared as something new was focused by the "youth revolt" in the late 1960s and early 1970s. Kenneth Keniston, author of the incisive studies of *Young Radicals* (1968) and *Youth and Dissent* (1971), advanced the notion elsewhere (Keniston, 1970) that societal changes had brought about a "new" stage of life following what had been conceived as adolescence. He pointed out that several factors, such as rising prosperity, prolongation of formal education, increased educational demands by a highly industrial and technological society that had contributed to creating an "adolescent society", also in a post-industrial society were behind the emerging youth phase following adolescence. Those who could be identified as "youth" were far from a majority but consisted of "forerunners" (Yankelovich, 1974) generally in college or university. Since then a growing number of post-adolescents are characterised by the fact that "they have not settled the questions whose answers once defined adulthood: questions of relationships to the existing society, questions of vocation, questions of social role and life-style" (Keniston, 1970, p. 634). Keniston's thesis is that "we are witnessing today the emergence on a mass scale of a previously unrecognised stage of life", a stage between adolescence and adulthood (p. 635). He tries to define youth by defining the major themes that dominate young people's consciousness, behaviour and development during this age and the specific changes in various domains (moral, intellectual, sexual, etc.) of development that occur during this stage. The youth as well as the adolescent is struggling with the identity problem.

Keniston's definition of youth is evidently influenced by the experiences of the youth revolt of the 1960s and his psychoanalytic background. Nevertheless, there is a growing recognition that something fundamentally new is occurring in the transition to adulthood. An increasing fraction of young persons, having passed through adolescence, experience a period during which neither work nor family is settled, a kind of moratorium before assuming the responsibilities of adulthood.

There is, however, a rather blurred borderline between youth and adulthood. This blurring arises because age corresponds less to lifestyle than in the past. There are certain ways of living, dressing and working that are characteristic of youth but that also continue among some people well into their 20s, 30s and 40s – and for a small fraction, indefinitely. Commercial interests try to maintain a "youthful" lifestyle which can be lucratively exploited. Many young adults with quite good incomes but with relatively few financial responsibilities can continue a youthful lifestyle.

There is also an inconsistency in the roles a young person has to play, which may lead to uncertainty about self-identity. Roberts (1983) quotes Kitwood (1980) who maintains that the problem for many young people is not to establish but to protect self-identity. "Their lives, like many adults in contemporary societies, are fragmented. They receive different treatment from different teachers, parents and employers. Sometimes they are assumed to be sexually experienced and politically aware: on other occasions they are treated as naive innocents" (p. 39).

The lack of perfect correspondence between age and lifestyle means that the "something new", although identified with youth, is not defined by age alone. It appears that what is new is a longer period, for the average young person in modern society, before assuming the responsibilities and the authority associated with adulthood, together with increased possibilities of deferring those responsibilities indefinitely. A part of this report will be addressed to the question of what conditions are responsible for this change, and what are its consequences.

One aspect of what is new is the growing youth unemployment which, as pointed out above, in present society appears to be endemic. The youth unemployment rate has steadily been going up in most OECD countries. A growing number of young people, in school or out, find themselves non-persons with empty roles. It has been said that they have to enter "a maze of waiting rooms where they can only mark time". Another metaphor signifying the same dilemma is to speak about "holding operations" that keep the young "in circulation" until they are permitted to "land" on their first occupation. This imagery, however, partly prejudges the matter by seeing youth as wholly passive, and the causes of youth employment problems wholly in the structure of opportunities. We shall try to leave these questions open at the outset.

There is ample reason to raise the question: Why does youth unemployment appear to be endemic? We shall attempt to address that question in the report, although the answer depends in part on research yet undone. There is equal reason to ask whether education, in its present form or another, can make this problem less endemic. There is a growing suspicion that the troubled transition to adulthood depends upon the inability of existing institutions, particularly the regular school, to cope adequately with what has become a greatly enlarged socialisation task. If this is so, then assigning an extended role to existing educational institutions, without modifying them to perform these new functions more adequately, must be brought into question. Therefore, it appears essential to identify institutional failures, along with the conditions behind them.

The *problematique* of young people in today's society is reflected in several paradoxes. Never before has the average young person had as wide a range of opportunities, but never before has competition for these opportunities been greater. A high percentage of students in the age range 13-16 take a negative attitude towards schooling, but a high percentage of those with such attitudes nevertheless continue formal education beyond the mandatory school leaving age. At a time when young people know less concretely what they are learning for, they are induced to go on learning for a longer time than ever before.

In preparing the report we have been drawing upon input, not least in terms of criticism, from several colleagues. We want particularly to thank Professors Erik Allardt and Ulrich Teichler as well as Dr. Kenneth Roberts with whom we met several times.

James S. Coleman Torsten Husén

Part I

CHANGING CONCEPTIONS OF YOUTH AND TRANSITION TO ADULTHOOD: AN OVERVIEW OF TRENDS AND ISSUES

1. A BRIEF HISTORICAL OVERVIEW

To be acknowledged as a full-fledged adult varies considerably depending upon the sociocultural setting. The very idea of a period of "adolescence" is rather new. The monumental *Adolescence* by G. Stanley Hall published in 1904 in two impressive volumes and based on questionnaire data was the first major empirical study of this stage of development. But even then, the picture he obtained was rather selective, reflecting at best views and reactions of adolescents of middle-class background. At that time less than 10 per cent of young Americans entered high school and very few of them went on to college.

Hall accepted as a biologically given fact a long, troublesome transition period characterised by "storm and stress". It was not until anthropologists began to study this period of life in primitive societies that it began to be apparent that this was an ethnocentric portrait of that stage of development among young people of middle-class background in Western societies[1]. In some "primitive" societies the transition from childhood to adulthood occurs in a brief period following the arrival of physical sexual maturity, often including a short period over a few days or weeks of "testing" or initiation rites. The latter signified that the young person was ready to shoulder adult responsibilities and to become familiar with the religious and mythical notions of the tribe, and thereby adopt a *Weltanschauung*. Some initiation ceremonies indicate that the individual passes from the position of being an economic liability to that of an asset, although children had economic functions in a subsistence economy, such as taking care of younger siblings or looking after cattle. In some cultures the initiation ceremonies are centred around the boys, which signifies that they play a more dominant role in the economy, whereas in others the ceremonies are centred around the girls. Confirmation in, for instance, rural Sweden was a remnant of such rites. At the age of 14-15 the young people went through a period of religious instruction by the parish priest and were examined. The boys were thereafter supposed to wear long pants which emphasized their adult status and were expected to deal with age mates from the other sex.

Thus, in many pre-industrial, "primitive" societies no period is singled out as being one of "youth". It is by no means coincidental that adolescence as a distinct period of development with its own characteristics and recognised as a period in its own right gains attention during the late 19th and the early 20th centuries. In Western society prior to that time, attention was paid only to the short transition from the stage of immaturity to one of mature adulthood.

There is no need to go back to "primitive" societies in order to find variations in the dynamics of adolescent development. When Stanley Hall in the 1890s studied adolescent crises in America he found that these in many cases were resolved by religious conversion. Studies conducted less than half a century later in America showed that the preoccupations were quite different. In their investigation of *Middletown in Transition* the Lynds (1937) showed how the juxtaposition of old and modern values and the ensuing inconsistencies created problems about identities and roles among young people.

Rapid social change tends to lead to tensions and eventual conflicts between the young and adults. This can be found in traditional societies where modern media bring young people into contact (via movies and TV) with Western values. Old authorities are challenged and conflicts easily occur. This has been observed, for example, in Moslem countries where young girls, who are especially constrained by traditional values, defy their families under the influence of Western values as mediated, for instance, by Western films and television.

The degree of complexity in the social structure also affects the tension between generations and contributes to enhancing adolescent conflicts. In highly urbanised societies with highly specialised production and formalised social relations, young people tend to come in contact with other groups with different value-orientations. During the period of adolescence and youth, the conflicts between family values and these values from outside can be especially severe.

Historically, and particularly in Europe, an apprenticeship system has served as a means of learning a craft or a trade. The guilds of senior craftsmen prepared elaborate rules of how to take on apprentices, how and for what length of time they should be trained and examined. Rules were set for adopting *Gesellen* who had to find opportunities for further training by wandering for a long period from one place to another before being regarded as qualified enough to open up their own shop. The apprenticeship system almost disappeared in some countries when mass production enterprises with routine jobs were established. There was little opportunity to learn a complex craft or trade. Young people who began to work in these enterprises were taken on as handymen or errand boys. But the apprenticeship system as a means of socialising young people into working life has persisted in some countries, such as Austria, the Federal Republic of Germany, Switzerland and Denmark, even in middle-sized or large enterprises. In such cases the learning of various industrial jobs has been achieved by a combination of working on the shop floor and attending built-in vocational schools.

It was during the Enlightenment that adolescence began to be conceived as a separate developmental stage in its own right. It was Rousseau who, not so much with *Emile* as with *Le Contrat Social,* founded the new way of thinking about adolescence. The "common ego" *(moi commun)* that emerged from the social contract gave rise to a society that was lifted from the raw natural to a cultural stage; young people were introduced into it by a "second birth", or rather, youth became constituted as a result of the second birth.

The breakthrough of a conception of youth as a stage with its own intrinsic value came through German romanticism. The very idea of the idealistic young person who was *himmelhoch jauchzend zum Tode betrübt* (heaven high shouting from joy and depressed to death) and the youth stage as a period of *Sturm und Drang* became generally accepted, not least in the literature of Romanticism. The expression was coined by Goethe in *Die Leiden des Jungen Werthers* and signified the romantic mood seizing on young people vis-à-vis adult society. A similar picture was developed by, for instance, Edward Spranger in Germany who typified his youth in the *Wandervögel* movement. This was a precursor of the conception of adolescence developed by Stanley Hall at the turn of the century.

Hall confesses in his autobiography (Hall, 1923), that Darwinian evolutionism was "music to my ears". The stages of individual development (ontogenesis) were according to the

Darwinian or Spencerian scheme seen as a repetition of the cultural development (which was paralleled with the biological phylogenesis). Thus adolescence could be seen as a parallel to the "heroic" stage of the historical development, the stage of ancient Greece.

The next important study of adolescence was conducted by Charlotte Bühler in Vienna. Her seminal study *Das Seelenleben des Jugendlichen* (1921) was important in two respects. She made a distinction between "puberty" and "adolescence", the first being the period of physiological maturity and the second the *Kulturpubertät* when the young person achieved the psychological and social maturity required for the culture to which he or she belonged. Her analyses of how young people reacted were mainly based on diaries written by the young persons themselves. Again, the subjects whose diaries were included came from rather sophisticated and articulate intellectual milieus in Vienna and could not be construed as representative of young people in general.

Altogether, the emergence of adolescence and youth as a *distinctive* developmental period is a product of the complex and specialised industrial society, where a long period of schooling and preparation precedes co-optation into adult society, and where a wide range of options and the lack of consistency between the preparatory functions of various institutions – family, school, workplace – make the establishment of an adult identity difficult.

Erikson (1968) regards the youth period as one of identity crisis, a disequilibrium of the individual's conception of the self, the trying out of the individual's conception of the self, and the trying out of various roles.

Before the 1960s there was seldom any mention of "alienation" or "youth revolt". The events of the late 1960s, particularly in 1968 at universities all over the industrialised world both in the West and in the East, gave rise to grave concerns about the relationships between generations. Kenneth Keniston's (1968) article on American youth, *Youth, Change and Violence,* appears to be rather typical. Soul-searching efforts were made in trying to explain the "radicalism" of the young people. But, again, "young people" in this context were still represented by an educated elite, in this case young college and university students. When studies of changing values among young people were conducted, such as the ones by Yankelovitch (1973), it was found that the variability in attitudes and values within the youth generation was considerable and even larger than that between generations of young people at various points in time or between them and their parents.

By the mid-1970s the focus of concern had shifted from revolt and alienation among young people in educational institutions to employment among those who had left those institutions. In spite of the fact that school enrolment had increased explosively, the unemployment rate, particularly among those who left school after completion of the mandatory minimum, continued to be high.

Doubts began to be raised about how the school "prepared" young people for life. Certain catch phrases were coined to characterise improvements, such as "career education". The social problems were moving into the schools, in particular the urban schools, and in the early 1970s there was talk of the "crisis in the classroom" (e.g. Silberman, 1970). Certain objective indicators pointed to the existence of undeniable problems, such as increasing student absenteeism, particularly at the secondary school level. In some schools in central cities in the United States, it began to be evident that between one-fourth and one-third of the students in the 13-16 year age range were absent on a typical day. Vandalism had grown and could objectively be measured by the soaring repair bills. Teacher turnover had increased and – in spite of a tougher competition on the job market – teacher training institutions were – and are – increasingly having difficulties of attracting high quality teacher candidates (D. Kerr, 1983).

In the United States no fewer than five commissions, set up by public and private agencies, took a close look at the problems of secondary schools in the early 1970s (Passow, 1975). Ten years later, another set of examinations of high schools has occurred, this time focusing on "quality" of secondary education (reference should here be made to E. Boyer, Sizer, 20th Century Fund Study and National Commission on Excellence). In one of these, John Goodlad (1983) and his co-workers have published a comprehensive and careful study of what goes on in typical high schools in the United States.

What can we expect to be major concerns about youth in the next few years? No doubt youth unemployment will continue to be a major one. But there are other structural changes in society that have important implications for youth. In Part II of this report we will describe some of these changes and their implications. The rest of Part I examines some theoretical orientations to the period of youth and to the transition to adulthood.

2. THEORETICAL CONCEPTIONS OF ADOLESCENCE AND YOUTH

In order to gain a better sense of the periods of life termed "adolescence" and "youth", and their relation to the social structure, it is useful to examine several orientations to this period. The next section examines the period from a biological and psychological viewpoint.

But first, it is in order to clear up the terminology, which varies not only between language areas but within them as well. In both British and American literature there is increasingly a tendency to distinguish between two periods referred to earlier: "adolescence", covering roughly the age range 12 to 17 or 18 years, and "youth", covering the period beyond 18. Whereas earlier accomplishment of autonomy by "choosing" a vocation, adopting an adult lifestyle and coming to play a more defined role in adult society occurred for most persons before the age of 20, now large numbers of young people beyond this age still have not settled questions whose answers once defined adulthood. Many have most of the problems of transition still ahead of them at this age; and some still hold to lifestyles and play roles that once were regarded as "puerile" or at least typical of adolescent behaviour.

In "an analytical sketch" on education and work, Clark Kerr (1977) points out the usefulness of the term "young adulthood". "It suggests an in-between period which is different from adolescence and adulthood and for which the term youth does not fit because it carries the impression of being young and not in charge of oneself" (p. 137).

Throughout this report, we will use most often the term "youth", indicating both a focus on the period beyond adolescence (or at least its early years) and a focus on the post-adolescent problems of replacing school by work and replacing the household where childhood was spent by another.

Biological and Psychological Issues

Whatever term is used – puberty, adolescence, youth or young adulthood – the development stage we are discussing here must be defined in physical and behavioural characteristics and not simply in terms of age. The perspective depends to a large extent on the relative weight one puts on physical and behavioural characteristics. The physical ones are the most conspicuous. After several years of linear growth a rather sudden "growth spurt" sets in when sexual maturity is reached. An exponential growth in height occurs during a period of two to three years, which may begin as early as 10 for girls, and may end as late as 16 for boys, when the young person almost reaches his or her maximum standing height. The onset of puberty, as indicated for girls by the growth of breast buds and for boys by growth of the testes, is followed by secondary sex characteristics, such as pubic hair, among girls, and the deepening of the voice among boys. The behavioural characteristics, although noticeable, are not easily observed and assessed, because they depend to a large extent on the sociocultural setting in which the young people are growing up. Group and individual differences are considerable.

Both the biological and psychological development during the period of life that concerns us here, the age range from 12-13 up to well beyond 20, became the subject of empirical studies by pediatricians and psychologists about a century ago (Key, 1885; Stern, 1929). Sociologists and anthropologists began to study youth problems later. Time series of information on the menarche and standing height of young people from the end of the 19th century up to now indicate that puberty now occurs almost 2 years earlier than at the turn of the century (Tanner, 1962). This secular trend has occurred at the same time that the psychological preparation or maturity period, not least by staying in school, has been considerably prolonged. Thus, adolescence and youth nowadays cover a much longer period than before – the age span from about age 12 until the early 20s.

Charlotte Bühler (1921), as mentioned earlier, made some distinctions which influenced thinking about the psychology of adolescence on the European continent for several decades. In the first place she distinguished between puberty and adolescence *(Pubertät* and *Adoleszens)*. By puberty she referred to the biological-sexual and social-psychological development covering the period from 12-17 years. But within puberty Bühler distinguished between two processes which are very different in length, which she termed the physical and psychological puberty, respectively. On the basis of her observations, most of them from diaries kept by young people, Bühler concluded that youngsters during puberty were dominated by a general feeling of uneasiness expressed by oscillations between, on the one hand, open defiance and temper tantrums, and on the other hand, indolence. On the whole, sensitivity to the social environment is high.

The second main phase of the maturity process Bühler labelled *Adoleszens*. The demarcation line between the two phases she set at the age of 17. After the "real" puberty followed the adolescence, with a switch in the basic attitude to life. While puberty is dominated by negativism and defiance, adolescence, as she saw it, is characterised by a more outward and positive attitude to life. The changeover from a more introvert to a more extrovert orientation, from the negativism of puberty to the more positive basic attitude during adolescence did not, in her observations, take place overnight but over a period around ages of 16 to 18 years.

These attempts to delineate the transition period between childhood and adulthood and to characterise it were made more than 60 years ago and were based on observations of young people who were in most cases from well-to-do families with advanced education, yet Mannheim (1928) and later Rosenmayr (1976) have emphatically pointed out that young people not only belong to different social classes but to different generations as well, which strongly affects their consciousness. Youth has to be understood as reproduction because it is a product of socialisation, but it must also be understood as transformation.

Generations are to a great extent defined by significant historical events that young people of the same age have experienced together. This means that criteria of adulthood, as pointed out earlier, vary not only between cultures but also from one historical situation to another within the same culture. Charlotte Bühler's observations about adolescence appear dated precisely because of the historical changes in this period of life over the past 60 years.

There has been a strong tendency to focus on the conflict between youth and the generation to which their parents belong. More recent research has shown that the "young radicals" in today's society have internalised their parents' values to a larger extent than traditionally has been assumed. The conflict perspective which has dominated youth psychology has been overemphasized due to the attention that young idealism and militancy, especially in the youth revolt of the 1960s, has attracted. Young people are caught, often more intensely, by the same conflicts that beset society at large. Therefore, the "youth problem", or the transition problem, must be analysed within the framework of the societal problems and conflicts which tend to permeate its institutions, including the school.

Socialisation Issues

Conditions for the socialisation of young people into adult society have in modern, individual society changed considerably at all levels, as can be seen from the following items:

1. *At the individual level:*
   - Improved standards of living.
   - Improved health care (earlier puberty, increased height).
2. *At the interactional level:*
   - Urbanisation. Children are increasingly growing up in agglomerations of apartment houses in complex physical environments with many physical constraints and many "taboos".
   - Families have become small, with growing numbers of one-parent families and in most cases with only one or two children. No three-generation families any longer.
   - Increasing numbers of working mothers.
   - Peer groups of greatly increased importance in the socialisation process.
   - Contacts with adults are fragmented, in terms of both continuity and type of contact.
3. *At the institutional level:*
   - An increasing number of years are spent in institutions (day-care centres, kindergartens, regular schools), beginning at an earlier age and ending at a later one.
   - An increasing fraction of each day is spent in institutions (day-care, school, youth centres).
   - An increasing fraction of jobs are in large organisations.
   - Mass media, particularly TV, absorb a major portion of time, and constitute an increasingly important agent of socialisation.
4. *At the societal level:*
   - The economy changing to a post-industrial service or "client" society.
   - The social order becoming more meritocratic with formal education becoming more important for social status.

These changes in the "ecology of education" (Cremin, 1980) constitute the background within which questions about the effects of variations in socialisation patterns can be examined. Any such examination should begin with the immediate pattern of determinants that influence the individual child. This begins with the mother-child relationship, then the family constellation (e.g. the completeness of the family), the nature of the neighbourhood, and the circle of friends and age-mates in the immediate surroundings.

Beyond this are institutions established separately from the family for socialisation purposes, i.e. for the education of children and young people in our society, such as kindergartens, regular schools, institutions for handicapped, delinquents, etc. These separate socialising institutions have, as pointed out earlier, grown in number, enrolment and importance over the last few decades. The socialising effect of various other public institutions and service organisations and agencies, such as mass media, churches, business enterprises, voluntary associations, can be seen as supplementing those institutions that specifically have been set up for socialisation purposes.

It should be kept in mind that socialisation institutions influence the social development of young people, not only by the explicit and intended educational actions but also by the way they are organised and actually operate. These latter influences are by-products, sometimes of an unintended character. For instance, high turnover of staff in day-care centres has been proven to adversely affect the children who, after repeated disappointments and ensuing frustrations, are reluctant to invest emotionally in the staff members taking care of them.

A typical feature of our highly specialised and complex society is that many socialisation agents or systems operate simultaneously and parallel to each other and without co-ordination, even to the extent of being almost perfectly partitioned, each institutional bureaucracy having carefully delineated its sphere of influence and prerogatives. The operations of the various agencies or organisations tend to be "vertical" (Wirtz, 1977) and act from the top to the individual at the bottom without any "horizontal" contacts with other agencies or organisations which are responsible for other limited aspects of the same individual. This means that an agency can create side-effects that are contradictory to the objectives of another agency. In the field of social welfare, lack of co-operation among school, police, home, local welfare board, easily leads to outcomes contradictory to professed goals.

The structure of society at large – particularly its industrial and technological development – sets the outer framework for the socialisation efforts of agents at the lower levels. The overall educational policy and the labour market policy influence the functioning of public institutions through the curricula they establish and through the budgetary appropriations. One illustration of attempts at the national level to influence the school is the Elementary and Secondary Education Act passed by the United States Congress in 1965 as part of the War on Poverty of the Johnson administration. Through this Act, policies were instituted on the assumption that the employability of people from poor backgrounds would be greatly improved by means of massive "compensatory" measures in education. In signing the Act the President was quoted to have said: "We are going to teach them out of poverty".

In the late 1950s, sociologists began to examine high schools as arenas of socialisation in which peers were especially important (Gordon, 1957; Coleman, 1961). The idea of an "adolescent society" or a "youth culture" as an important agent of socialisation arose, with replications of American studies carried out in Europe, for instance, by Andersson (1969), and by Kandel and Lesser (1972). But there was extensive controversy among sociologists about the extent and importance of such a youth culture. (See Gottleib, 1963, for a discussion of the controversy.)

This controversy focused on an important question in the socialisation of youth: To what extent in modern society are adults the agents of socialisation for adolescents and youth, and to what extent are their friends and peers the agents of socialisation? On the one side are those who contend that the youth culture is exceedingly important, and that the directions in which this culture focuses the attention of youth can seriously impede or modify the process of transition to adulthood. On the other side are those who see the youth culture as merely an epiphenomenon, and as closely derivative from, and reflecting the values of, the adult culture.

The relative importance of peers and adults as agents of socialisation of course differs for different young persons, and it differs among countries. But recognising these differences, there remains the overall question.

A part of the source of the controversy may lie in two different meanings attributed to the notion of adults as socialising agents. Those sociologists who contend that the adolescent and youth culture are especially important are contrasting this agent with parents and teachers, whose influence as socialisation agents, it is argued, has waned. Those who see a continuous fabric connecting youth and adults, with youth culture merely reflecting the values of adult culture, may be seeing adults not principally in their roles as parents and teachers, but as producers and consumers of commercial entertainment and accoutrements of leisure. Seen in this way, there is a significant portion of adult culture which differs little from the commercialised portions of youth culture.

Assuming this difference in the adults to which the youth culture is compared to be a major source of the controversy, there is basic agreement on a change in the agents of socialisation: change away from socialisation by parents and norms of the local community (which were in part mediated by peers of the youth themselves), toward a socialisation process in which the adolescent and youth culture are the direct agents, but this culture is itself shaped by adults – not in their roles as parents, but in their roles as producers and purveyors of commercial entertainment, commercial fashion, commercialised leisure.

The initial controversy over a separate adolescent or youth culture preceded the youth revolt of the late 1960s and early 1970s. Whatever can be said about the absolute importance of youth culture as a socialising agent, its relative importance, compared to earlier periods, increased sharply throughout Europe and America in the late 1960s. A major source of that increase was the coming of age of the postwar baby boom generation, which brought greatly increased numbers of youth into an institutional structure designed for fewer people, thus forcing youth in on itself.

In America, where the youth revolt first arose in the middle 1960s, five panels or commissions issued reports in the 1970s on secondary schools and youth socialisation, all beginning from the premise that the larger society's principal institution of socialisation, the school, was not functioning properly (Panel on Youth; Brown 1973). These reports all pointed to an increased malaise among youth brought about through extended schooling in an academic mould. In general, these reports argued that some experience with adult roles in settings that were age-integrated was important as a supplement to the extended institutionalisation of youth. The reports called for a diversification of education for those in extended schooling, with more "experiential" types of learning, including work or other productive activity.

But as occurred a decade earlier over the conception of an "adolescent culture", this conception of a "youth culture" separated from adults by a generation gap reinforced by institutions was challenged. Timpane *et al.* (1976) brought this conception into question, arguing that there was no basis for changes in the structure of secondary education. Changes have, however, occurred in the form of "alternative schools" in America and their counterparts in some other OECD countries. In some cases, these changes have constituted an *enrichment* by adding experiential modes of learning; in other cases, they appear merely to have constituted a *dilution* of the academic content of the curriculum.

The period since the late 1960s has involved extraordinary institutional change affecting youth. This raises anew the question concerning the thesis that youth are segregated from adults. What is meant by age segregation includes, however, two components: objective contacts with adults and subjective values or attitudes. Using these criteria, what evidence can be advanced for or against the existence of age segregation?

In presenting the thesis that an adolescent society is important in socialising the young, Coleman (1961) points out that in a survey of high school students, only a small minority of young people chose the occupations of their parents, perhaps at most 10 per cent, whereas in the pre-industrial society most of the occupations meant working "at home", the major categories of work being agriculture, cattle-raising and artisan crafts. The family was a unity of work and consumption, with the children working along with the parents from an early age. In the present-day urbanised society most of the fathers work outside the home, and their children often have only a vague idea about what their fathers actually are doing during their absence. In most industrialised countries there has more recently been a drastic increase of working mothers, since the availability of two salaries has been seen as a prerequisite for a sustained or increased standard of living, given the increased costs of raising children. This means that children in a majority of the homes are not only out of touch with the sphere of work which occupies their parents most of the day, they are increasingly taken care of by institutions, such as day-care centres, nurseries, and schools. This has certain consequences in terms of shifting the responsibilities for upbringing from personalised care in the family to impersonal agencies.

The most basic and undisputed evidence which can be advanced to support the contention of age segregation is prolonged formal schooling. In 1945 some 80 per cent of the 14-year olds in Western Europe left school and joined the work force. This was at a time when the apprenticeship institution in some countries still was predominant for introduction into the work force, and young people, at very modest salaries, were introduced into the work places by being assigned simple tasks of the errand-running and handyman character. Today, in most OECD countries, some 80-90 per cent of the 14-17-year olds are in full-time schooling. Mandatory school attendance in most countries lasts until 16 and the majority opts for upper secondary school for some additional years. Thus, whereas most teenagers of the 1940s and 1950s were to be found at the work places with adult workers, their counterparts of the 1980s are to be found in the schools with adult contacts limited to teachers, and with quite another role than that of young people of their age some decades ago. Work participation, even though one's status was at the bottom of the hierarchy, was of a functional character. In principle, however, youth were involved in the work process along with full-fledged workers and gradually had to take responsibility for their part of it. But the role of a school pupil means being a client, being taught and examined by somebody who plans the teaching-learning process for both partners and without sharing functional responsibility of the work process.

Yet the matter is not so simple, as pointed out by Timpane *et al.,* (1976), in criticising the idea that there is age segregation. The critics refer to United States statistics according to which young people in the age range 16 through 24 since the early 1950s have slightly *increased* their participation on the labour market, both full- and part-time, in spite of increased participation in further education. The 1975 National Longitudinal Study found that in the early 1970s 70-80 per cent of those in senior high school held summer jobs. Although the proportion of youth who work while in secondary schools is less in other OECD countries, a similar trend can be seen in these countries.

These trends suggest that in *some* respects, at least, there is a reduction in the age segregation and an increased contact of youth with the world of work, contrary to the assertions of the 1970s analyses of youth and secondary education in Europe and the United

States referred to earlier. Yet again the matter may not be so simple. The increased interest of youth in part-time work while still in secondary school appears to be due to two factors. First is reduced attention to and interest in schoolwork, a reduction suggested by the decline in achievement levels of high school students through the 1970s. Second is an increased interest in disposable income that can be spent independently of parents' approval: on entertainment, cars, records, clothes, cigarettes, and in some cases alcohol and illicit drugs. Thus it may well be that the part-time work of youth who are in school not only shows an opportunity for reduced age segregation through work contacts, but also shows a motivation toward distinct leisure activities.

Quite apart from the motivations of youth for finding part-time work while still in school, schools themselves have instituted programmes to increase contacts with the world of work. This occurs in some countries by means of what in Sweden has been referred to as "practical vocational orientation". Students have the opportunity for a short period of a few weeks full-time to be spent at various work places. The meaningfulness of this could be questioned, not only because of the limitation of time but foremost because of the absence of "real" responsibilities. There is ample reason to raise the question whether short visits to work places are of greater value than the insights obtained through various other channels, for instance, by means of television.

Altogether, the increase in length of schooling creates the important and undisputed fact that the majority of teenagers are not any more at the work place but in school, assigned the "client" role.

What about the subjective criteria of "disconnection" or "isolation", which is part of what is meant by age segregation of youth? We will turn to that in the next section, after further examination of research on agencies of socialisation.

Another body of work on youth socialisation had its origins in central Europe, particularly among sociologists in Germany, shortly after World War II. The studies were to a large extent inspired by the shattering of the previous values and ideals by the German defeat in the war and subsequent revelations of atrocities. Much of the research in this field was referred to as "socialisation studies" under the influence of the critical philosophy of the Frankfurter School of social research under the intellectual auspices of Horckheimer and Adorno, later Habermas. The major concern was to map out how young people perceived their situation and what attitudes they were heeding. But socialisation in terms of the learning of social roles, internalisation of norms and beliefs and the learning from social skills useful in the existing society also came into focus. We shall here examine one among several major surveys in some detail.

The German Shell Youth Foundation sponsored a study of the situation characterising German youth, with two reports published in 1954 and 1955 respectively, *Jugend zwischen 15 und 24* (Youth between 15 and 24). Some 20 questions given to the sample of young people in 1953 were repeated in an interview questionnaire a decade later, and a report was published in 1965 under the title *Jugend, Bildung und Freizeit* (Youth, Education and Leisure).

This work continued with two major studies in the 1970s: The first was *Jugend zwischen 13 und 24* (Youth between 13 and 24) with the subtitle *Vergleich über 20 Jahre* (Comparison over 20 years). The survey on which this study was based included a subset of the questions given to youth in 1953 and again in 1963. The second study was based on fairly representative samples of young people at 12, 17 and 22 from the Federal Republic of Germany, France and the United Kingdom and was thus cross-national in orientation. The title of the ensuing comparative publication was *Jugend in Europa: Ihre Eingliederung in die Welt der Erwachsenen* (1977) (Youth in Europe: Its integration into the world of adults). The last two studies were rather explicitly oriented towards an elucidation of the socialisation of young people.

In the second study, three domains of socialisation were explicitly investigated:

1. *Primary socialisation* occurring in the family, kindergarten and among playmates with the goal of achieving emotional stability, working affective relations, co-operative abilities with peers and respect for non-parental authority. This was studied with a cohort of 12-year-olds.
2. *Secondary socialisation* brought about by the various stages of formal schooling, vocational education and leisure time activities with the goal of achieving cognitive competencies, readiness for continuing learning, political socialisation and work discipline (studied with a cohort of 17-year-olds).
3. *Tertiary socialisation* conducted by various legal and religious organisations, voluntary associations, mass media and by the political system with the goal of social integration, social recognition, readiness and capacity to assume new roles and to achieve responsibility and vocational competence (studied with a cohort of 22-year-olds).

This study was unusual in focusing upon socialisation over a broad age range, and on different agents of socialisation over this range. The authors advanced a socialisation model according to which the various socialisation conditions were seen affecting the interaction between the young person and the agent of socialisation.

This is only one example of research carried out on youth socialisation in a European context. Socialisation research has also been conducted over a period of time in Scotland by Gow, McPherson (1980), Germany and Austria by Kreutz (1980), by researchers at the Max Planck Institute for Educational Research in Berlin (Oevermann, 1976) and by Hurrelmann (1980)[2].

Finally, reference should be made to a Symposium conducted by the Dutch *Jeugdprofiel* in 1975. A group of leading social scientists were invited to deal with current youth problems in papers that served as inputs for the Symposium whose proceedings were edited by Hill and Mönks and published in 1977 under the title "Adolescence and Youth in Prospect".

The *Bergedorfer Gesprächskreis,* a politically and economically independent forum which contemplates and discusses new ways and new initiatives for the development of free individual societies with participation of intellectuals from universities and government agencies as well as representatives from industry and business, devoted its 63rd meeting conducted in June 1979 to the problem of generation conflict. It is published under the title *Jugend und Gesellschaft: Chronischer Konflikt – neue Verbindlichkeiten?* (Youth and Society: Permanent Conflict – New Commitments?).

The participants in the Bergedorfer Symposium seemed to agree that the "generation conflict" – at least as perceived on the Central European scene – is not primarily a conflict between youth and adult society or something that could be conceived basically as a youth problem but should be understood as unsolved problems and tensions *within* a given society at large. However, it is not just a matter of young people reacting to the dynamics of society at large. There is an interaction going on both at the individual and collective level, or something that involves at least part of the collectivity. There is on the part of young people an intrinsic drive for self-assertion which is part and parcel of their psychological and social maturation. They are not just influenced but have an inner urge to influence and be part of the action in the larger context.

There is in modern society an important factor behind such a self-assertion. Technological changes promote a polarisation of generations by fuelling an innovation drive. A new generation tends to identify itself with the emerging new devices, habits and values and tends to support these in order to achieve visibility and power.

In an age-stratified society the creative role of young people in innovative tasks is easily overlooked. By starting with a "clean slate" the young are less burdened with old ideas, more receptive to new ideas and approaches, and therefore more innovative.

Rosenmayr, an Austrian sociologist who has specialised in youth socialisation, points out in his preamble to the discussion at the Bergedorfer Symposium that there is a tendency to *Selbstausschliessung* (self-exclusion or isolation) on the part of young people as a kind of introvert reaction. He referred to several international studies from the early 1970s showing how articulate young people rejected the growth-oriented society and became involved in ecological movements to protect nature from exploitation. Another example is their involvement in the peace movement. These reactions can be regarded as post-materialistic promotion of expressive (i.e. self-fulfilling) values, but they can also be regarded as a reaction against powerlessness vis-à-vis adult society.

The young today reach physical maturity as early or even earlier than previously but are kept outside decision-making, particularly in economic affairs, and are confronted with great difficulties in being more permanently "admitted" to responsible positions in the world of work.

The young people who are with particular intensity, not to say aggressiveness, promoting the new post-materialistic or expressive values are usually not so much reacting to their parents. Studies conducted in Germany, Austria and the United States (Barnes, 1979; Ingelhart, 1977) indicate that young protesters tend to come from families where the parents have a liberal bent and sometimes heed utopian value components. Thus, the emotional relations with the parents are often not troubled by major political differences.

Rosenmayr (1979) sketches four types of self-exclusion: apathy, sub-cultural ghetto-formation, sects and – a reaction which at that time occupied the front-page in German newspapers – the armed self-exclusion, pure terrorism.

Apathy is a reaction among young workers determined by the fragmentation, monotony, insecurity and stress of the industrial work process and by the lack of empathy with other individuals at the work place. The French sociologist Jean Rousselet has in extensive studies referred to an *allergie au travail* which is a cynical reaction to the situation with which many young workers are confronted. The apathy is expressed in, for instance, desertion of the youth centres and non-voting in general political elections. It also tends to lead to a reduction of the ability to express oneself to the extent of ending up in a "cultural aphasia".

Sub-cultural ghetto-formation takes various forms, such as the establishment of communes or the setting up of small businesses that will provide the bare subsistence minimum. The tendency to set up sects around certain religious beliefs that focus on the inner life is another tendency to seal off oneself from the outer world and escape to one's inner world.

Finally, the most aggressive form of self-exclusion is the terrorism where adult society as it exists has been totally rejected. This form of total self-exclusion is found in a small, armoured kernel of protesters. But it is not entirely isolated from society at large. It has partly psychological, partly real support from those groups that in various sub-cultures live in partial exclusion from society at large.

3. CHANGES IN VALUES AND ATTITUDES

Youth can be conceived as a period of both reproducing society and transforming it. Young people become socialised by their family, school and peers. By and large they internalise prevailing values of these agents. The preponderance of one of them over the others varies between generations and social groups. But the influences of the commercially spawned culture outside the local community, operating largely through the agency of peers, has somewhat changed the balance among these socialising agents in recent years.

Since there is an excess or diversity of values projected toward youth from the preceding generation, there is always a margin of freedom in the *selection* of values which guide the socialisation process. This is noticeable among the more articulate young people leading the youth movement. They are described by Yankelovich (1974) as "fore-runners" who herald new paradigms of thought and new valuations. As can be seen from surveys of American youth, the new values then spread to the masses of youth. For instance, in the late 1960s the challenge to the traditional work ethic was confined almost entirely to campus youth, whereas in the 1970s, when the work ethic became more strict on campus, it tended to slacken among the rest of the youth generation.

The fore-runners, according to Havighurst's (1975) estimate, make up some 20 per cent of the age group 15 to 25. The "dissenters" and "young radicals", in Keniston's terms, constitute a sub-group among the fore-runners. They are oriented toward expressive values and want to use their education as an instrument for criticising society rather than as a means of perpetuating a growth-oriented, technological society. A change towards expressive values can be seen behind the quest for improved quality of life and the increased emphasis on spiritual and humanistic pursuits in opposition to pragmatism and technology.

The practically-oriented youth, on the other hand, are, as Havighurst puts it, "the apprentices to the leaders of the technocratic, production-oriented, instrumental society". They endorse values of productivity, achievement motivation, materialism, and social responsibility (law and order). They comprise in his estimation some 60 per cent of the youth of 15 to 25. The remaining 20 per cent are less educated, less articulate, and, in our terminology, recruits of the "new underclass".

It should be emphasized that we are here dealing with Weberian ideal types. There is, of course, in reality no dichotomy between "fore-runners" and "practically-minded" or expressively-oriented on the one hand, and instrumentally-oriented on the other. But the value characteristics tend to cluster in the two syndromes described above, which makes a distinction between the two types heuristically useful. It helps us to understand the reactions among young people toward education, the way their performances are evaluated and what they expect from their working life.

Changing Attitudes to Education and Work

The rapidly increasing availability of further education beyond the mandatory age has led to rising expectations and hopes. These include hopes about the status and the nature of the jobs to which additional schooling and ensuing credentials will entitle those who "stick it out".

Young people and their parents are keenly aware of the decisive role played by formal education in determining future social roles and job status. Therefore, even those who hold quite negative attitudes to the content of education offered by the institutions they attend are ready to embark on long courses of study in order to pull themselves as far up as possible on the ladder of general education with an academic bias. They realise perfectly well that the amount of formal education they have absorbed according to conventional scholastic criteria will decide the place they will occupy not only in the line seeking entry to the next level of education but in the line of job seekers as well. They are also aware of the fact that, in spite of much talk about "over-education" and unemployment among university graduates (Freeman, 1976), holders of basic degrees are several times less likely to become unemployed after leaving school than are secondary school leavers or drop-outs.

In the six-subject survey conducted by the International Association for the Evaluation of Educational Achievement (IEA) student attitudes towards schooling in general as well as towards the importance of success in school were assessed (Husén, 1979). Students in industrialised and affluent countries, such as Sweden and the Federal Republic of Germany, tended to score rather low on the Like-School Scale, whereas students in developing countries (often with miserable school facilities) scored very high. One possible explanation is that there are so many other agents in affluent countries which compete with the school, such as media, sports and other leisure time activities. But since the negativism among students from affluent countries tends to increase during the last stage of mandatory school, another reasonable explanation is the growing awareness among people of circumstances that frustrate their hopes and aspirations.

A second attitude measure in the IEA survey was the School Motivation Scale. It was based on questions, such as "Do you consider it important that you perform well in school?" The mean score on the School Motivation Scale *increased* as students moved up through the grades, while the general attitude toward schooling went *down* over the same period. This shows the bind in which some students in developed countries find themselves as school progresses: they like it less, but they see it as more important to their future. For these students, the "voluntary" continuation in school beyond school-leaving age – a path pursued by the majority of students in the developed countries with low like-school scores – can be seen as a continuation under duress. The awareness of the importance of formal education for selection to the job market and the career prospects that go with it spurs extrinsic motivation to "stick it out". Yet the duress shows up in low levels of effort and high levels of truancy. A minority of students, who fail and give up at an early stage, perceive mandatory schooling beyond the mid-teens when they are physically able to work as an "extended jail term".

Young people are aware of the increased competition in school and the world of work, and of the role played by the school as a sorting and sifting device. They are also aware of the importance of marks and examinations as part of the selection game they have to go through in order to get as far as possible on the school ladder.

In spite of rhetoric to the contrary, credentialism seems to gain ground (Teichler *et al.* 1976; Dore, 1976). The classic, ideal picture of young people, who after careful guidance and realising their "real abilities and bents" progressively approach the occupational domain and the specific job of their choice, contrasts with the grim reality facing many youngsters, namely, that of a negative choice made for them by school at an early stage.

The intense, at times aggressive, debate in recent years about school marks, examinations, and *numerus clausus* reflects a reaction against credentialism among some articulate and vociferous students. In several European countries student leaders at both secondary and post-secondary level have come out strongly against norm-referenced tests, the system of relative marking in which students are compared with each other on the basis of norms derived from the distribution of performances in the entire student population. In many cases the reaction has been against all kinds of individual examinations, with demands for group examinations instead. The quest for so-called criterion-referenced tests and examinations, where some kind of absolute mastery has been defined, and where the ranking of students is replaced by a categorisation of passes and failures, is in part a response to the defects of a ranking procedure, though it in no way eliminates competition.

A similar reaction against school marks and the credentialism made possible by them occurred, and passed, a decade earlier in America. Both in America and in Europe, the reaction can be seen as a dissatisfaction with the school's role as a selection mechanism.

No doubt, the emotional heat of the debate on school marks in countries which have gone through an enrolment explosion is fuelled by the fact that marks are instruments of competition. The fight against them by a minority of students is grounded in a change in values that has had repercussions on a broad range of issues in the political domain. Marks are conceived as instruments of the capitalist society, and the educational system is regarded as an instrument for educating young people for the "slots" in the employment system, without consideration of intrinsic educational values. The sorting and sifting performed by the school is seen as subservience to the forces of the labour market. The marking and ranking of individual performances are also incompatible with the values expressed in certain curricular rhetoric on education for co-operation, self-fulfilment and personal initiative.

It should, however, be emphasized that opinion on the marking and ranking system among students is highly divided, and fluctuates greatly over time. There is, for example, little such reaction in the United States now, while there was some years ago. There are indications, to which for instance the Swedish National Board of Education has referred, that if asked about the matter the majority of students in Sweden would prefer marks to other types of individual assessment. One could venture the diagnosis that a degree of pragmatism goes into the attitudes of the less articulate majority, since such assessment gives greater equality of opportunity than the more particularistic selection procedures based directly on social background, which it has replaced. They have more to gain by accepting the system than the upper-middle class rebels.

The reality of meritocracy is there, and the school is an important part of it, reducing the dependency of opportunity on family background, but increasing its dependency on educational credentials. In the book *Talent, Equality and Meritocracy,* Husén (1974) raised the question just how inherent a certain meritocratic element is in the social fabric of the high technology of industrialisation, and whether the drawbacks of meritocracy constitute the price that must be paid for the benefits it brings in technological development and a growing economy.

Lack of systematic evidence, particularly comparative evidence, covering a long period, makes it hard to judge to what extent changes in attitudes toward education and work have taken place among different categories of young people. Student activism referred to above on questions of marks, examinations and participation in decision-making is a symptom of such a change. Confrontations between students and bureaucracies are other indications. Altogether, the changes could be described as a revolution in values related to work.

There appears, however, to be more than one direction of new values competing to replace those of the preceding generation. One set of values, which can be traced from the period of the youth revolt beginning in the late 1960s, has already had a strong impact. Another, which shares some elements of the first, but conflicts with it in other ways, has emerged more recently. We will describe the first in some detail, and then sketch some of the elements of the partly conflicting values that have arisen more recently.

A first set of new values: six salient features of the first set of new "expressive" values are listed below:

1. The notion of a successful career as continuous advancement and promotion in a process of selection and competition, where the able and ambitious succeed and the others fail, is rejected. More emphasis is placed on self-fulfilment, security, and development of rewarding leisure time interests.
2. There is a belief that education should attempt to attain a wider range of objectives, of the sort that are espoused in the rhetoric of education bills, preambles to curricula and graduation speeches. Studies are seen to be not for narrow vocational goals only, but for the development of the whole personality. Reforms of higher education in Sweden and France, for instance, aimed at making university programmes narrowly defined in vocational competencies, have been strongly rejected by students who resent being prepared for vocational "slots". The resentment is strong against programmes which lead to specific occupational competencies, because of the feeling of being locked into a particular occupation with limited leeway for a career change.
3. There is a growing view, among young people who are actively reacting against competition and credentialism, of education as something that in the future in different ways can be "mixed" into the career pattern. The growing number of students who do not take their entire formal education *en bloc,* but "stop out" for a certain period before proceeding with, for instance, university studies, is an important symptom of changed attitudes.
4. Concomitant with the rapidly rising level of formal education among entrants to the world of work, and to a large extent an outcome of it, is a profound change occurring in the perception of what work means. As is shown by Yankelovich (1974) surveys, middle and upper-class youth tend increasingly to strive for intrinsic rewards, whereas the lower class and less educated more favour instrumental ones – but the gap between the two groups is narrowing, which is another indication of new values spreading from privileged and more articulate groups to the entire youth generation.
5. There is an unwillingness to accept supervisory authority. Young people increasingly want to hold jobs which allow them freedom of personal initiative, and jobs subject to minimal authority. They want to have more say about the working conditions and planning of the work process. They do not accept the drudgery and boredom which their elders had to put up with, unless it is self-imposed. The better educated they are, the more they loathe taking repetitive and routinised jobs in the manufacturing industry.
6. There is today a marked tendency among young people in highly industrialised and affluent societies to push upward the age when they feel it appropriate to "settle down". One leading expert of vocational guidance, Donald Super (1957) who has conducted an intensive follow-up study of a group of youth from the beginning of high school through the 20s, writes about a stage of "floundering" in the development of vocational maturity. Whether the extension of a pre-settling-down period is seen as floundering or as a desire to defer adult responsibilities, it is characteristic of youth in developed societies.

These values appear to be most characteristic of activist youth and disadvantaged youth. However, under the title *Education and Work: The Views of the Young* (OECD, 1983) CERI has recently published a study based on a wide range of sources of information on how young people in 15 Member countries viewed their situation vis-à-vis the institutions – family, school and work – that most closely affected them in a society with low economic growth and high unemployment. This study covers values not merely of activist or of disadvantaged young people, but of the whole range of youth. The view that emerges from the value study differs in several respects from that presented above. The overall picture, however, is one of insecurity and confusion in the majority of young people in an era of job scarcity.

A second set of new values: The second, more recently emerging set of values which partly conflicts with the first can be seen as an outgrowth of the technological explosion that is currently in progress.

This technological explosion is in some part youth-driven. Nowhere is this more evident than in the micro-computer industry. The industry originated among young hobbyists, developing both hardware and software in basements and garages. Hundreds of very small firms, begun by young persons, have sprung up, and the industry is largely made up of the young. The mass manufacture of the products of this technology is not particularly an activity of the young; it is highly mechanised, and the assembly tasks are little different from those of other electronic or electrical appliances. But all other activities, including hardware and software development, management and marketing have absorbed a disproportionate number of youth.

The characteristics of this activity share a number of elements in the first set of values. The work has a high component of intrinsic satisfaction, the supervisory authority is minimal, and success does not depend on attaining an extensive set of educational credentials, although the acquisition of skills is vital. The work is not so much rule-governed as goal-governed. In this period of rapid expansion – and perhaps later as well – it does not require long-term commitments to a single organisation or to a narrow task, and thus does not require early settling down to a fixed future.

But this kind of work entails values tnat differ from the ones described earlier. It is not anti-competitive, but extols competition. The youth involved are not oriented to security, but to immediate success, even at the cost of security. They are not interested in an increase of leisure, but are intensely involved with their work. They are not interested in an education focused on a wider range of objectives, but are interested in deeper and deeper involvement in a narrowly focused activity.

In other words, this set of new values reflects a more optimistic note. For instance, the Social Science Research Institute of the Konrad Adenauer Foundation in the Federal Republic of Germany has recently presented a preliminary report of a sample survey of 2 000 young people between 14 and 21 on their values. The information was collected in December 1983 using the same questions posed in an earlier survey of 1979, which made it possible to measure changes that had occurred over the last few years.

The report points out that the prevailing concerns aired by vociferous young people are unemployment, environment hazards and armament, creating frustration leading to drop-out and drug abuse. This portrait of youth has dominated the media. The present report is an attempt to give voice to the "silent majority". Two types of findings stand out. In the first place, young people appear on the whole to be more satisfied with life than is reflected in the media. In the 1983 survey by the Konrad Adenauer Foundation, 3 per cent were "satisfied", 49 per cent "somewhat satisfied" and only 11 per cent "not satisfied" with their present life conditions. Young people appeared by no means euphoric, but the picture was considerably brighter than that frequently conveyed to the general public. A majority, albeit narrow, seem to look to the future with confidence.

The second striking feature was how rapidly values and attitudes change, in this case from 1979 to 1983.

It is not clear just how much growth there will be in occupations compatible with this new mix of values, which combine the work ethic of the old values with the anti-authority, anti-credential, pro-self-fulfilment ethic of the new values. What is important from the present perspective, however, is that this new mix of values has come to characterise large numbers of youth in highly developed countries. The set of values indicates that the potential for extensive economic growth and social change exists if restructuring of school and work provides opportunities for capitalising on this new mix.

4. SCHOOLING ISSUES IN AN ACHIEVEMENT-ORIENTED SOCIETY

After World War II hopes ran high about what the school would be able to accomplish in improving the conditions of man, as society entered a period of spectacularly increased material standards of living. Opportunities for formal education were multiplied and secondary education became universal. Education in the 1960s doubled its share of GNP over hardly more than a decade. Equality of educational opportunity was expected to lead to equalisation of life chances. The school was expected to play an important role in the social education of young people, to promote co-operation, mutual understanding and social integration in an increasingly equalized society. But rhetoric and reality in hindsight turned out not to be quite congruent. It has begun to be apparent that the school of today is beset by built-in goal conflicts that cannot be removed by sweeping rhetoric.

Before discussing some of these conflicts we should point out that the school has to some extent fallen victim of its own spectacular successes. Within a short period enrolment has expanded greatly. Children from social strata who in earlier times were practically barred from further education no longer have economic barriers to advanced education. Schools now take much better care of the children, as exemplified by the widespread availability of school lunches and health care. Young persons have a broader spectrum of learning opportunities in addition to the traditional ones which in a one-sided way aim at cognitive competencies. The school, with its extra-curricular activities, plays an important role as a custodial institution in countries where in recent years mothers have gone out to work outside the home to an unprecedented and unforeseen extent.

Yet today the school – particularly the secondary school with universal enrolment – has become a troubled institution, a target of much dissatisfaction and criticism. There are objective indicators that quite unequivocally point to a malaise, such as absenteeism, vandalism, dropout, and turnover of teaching staff. Attitude surveys, such as the ILEA survey referred to earlier (Husén *et al.*, 1973, Husén, 1979), have found a high proportion of young people in the early years of adolescence, particularly in the age range 14-16, who say they are fed up with school and heartily dislike it. As we shall see later, many hopes about what reformed schools with widened access would achieve have been frustrated, further fuelling the criticism.

It seems to us that a deeper diagnosis of the situation is called for in order to avoid remedies of a cosmetic character. Quick and easy cures cannot be prescribed for school conditions which reflect problems at the core of the society. There are reasons to underline this when reading the recent reports on the state of the American high school. Some of these reports, such as the one by Goodlad (1984), although they focus on what is happening in the schools, recognise the roots of the school's problems in problems of the larger society.

A deeper diagnosis of the present problems besetting the school as an institution should be guided by the following perspectives:

1. Schooling has become increasingly formalised and bureaucratised, and is often conceived of in terms of the product-oriented manufacturing or service industry. This has led to an organisational crisis;
2. Modern society has become increasingly meritocratic with schooling being the main vehicle for achieved social status and job career. This has led to a selection crisis.

The bureaucratic and meritocratic syndromes are closely related. Increased credentialism enhances bureaucratic formalism and vice versa. Both syndromes are inimical to the flexibility and innovative spirit which are prerequisites for achieving the genuine purpose of school education, what Sizer (1984) calls "to train the mind".

The Organisational Crisis

Schooling has increasingly become a processing industry with the specialisation, formalisation and bureaucratisation that go with mass production. In those respects the school has developed along the same lines as institutionalised health care. The school was from the outset an outgrowth and an agent of the immediate surrounding community. It supplemented the education provided by the family and the community. Its teachers acted *in loco parentis.* The school operated on the principle of a relationship between partners on equal footing. This relationship is in jeopardy in an "asymmetric society" where the individuals have to deal with powerful corporate actors. Parents confronting a school are increasingly faced with an impersonal bureaucracy.

As the school has become bureaucratic, it has become an instrument of the State, that is to say, an agent of a corporate conglomerate. Schooling has become a complex state-wide system with large districts and a hierarchy of administrators. Teaching staff are assigned to their jobs according to elaborate legislation and rules established by collective bargaining.

In most countries the size of individual schools has grown considerably. In Western Europe and America the typical 10-year-old by the early 1970s went to a school with 400 students, the typical 14-year-old to a school with some 700, and the typical 18-year-old to a school with 1 000 students (Husén, 1979). Secondary schools with their diversified curricula calling for specialised teacher competencies and schooling facilities, such as special classrooms, tend by their nature to be large. But size brings with it organisational problems.

The large school with one thousand or more students is conducive by its sheer size to a formalisation of social control. In the small setting, where individuals are familiar with each other, social control can be exercised in an informal way. Those who break the rules can easily be recognised and subjected to proper sanctions. They cannot escape into the anonymity of the large collective. They have to suffer the embarrassment that goes with rulebreaking. But in the large setting of a school with high enrolment anonymity facilitates deviance, and social control becomes procedurally complex.

The size of the school enterprise and its formalisation leads to a fragmentation of children's contacts with adults. Instruction tends to be divided between a large number of teachers with specialised competencies. Non-teaching staff, composed of psychologists, janitors, nurses, social workers, cafeteria personnel and librarians, has grown in size. As a matter of fact, a considerable portion of the increased unit cost (per student and year) is due to the relatively rapid increase of non-teaching staff. Different aspects of the child, particularly a child with problems, are partitioned out to various specialists, just as in hospitals where the tendency is to treat the disease and not the patient. This also applies to child care in society at large. Various agencies, sometimes jealously guarding their respective territories, deal in a disconnected way with different aspects of the child. One only needs to observe how problems of juvenile deliquency are handled by the various public agencies involved: the police, the courts, the school and social welfare agencies.

A particular problem of size is offered by the rapid growth of public agencies in most OECD countries over the last 10-15 years. In some countries the number of employees in the public service sector (school, welfare, and health services) has almost doubled. The sheer number of people has made communication and co-ordination more difficult and has exacerbated the problem of fragmentation of the care of an individual child.

At the core of any genuine educative experience we can identify an interaction between two individuals: the teacher and the student. Teaching often is misconceived as simply being transmission of knowledge from a more informed to a less informed person. But the interaction between teacher and student implies that the former motivates the latter and can serve as a role model. This explains why as a rule teaching cannot be replaced by technological devices, however information-rich in a mechanical sense they may be.

Contradictory demands are posed on the school when stated goals in legislative documents and curricula are in obvious conflict with the administrative regulations. Professed goals, such as the humanistic ideal of self-realisation and child-centred instruction, clash with the hierarchically structured administrative machinery with directions from above. This affects what goes on in the classroom in terms of instruction and evaluation of student progress. In many curricula teachers are formally given a large margin of freedom of implementation, but centrally approved textbooks and other teaching materials decisively influence the actual teaching. Centrally prepared examinations and tests operate in the same direction. Co-operation, both among teachers and students, is often emphasized in the curricula. But central regulations and collective bargaining agreements are stumbling blocks when attempts are made to establish co-operative instruction. So is the hierarchy competition in the system, since co-operation presupposes partnership on equal footing.

The Selection Crisis

Modern, highly technological society on the verge of post-industrialism has tended to become diploma-ridden, credential-oriented, and meritocratic, with formal schooling increasingly a principal determinant of status. Competition for employment increasingly occurs on the basis of school records. Several studies have shown a strong tendency for employers to use an amount of formal schooling as the first criterion of selection among job seekers (Teichler *et al.*, 1976). The more formal the schooling the better the chances of becoming employed.

The rising credentialism has strong repercussions upon what goes on in the school. Marks, tests and examinations tend to occupy the focus of attention since they are the main instruments of the sorting and sifting process which begins at an early stage. Learning tends to be rewarded by extrinsic rather than intrinsic rewards. B.F. Skinner puts it this way: an American child who with perfect French grammar and pronunciation can say in French, "Please, pass me the salt" gets an A. But a French child who says the same, gets the salt! The perverse effect of the scrambling for marks, examination scores, etc., has been dealt with in L. Husén *et al.* (1959).

Although those who are successful in the competition for marks can come to be motivated by intrinsic rewards, those who lose at an early stage and lag more and more behind tend to give up. They represent what one of us (Husén) has referred to as the "new educational underclass". They are poor readers and their deficient reading ability affects the whole array of cognitive competencies in various subject areas that the school is expected to transmit. In most OECD countries these students are promoted to the upper grades, in spite of poor performance. They show a high absenteeism rate. At the end of mandatory schooling they tend to drop out, often being pushed out by well-meaning teachers.

Concurrent with the development of the new educational underclass is the ability of those with educational and status advantages to pass them on to their children, using the selection criteria of the school to do so. It is striking to note how the meritocratic prerogatives are passed on from one generation to the next. The classical liberal conception was that talent should have its way, *freie Bahn den Tüchtigen.* This also meant that those from privileged backgrounds but with a limited capacity would slip down on the social ladder to positions corresponding to their "real" ability. There would be a "just reshuffling" between generations. But the notion that social classes would be reshuffled in the next generation has not been supported by existing empirical evidence. Within widely different social orders it is found that those who have "made it" to advanced positions, not least by advanced education, tend to pass on their status advantages to the next generation. This is shown, for example, in Poland, in empirical research reported by Adamski (1980, especially Table 4). In societies where inheritance of material wealth is nil or close to nil the best thing parents can do is to use their favoured position (with their good education, influence and contacts) to get their children into high quality, prestigious educational institutions.

Existing evidence tells us that formal education in a technologically sophisticated and administratively complicated society cannot serve to produce equality of result. Formal education imparts competencies which determine subsequent career and therefore of necessity creates differences. Thus, the school cannot bring equality of results and at the same time be an instrument that develops and legitimises individual differences. What it can do in principle, but has only partially succeeded in doing, is to equalise the opportunity for children from different social backgrounds.

One of the implications of what has just been said is that the quest for equality easily comes into conflict with the demand for quality. Equality, that is to say, quantity of enrolment, can easily be bought at the cost of quality, an experience that many public colleges in the United States have had. But this is not an unresolvable conflict, as is evidenced by comparative studies of the standard of the elite in comprehensive and selective systems of secondary education (Husén, 1983).

An example which illustrates that there is some reshuffling, that the new underclass recruits from children of all backgrounds, is provided by an item in the Swedish newspaper *Dagens Nyheter* on 17th July 1983. It reported an interview with a 19-year old girl from Lidingö, an upper-middle class suburb of Stockholm who had taken a job as a janitor. "To hold a job and to have a regular income gives me enormous freedom", says the girl who dropped out from the last grade of the upper secondary school. "It gives me a wonderful feeling of being somebody when you are able to pay something for food and room at home. And to be able to travel freely during the vacation. Buying a Euro-rail, a 'train-tramp' card, for travel all over Europe".

She could not stand the competition in school. The marks one obtained decided who could enter the university directly from upper-secondary school. According to the admission rules set by the government until recently only one-fifth of the places at the universities were reserved for the so-called "direct transfer" quota, whereas the majority of places were reserved for other quotas, mainly for those with a certain number of years of "work experience", many of the adults with full- or part-time jobs. The competition is particularly keen for places in high-prestige studies, such as medicine, but also for studies where the chances of getting a job after graduation are rather good, such as teaching and nursing.

The girl who left school to become a janitor was in search of freedom and independence she could not find at school. "The quest for independence easily makes adolescents perceive the status of the teacher as threatening. The threat is magnified by the authority given to the teacher. He is entitled to evaluate student performance and serve as a kind of gate-keeper to high status positions." (Husén, 1979, pp. 120-159). In order to cope with this situation students tend to play a "withdrawal game". The more sophisticated do so by making themselves more faceless, which enhances formalism and rules. The less sophisticated simply drop out, thereby sometimes being helped by the school.

Two Illustrative Cases

Hurrelmann (1983) has embarked upon a longitudinal study of how "children of the educational explosion" *(Kinder der Bildungsexpansion)* perceive their opportunities of access to further schooling and the job market. The preliminary findings so far published illustrate the reactions of young people in an achievement-oriented society with an enormously increased secondary school enrolment and a shrinking labour market.

Children born in the early 1960s in Germany grew up during a period of educational explosion which came somewhat later than in some other European countries, and much later than in North America. The percentage during the 7th year in academic secondary schools *(Gymnasien* and *Realschulen)* rose from 26 in 1960 to 50 in 1980. Among the 15-year olds the proportion who took lower secondary school certificate *(Mittlere Reife)* rose from 20-41 per cent, whereas those who completed upper secondary school with leaving certificate *(Hochschulreife)* rose from 7 to 22 per cent. The enrolment in absolute numbers doubled from 1965 to 1976. Attempts to "siphon off" the pressure on academic secondary schools by establishing comprehensive schools or by trying to make vocational programmes more attractive were not very successful in terms of size of enrolment.

Thus, by 1980 the objective situation was that about a quarter of the age group had qualified for university entry and about half of it had obtained at least a middle school certificate which traditionally had guaranteed entry into qualified occupations. Since the latter had increased only moderately, the result was an inflation of entry requirements. Just as a high school diploma had earlier become important for employment in the United States, *Mittlere Reife* has become almost a necessity in the Federal Republic of Germany. The *Realschulen* and *Gymnasien* are no longer elitist institutions.

Hurrelmann's interviews with the students in their 7th school year showed that schooling was seen as highly instrumental, as a sifting and sorting institution for occupational positions and social status, without particular value in itself. The certificate was the important and tangible value derived from spending some extra years at school. The young person developed an "optimisation strategy" with the goal of achieving most efficiently a good position in the line of job seekers.

Those who leave school with the statutory minimum at the age of 15-16 would have to scale down their aspirations, given their bad starting point in the scrambling for jobs. In order to protect his own identity the young person is forced to interpret his situation as a social order he must accept. "Children of the educational explosion" do not perceive the objectively spectacular widening of educational opportunities far beyond what was available to their parents as progress. Improved educational opportunities become a liability in a meritocratic system where failure is regarded as self-inflicted. "You had your chance and failed to take advantage of it."

Those who attend upper secondary school develop long-range strategies of optimisation. Their background is such that they are both successful in their school achievements and in relating these to their coping strategies for reaching more advanced levels in the educational system or in gaining entry to highly qualified occupations. The postponing of transition from school to work allows a sequencing of strategies, with a higher likelihood of success, not least because social problems of adolescence do not coincide in time with the change-over from school to "adult" status in working life. The school for these young people with prolonged education serves a "cooling off" function.

One result of this sorting process is that school, together with an excess supply of youth relative to demand, facilitates a queueing process according to educational credentials. Those who have not been able to "make it" up to a secondary school certificate are outcompeted by their age mates with such certificates in applying for jobs in business, administrative and service occupations. The rising general level of schooling cuts those with low (mostly statutory minimum) qualifications out from major segments of the labour market. Shrinking employment opportunities due to recession and attempts to rationalise production by saving labour results in rapidly increased unemployment among young people with minimum schooling. Attractive jobs traditionally occupied by young people with some further schooling are taken by those who have been able to proceed up through the next level. Unemployment rates among people reflect a pecking order according to educational credentials.

An examination of questions related to those studied by Hurrelmann has been carried out in Sweden. The Minister of Education in 1982 appointed a task force to inquire into the problem of students who with a euphemism take an "adjusted course of study" in the last grade of the 9-year comprehensive school. They belong to the category which sometimes, again euphemistically, is referred to as "book-tired" or "school-wearied". Instead of taking all the classes at school, these students spend most of their time at workplaces formally under the supervision of the guidance teachers.

The students with an "adjusted" course of study in Sweden – since the mid-1970s some 5 per cent of the 9th graders – are in the Swedish debate often referred to as being "knocked out" from school. Whether this is an adequate metaphor or not could be questioned. But evidently many of them are "pushed out" and thereby at least partially removed from the purview of the school. Most of these students have a rather deprived social background. From the beginning they tend to be slow learners. They lag increasingly behind, particularly in acquiring basic skills, such as reading.

The Swedish task force on the basis of observations and interviews with students, teachers and school administrators identified two typical school environments, one with less than 1 per cent and the other with a much higher percentage of students with "adjusted" course of study. The former type of school was characterised by more informal and direct relationships between the various partners involved in dealing with difficult students. In the latter type these relationships were rather formalised and impersonal.

Adherence to formal rules, great social distance between strata, relationships being expressed in terms of authority, and motivation by extrinsic instead of intrinsic rewards are characteristic of bureaucratised educational systems.

The Emerging "New" Educational Underclass

The system of common public school during the entire period of mandatory school attendance prevailing in many industrial countries was created to serve as an instrument of egalitarian policy. Everybody is expected to attend the basic school for an equal number of years. The school served all walks of life and drew its enrolment from what was its natural catchment area. But within the framework of this system a new differentiation that leads to a new social stratification is emerging. A bottom layer, a "new underclass", is being formed. It is "new" in three major respects.

1. It consists of a minority of, say, 10-20 per cent of the student population, whereas the majority stands out as the mainstream that is able and motivated to take advantage of the increased public offerings. The "old" underclass consisted of a majority which – not least thanks to its own articulate spokesmen – organised itself politically and can claim partial credit for the immensely expanded opportunities for schooling that are now formally open to all young people. Before formal education was "democratised" and made available to everybody on a formally equitable basis, the structure of the school system was much more differentiated and highly stratified. It served the various social strata with parallel paths. Ever since a common basic school was set up for all at the bottom of the educational ladder, further institutions of schooling were socially highly divisive. The parallel paths of schooling, one for the select few and one for the broad masses, were serving what sociologists refer to as an ascriptive society. Many were born into their future social status as well as into the type of education that led to their social destinations. Formal education beyond the mandatory minimum provided in the basic, elementary school was beyond the economic means as well as the aspirations of young people of lower class background.
2. The "new" underclass consists to a large extent of young people coming from culturally and educationally rather than economically disadvantaged homes. The "old" underclass came almost entirely from economically deprived circumstances. In comparison with the "old" underclass many in the "new" one are well off. But a high percentage of the children in the "new" category come from homes that in various ways are affected by pathologies and hardships. Many grow up in one-parent homes. Others are handicapped by being immigrants and having difficulties with the language of their new country, a phenomenon not uncommon in these days in Europe with its high percentage of school children whose parents are "guest workers".
3. The "new" underclass, unlike the "old" one, is in many countries formally offered the same opportunities as their more privileged age mates. They do not have to pay school fees or pay for books, other learning materials and school lunches. Various social services are provided free of charge. But theirs is the handicap of not being able to take advantage of the opportunities to the extent that mainstream families and their children do. They often do not hold the aspirations that others hold. And they tend to be "pushed out" of school because they are perceived by the teachers as troublesome and as having low motivation. They show evident signs of trying to evade school by being absent and by disturbing what goes on in the classrooms.

How is it, then, that in an era of egalitarianism – especially in education – that has swept the highly industrialised countries, a new stratification is emerging? Before trying to advance an explanation, it would be useful to point to two paradoxes which bear on the question.

Paradox 1: Never before have so many places in further education been available, relative to the population of youth. The "explosion" referred to in the 1950s through mid-1970s consisted in increasing the number of places in secondary (particularly in upper secondary) education, that in some countries in practice has become almost universal, and at the universities. Instead of admitting some 10-20 per cent of the age cohort, which was the general practice in Western Europe by 1950, all are taken into lower secondary school which in practice has become part of the basic school system, and some 20-25 per cent go on to higher education. But in spite of the fact that capacity quadrupled over a couple of decades, the competition for places has become much more intense than ever before. Competition for university places has tended to have strong repercussions on the lower stages and creates what the Germans refer to as a *Leistungsdruck,* achievement pressure. It has been said that parents in Japan try to get their children into the "right" kindergarten in order to get them successfully into the "right" elementary school and the "right" high school in order finally to get them into the "right" university! Why this intensive competition in spite of the enormously widened opportunities?

Paradox 2: Asked about how they like school (which occurred in, for instance, the IEA comparative surveys in the 1970s) a strikingly high percentage of students, particularly at the age range 14-16, answered that they disliked school and many would prefer to leave as soon as they had reached the end of mandatory school attendance (Husén *et al.,* 1973, Walker, 1976). But it turns out that many of those who say that they "hate" school in spite of this opt for further schooling beyond the mandatory stage. For example, in Sweden where a rather high percentage of 14- and 16-year-olds indicated that they disliked school, more than 80 per cent decided to apply for admission to a *gymnasium* programme after completion of the nine-year compulsory school. Why this inconsistency – or is it an inconsistency? Is their behaviour basically rational?

The overriding explanation of the two paradoxes appears to be that formal education has tended to become a major stratifier in an increasingly meritocratic and socially more mobile society. In order to "make it" in life one has to "make it" at school, get the formal credentials that can serve as the first stepping stones in a career, not least to make it in acceptable and stable jobs. We have been entering a society beset by what Dore (1976) calls the "diploma disease". In the first place, a case can be made for maintaining that more formal education is required for most of the jobs and certainly for citizenship in the more complex and technology-oriented society of today. Secondly, there is a tendency to use formal education as the first criterion of sorting job applicants and to demand qualifications that in many cases are irrelevant both to the vocation and the personal life of people. Economists (e.g. Arrow, 1973; Thurow, 1972) have advanced what has been called a "job-competition model" in accounting for increased importance attached to formal education in working life. It is assumed that formal education serves as a kind of "filter" by means of which the less motivated, the troublemakers, the less disciplined and less docile are sorted out. The higher up in the system that the job applicant has reached the more he has been able to "stick it out", the more motivated and docile he is supposed to be. In addition, by obtaining a certain breadth of general education he can more readily absorb the specific skills required in a particular job. He is easier to train.

Teichler *et al.* (1976), as indicated earlier, have convincingly demonstrated that there is a growing tendency in the employment system to use the number of years of formal education as the first criterion of selection. The job seekers are symbolically lined up according to the amount of formal education completed. Those who have been successful in making it to the higher stages tend to outcompete those who got stuck at the lower grades or stages. This means that those whose credentials include only the mandatory minimum of schooling have great difficulties in getting jobs because they are further back in the line. The youth unemployment statistics bear witness to this. Among the 16-19-year-olds in Sweden with only basic school about 10 per cent were out of jobs in 1979 as compared to 6 per cent who had some upper secondary schooling. In the total work force above the age of 25, 3 per cent with elementary school were unemployed and less than 1 per cent of those with post-secondary education. A similar picture is obtained from other countries with even more striking differences between the various levels of education.

Young people are keenly aware of the hiring practices and of the fact their chances on the labour market are closely related to how high up on the educational ladder they have been able to climb. In a study in Australia of 18-year-olds by Wright and Headlam (1976) it is pointed out that there are strong forces within our industrial and technological culture to press for more and more education, something that has been assumed to be wholesome. But the result has been prolonged initial, formal schooling with higher academic qualifications. The trend towards higher entrance qualifications to many occupations has tended to put heavy pressure on young people "to stick it out". This has resulted in serious flaws in the socialisation of young people. The Wright-Headlam report says: "While providing greater educational opportunities, in fact our society has created pressures and demands upon young people which run counter to some of their basic developmental needs, particularly those to do with establishing their own growing identity ...". The report finds a "serious dissonance" between what the young people feel that they should get from education and what they perceive as valued by authorities, such as marks and examination scores.

This is not the place to dwell upon the dynamics of a meritocratic society and the role played by education in social promotion and mobility (cf. Husén, 1974). We are in this section mainly concerned with the losers in the meritocratic sifting and sorting process. As long as the economy was expanding, the "products" of further-going educational institutions could become easily employed in the public sector and also in the education-dependent service industry. The upward mobility by means of educational credentials was matched by expanding job opportunities. But also the failures could be taken care of by means of compensatory education and other means in a welfare society where there was a growing awareness of the responsibility for those who were lagging behind of disadvantaged background. But with the economic slowdown and even recession the competition for resources and for jobs became tougher. Not everybody could be accommodated in stable jobs. The number of losers increased, not so much in the schools as on the job market. The adults tried in the first place to secure and protect their own jobs. It has become increasingly difficult to accommodate the school leavers, particularly those at the age of 15-18 who finish school with the mandatory minimum. The consequence has been a rapid increase in the rate of unemployment among low-educated young people, an increase that has been much more rapid than among adults.

What are the characteristics of the "new underclass"? Martin Trow (1979) in dealing with youth problems in a Carnegie project has advanced a two-dimensional typology where he distinguishes economic-material from educational-cultural advantages and disadvantages. He then comes up with a fourfold table where advantages and disadvantages in both these two respects are distinguished. The four categories are: deprived, disadvantaged, advantaged and privileged. The "new underclass" consists essentially of the deprived but also to some extent of the educationally disadvantaged, those who are victims of the common pathologies of the nuclear family of today: divorces, alcoholism, parents spending most of their time without contact with their children. As pointed out above, the "new underclass" is identified more by socio-cultural than by economic handicaps. This has been more closely examined by the Commission of Inquiry into Low Income families in Sweden (Johansson, 1971).

The following characteristics of those who are unsuccessful in school appear to be especially important.

- Lack of "school readiness" at the time of entering the first grade (see, e.g. B.A. Johansson, 1965).
- Early failures in school.
- Weak parental support of school learning (no participation and contact in PTA activities).

- Failure to learn to read during the first years of schooling which is aggravated by automatic promotion.
- Absenteeism.
- Disciplinary troubles (aggressive and disrupting behaviour in the classroom, vandalism, etc.).
- Low marks and grade repeating.
- Low motivation, no aspiration to take advantage of opportunities beyond mandatory school.
- Tendency to leave school early. "Adjusted" courses of study.
- High unemployment rate immediately after leaving school.

5. ECONOMIC AND LABOUR MARKET ISSUES

By the mid-1970s youth unemployment was becoming a matter of concern in all highly industrialised countries. A combination of an economic recession and the baby boom generation reaching the labour market rapidly increased youth unemployment, which in most OECD countries doubled during the latter part of the 1970s. The enrolment explosion in secondary education and the deliberate policy in some countries of using schooling as an instrument of coping with unemployment did not offset the surplus of youth in the labour market.

The OECD has since the mid-1970s made studies of transition from school to work and factors related to unemployment among young people a priority area of study. A task force chaired by Clark Kerr submitted a report on policy issues with recommendations in 1975 (OECD, 1975), a report which discusses the matter of transition from school to work at some length. The OECD has conducted several studies of youth unemployment (OECD, 1980 and 1981a), and has collected available statistics in order to elucidate causes and consequences of the problem. In the United States and Japan the majority of young people in the age range 15-18 was already by the early 1960s in full-time education. In Western Europe most of them were either entering the job market or were already employed. Since then most West European countries have caught up.

As all this indicates, the movement from school to work is one of the important components of the transition of youth to adulthood, and a component which has undergone extensive changes in recent years. Some economic theory is directly relevant to this component of the transition. The difficulty in using this work lies, as in the case of work originating in the other social sciences, in theoretical disagreements. Nevertheless, a summary of the theoretical positions will give an indication of the basis of the disagreements.

A major theoretical orientation to the functioning of labour markets is the theory of human capital which derives directly from neoclassical economic theory (Schultz, 1961; Becker, 1964; Mincer, 1974). According to human capital theory, young persons will remain in school so long as the present value of the expected gains due to increased education exceeds the cost of education plus the income foregone by staying in school. When that is no longer so, the young person will leave school and search for a job.

This theory may be quite useful in accounting for certain observed phenomena. For example, when in the late 1960s and early 1970s, the large cohorts of the baby boom years were at the age of entering the labour force, with a demand for labour far from large enough to absorb them, there were large increases in university enrolment as occurs to a lesser extent whenever the demand for labour slackens. This is in accord with human capital theory, because when the probability of unemployment is high, the expected income foregone is less.

Yet the principal focus of human capital theory is not on the effect of unemployment in keeping young people in school. It is on the effect of school in making youth more productive, possessing greater "human capital".

A number of studies in various countries which relate years of education to subsequent income have shown a substantial rate of return to education for individuals. Challenges to these results have, however, argued that the causation is reversed: because of the selective character of schools, those who have higher productive potential will both remain in school longer and have higher subsequent earnings, without any effect of schooling on their productivity.

Even if it is accepted that the same young person will have increased income from staying in school, there are a number of questions this leaves open. First, the expected income gains from staying in school may not arise from an increase in productivity at all. Employers may use the educational credential as a signal of worker quality (see Spence, 1974; and Berg, 1971), and the schools may be carrying out a screening function, rather than an educative one (see Taubman and Wales, 1973). Thus it may be quite rational for young persons to act just as human capital theory would predict, and for employers to do so as well, even if schools were wholly ineffective in increasing productivty.

Human capital theory, however, goes beyond prediction of the choice between school and work. It assumes that the education obtained provides not only a private rate of return sufficient to make the investment worthwhile, but also a social rate of return equal to the private rate of return. This is in effect arguing that the economy's growth is limited by the supply of human capital, i.e. by the level of education in the society. We can call this a model of a "supply-limited" economy, where by supply is meant supply of qualified labour. The assumption is that the economy will grow at the rate of its educational growth. There have, in fact, been a number of analyses of countries' economic growth as returns to human capital (see Psacharapoulos, 1981, for a review).

However, other economic models, less completely in a neoclassical framework, assume a "demand-limited" economy (see, for example, Thurow, 1972). In this conception, the limiting factor is economic demand for labour, and the assumption is that variations in youth unemployment over time are merely indicators of fluctuations in demand, while differences among population groups at the same time within the same economy (for example, blacks and whites in the United States) are due merely to positions of members of these groups on employers' preference scales. This orientation, unlike the supply-limited theories, assumes that the choices of youth between education and work lead to an excess of education (see Freeman, 1976).

These two theoretical positions have very different implications for policy. The supply-limited human capital theories imply that expansionary educational policy will lead to economic growth, while the demand-limited theories see education largely as a consumer good, with the result that educational credentials are in oversupply, while economic growth depends on other factors.

It is quite possible, of course, that both these orientations may hold, but for different times and places. The fact that manufacturing has migrated rather rapidly in the 1970s from developed countries with high levels of education to less developed ones with lower levels (and that labour with low levels of education from less developed countries has been employed as "guestworkers" in various developed countries) implies that education of the manufacturing labour pool has not been the limiting factor to economic growth. Complementary to this, the extensive migration of highly-educated persons from less-developed countries to developed countries (sometimes termed the "brain drain") indicates that high levels of education among a portion of the population is not a sufficient condition for economic development.

On the other hand, those less developed countries which are undergoing most rapid economic development include those in which rates of literacy have been high for some time (particularly countries in the Far East and Southeast Asia). Thus it may be that while this education was not a sufficient condition for economic development it nevertheless contributed strongly.

This unresolved question about the role of education in economic development is part of a broader issue we are concerned with here, namely, the difficulties of youth in making the transition from school to work. Stated most provocatively, it is the question of whether the employment difficulties of youth are due to deficiencies in youth – which may be of a variety of kinds – or to defects in the economic system which create special barriers to youth entering employment, or fail to provide a sufficient demand for labour. Proponents of the first thesis point to the great difference in unemployment rates of youth with differing levels of education, to the extraordinary growth in employment of women while youth unemployment has been growing, and to the success of immigrants and guestworkers while youth show high levels of unemployment. Proponents of the second view point to extensive changes in the structure of the economy, changes in directions that are unfavourable to youth employment.

But to see the problem as either due to deficiences in youth or due to defects in the economic system may put the matter too simply, for the difficulties may arise from a mismatch which could be rectified either by changes in the youth or in the occupational system. Ordinarily, it is assumed that the characteristics of the system must be taken as given; but that assumption need not be correct, as the examples of work restructuring (e.g. Volvo and Philips) and flexitime (e.g. in many United States firms) indicates.

Similarly, the characteristics of work in the microcomputer industry, discussed in an earlier section, have a high compatibility with certain values expressed by youth: low levels of supervisory authority, intrinsic satisfaction from work, flexible working arrangements, short-term commitments. It is likely that some of the youth working in this industry would not be able or desire to hold a job in the industry if the organisation of work were more traditional.

The principal orientations of economic theory do not resolve the problem of why youth have special difficulties in getting integrated into the labour force. But they do clarify the major unresolved issues, as well as the different policies that are indicated, depending on the resolution of these issues.

Part II

CURRENT CHANGES IN FAMILY, SCHOOL AND WORK, AND THEIR EFFECTS ON YOUTH

There have been, during the period since World War II, extensive changes in those institutions of society that most directly affect youth. In particular, changes in the family, in school, and in the economy have been great, and in each of those institutions the changes show no signs of abating. This section will focus on recent and continuing changes in society affecting youth, with special attention to these three institutional areas, family, school and work.

1. CHANGES IN THE FAMILY

It is useful to distinguish three broad phases in the state of a family's social and economic conditions as they affect children and youth. The family's social and economic conditions can be described as bare subsistence or poverty, economic respectability, and affluence. These three conditions correspond roughly to phases of a society's social and economic development, ranging from a subsistence economy, through an industrial economy, to a post-industrial affluent economy.

We will outline each of these phases, and attempt to give for each a sense of the role of schooling in the transition of youth to adulthood. We distinguish these three phases because in each the family has a certain set of interests in its children that shape the way it acts toward them – and thus set the environment that the school confronts.

In seeing these three settings as phases of a society's social and economic development, it is important also to recognise that at any point in time, there are families whose social and economic level deviates from the societal average. In discussing each of the phases, we will mention the most important deviations.

Phase 1: The Exploitation of Child Labour

Phase 1 refers to subsistence level, and a Phase 1 society is one in which most households are at or slightly above a subsistence level. An economy based largely on subsistence farming is the most widespread example, though extractive economies in general, in which most occupations are in the primary economic sector, fit this phase, as do other village-based societies, or societies in the very early period of industrialisation. In such social structures, households directly produce most of what they consume; economic exchange and division of labour are minimal.

In Phase 1 societies, the labour of children is useful, both because in the diversified activities of the household, there are always tasks that children can carry out, and because the economic level of the household is sufficiently low that the effort of all is needed. The economic burden of children is offset at least in part by their labour. Families have many children, and exploit their capacity for labour, often with little regard for the impact of this upon the children's opportunities. Families have narrow horizons, are inwardly focused, and have little interest in or resources for extending their children's horizons broadly.

In an economic and social structure of this sort, the principal role of the school is in protecting children from exploitation by the family, and in providing a broadening influence beyond the family's horizons. The family constrains and limits the child; the school breaks some of these bonds and reduces the constraints. The school often stands, in such a setting, in an antagonistic position to the family, for the interests of the two often conflict. The school is the liberator of the child from the exploitative grasp of the family. If the school does succeed, it aids the child in providing social and economic opportunities beyond the horizons of that child's family.

A great number of children are at this stage regarded by the parents as an economic asset and a guarantee for old age. Yet nations whose economies and social structures are of this sort are the poorest, so that the economic resources necessary to provide educational opportunity in terms of formal schooling are most limited. The nation's capability of providing a strong school system to oppose the constraining force of the family is weakest. Consequently, it is in this phase that educational resources are ordinarily most unequally distributed between rich and poor regions. Educational opportunity depends largely on the opportunity provided by the family and the immediately surrounding area – and on the educational resources provided by the school, where it exists.

There are some families, in such societies, that are sufficiently free from economic distress that they have broader horizons for their children. Schools, in societies of Phase 1, tend to nurture and encourage these select children, for these are the few whose families reinforce and strengthen the schools' aims. The principal successes of the schools in Phase 1 societies are among these select children. These families are precursors of the majority of families in Phase 2 societies.

Phase 2: Children as Investments for the Family

Phase 2 is a post-agricultural, urban, industrial society, engaged largely in manufacturing and some commerce. Here the economy is an exchange economy, most labour is performed in full-time jobs, and the family's economic needs are provided mostly through the exchange of wages for goods. Children's labour is no longer needed for the household's economy, and there are fewer possibilities for productive work of children within the household.

In such a society, the family continues to have a strong interest in children, for a more long-range goal. Children are the carriers of the family across generations from the past into the future, and investment in children is an investment in human capital for the family's future. While in Phase 1, a large number of children was valuable as an investment in the future, in Phase 2, what seems important is high investment in each one of the few, to increase the status position, economic position, and social respectability of the family in the next generation. The family shows a great deal of interest in the transition of its young to adulthood, because the family's position in the next generation depends heavily on this transition. Parents are exceedingly concerned about daughters' marriages, and are exceedingly concerned about sons' beginning careers.

This change in the family's interest in children has many implications. One is a decline in the birth rate. Another is an increase in the demand for universal prolonged formal education and for equal educational opportunity. Still another is that great attention is paid by the family to the process of transition of its youth to adulthood.

The family is no longer the school's antagonist, but its most important ally. The family creates a strong motivation for schooling in its children, for the school's goals for the children coincide with the interests of the family. The school is a principal avenue by which the family can achieve, through its children, status mobility in the next generation.

High academic achievement is to be expected from children whose families are in Phase 2, and high academic achievement in the nation as a whole when the nation is in Phase 2. Family and school are reinforcing each other's actions toward high achievement and extended education. The family is the principal agent to guide youth into adulthood, and the school's role supplements this.

Phase 3: Children as Irrelevant

Phase 3 is an advanced industrial society (what Daniel Bell has called a post-industrial society), a welfare state with a high degree of affluence. In this phase, the family's central role in the economy has vanished, and the family itself has become a kind of appendage to the economic structure. It is an institution relevant to consumption, but no longer to production. Its functional role has been reduced to that of childrearing, and providing an emotional anchor-ground.

The family's central place in the economy and society has been taken over by large corporate bodies – industrial, commercial, and governmental. As the economic functions of the family are withdrawn to other institutions, the family loses much of its raison d'étre. It is no longer an institution spanning generations, but forms anew with each generation, so the family's interest in children to carry the family into the future declines. The stability of marriages (and thus of households) declines, as the multi-generational family is no longer present to restrain its members from individualistic solutions at the expense of the family.

As a society moves toward Phase 3, the State takes over many of the welfare and redistributional activities once performed by the family. This further reduces the family's functions, and the incentive of its members to achieve family goals rather than individual ones.

In such circumstances, we can expect that families would make fewer investments in children, would press less strongly toward academic achievement, and would support the goals of the school less strongly than in Phase 2, although differences between social strata with regard to family support for children's schooling can be expected to persist. Because the family no longer spans generations, its interest in the destinations of its young should be less strong, leading it to make fewer sacrifices to achieve desirable transitions of its young to adulthood. Its attention to the process of transition of its youth is less than in Phase 2.

Western developed countries are in Phase 2, and Phase 2 families represent the modal type. However, the most affluent OECD countries are approaching Phase 3, and in those societies a large number of families are already in Phase 3. In addition, the welfare provisions of the State move families more quickly toward the psychological condition of Phase 3 than their own affluence would dictate.

Because some OECD countries are moving into Phase 3, we should see evidence of the predicted reduction of interest in children and their future. The evidence supporting these predictions for Phase 3 is mixed. In the United States and some countries in Europe which are closest to Phase 3, there appears to be an even stronger demand for education, and more resources invested in education than in the earlier period of Phase 2. There is a strong professed interest of parents in their children's educational development.

On the other hand, families have shifted much of the responsibility for financing higher education to the government. Parents spend less time with children, and children less time with parents in whole-family settings, than in earlier periods in these countries. Leisure activities instead take place in age-segregated settings: cocktail parties and work-related gatherings for the adults, rock concerts and sports for the youth. Increasing numbers of children are abandoned, run away from home, or become addicted to drugs, and an increasing number of children of divorced parents are unwanted for custody by either mother or father. Abandonment, runaways, drug addiction, and unwanted children, of course, involve only a minority of children.

Our own assessment of trends in the most advanced societies is that there is, as predicted above, lesser investment in children and youth by families than was true forty or fifty years ago, and that the evidence will begin to show this more clearly. This means that the school will lose much of the active support it had during Phase 2, and with that the motivation to achieve which families imparted to their children. The school's task, in this condition, comes to be one of supplying not only the resources for learning, but also taking active responsibility for bringing about learning. The school, under these conditions, takes over some of the functions which the family once provided, but which it no longer provides. Whether the school is able to exercise these functions is another matter.

If this picture is a correct one, it accounts for an otherwise puzzling result: in less developed countries still in Phase 1, there is a relatively strong relationship between the tangible school resources in a region or locality and the level of academic achievement of the students in that region or locality. However, this relationship is sharply reduced in highly developed countries. The achievement attributable to the school itself in highly developed countries is weakly related to the level of tangible school resources provided by the community or the nation. The achievement is not independent of the way the school is organised, nor the disciplinary constraints it imposes on students, nor the academic demands it makes on them. But a school with excellent physical resources, laboratories, books, and teacher qualifications, a school with high per pupil expenditures, does not produce high achievement if these less tangible organisational elements are missing (see, e.g. Hanushek, 1981).

In Phase 3, the tangible school resources are in over-supply, not only in the school itself, but in the home, through television, and quite generally throughout the society. What is in short supply in the affluent Phase 3 are the intangible resources, such as the motivations that strong families, interested in investing time, effort, and attention in their children, provide. Achievement in school is highly related to those family characteristics that are most effective in bringing about interest in and attention to the child's achievement[3].

The schools that are most effective in this third phase are those that are able to supplement parental interest, attention, and demands to supply the intangible qualities that impel students to take full advantage of the opportunities provided by the tangible resources. The school, in Phase 3, is one of many elements competing for the attention and interest of children and youth, and what cannot be taken for granted are the motivational forces that direct the young person's attention and interest toward school learning, rather than toward other attractive competitors.

What also cannot be taken for granted is the attention given by families to the transition of its young to adulthood. As the family's continuity through generations breaks down, the society, and the school as its agent, takes an increasing share of responsibility for the transition from childhood and youth to adulthood. This includes not merely a transition from school to work, but a transition to adulthood in other ways as well: in the transmission of culture, and in the development of maturity of judgement, the school increasingly becomes involved in the transmission of tastes and leisure activities.

However, the shift of these responsibilities to the larger society does not take place without problems: the process of transition from youth to adult appears to be increasingly taken over by youth-oriented commercial interests: popular music, fashion in clothes, and for a minority of youth, mind-altering drugs. The interests of the society as a whole in the formation of the coming generation are hardly evident in the institutions which are replacing the family as guides in the transition to adulthood.

Implications of Phase 3 for the Process of Transition

In Phase 3, both the family's capacity for guiding its youth in the transition to adulthood, and its interest in doing so, are reduced. This thrusts on the society as a whole, and on the educational system in particular, a task for which it is presently poorly prepared[4].

The implications of this change in the family's role are both direct and indirect, and they pervade all aspects of the transition to adulthood. We will indicate briefly what some of those implications are.

Finding a Job and Beginning a Career

It has traditionally been true that the most common means of finding a new job is through family or friends. This is not merely a carryover from rural society; it functions in industrial settings as well. This, of course, partly results from the fact that informal channels of communication remain important in most areas of life. In part, however, this is due to the explicit care and attention of the family – and in ethnically diverse settings, of the ethnic group as the family's extension – to the occupational success of its coming generation. With the decline in the family's strength and in its interest in family continuity over generations, this attention to the occupational careers of its youth can be expected to decline. This implies greater difficulties in the job search process, and longer periods of unemployment for youth. Such changes may already be in part responsible for the extensive growth in youth unemployment relative to that of adults in the past two decades (see Table 2).

It is evident that if the difficulties are not to increase, society as a whole must, especially through its educational institutions, devise new mechanisms for aiding youth entering the labour market. As the possibility of finding a job and beginning a career through the old channels declines, the capacity of schools to take over a larger portion of this task becomes more important. The specific ways this may be done will differ at different educational levels and in different social settings; but the problem is similar across educational levels and in different developed societies.

Children and Youth as an Increasingly Disadvantaged Minority

Another change has occurred for children and youth as societies have moved from Phase 1 toward Phase 3. The economic position of children and youth relative to that of adults in society has been lowered.

In Phase 1 society, the households are multigenerational. As a first approximation, we can assume that all dependants are contained within the household and all households have the same number of dependants, including grandparents, parents, children, and other kin. Although there is not equal distribution of income among the household's dependants, we can speak of the per capita income of each person in the household as the household's income divided by its size. In such a society, the per capita income of a child is equal to that of an adult.

In Phase 2 society, grandparents no longer live in the household. Children leave the household, form their own nuclear families, and have their own children. Assume that each person spends half his life in a household without children, either before having children or after his own children have grown. Then if the income to households is equal, and all households with children have four dependants (two adults, two children), while households without children have two (both adult), the per capita income in the households with children is half that of the per capita income in households without. Altogether, children have two-thirds the per capita income that adults do.

In Phase 3 society, each adult is a wage-earner, except for some mothers with young children. Assume that a quarter of the population chooses not to have children, which concentrates the children in a smaller fraction of households. If each person who does have children spends, as in the Phase 2 society, half his life in a household with children, and each who does not spends only a quarter of his life there, then 7/16 of the households have children, with about three each necessary to maintain the population level. If mothers work about half the period that a child is in the household, then the average household with children has five dependants and one and a half incomes. The household without children has one income per dependant, since all adults except some mothers work. In this society, the per capita income of a child or youth is only 0.3 that of adults without children in the household, and only 0.43 that of adults as a whole.

The overall comparison of relative incomes of adults and children in these prototypical Phase 1, Phase 2 and Phase 3 societies is shown in Table 1. The table gives both the per capita income in households with children relative to that in households with no children, and the derived per capita income of children relative to that of adults.

Table 1. **Overall comparison of relative incomes of adults and children**

	Phase 1	Phase 2	Phase 3
Proportion of households with children	1.0	0.50	0.44
Per capita income in households with children relative to that in households without	–	0.50	0.30
Per capita income of children relative to that of adults	1.0	0.67	0.43

What makes this exercise with hypothetical societies of interest is the fact that developed societies are moving from Phase 2 to Phase 3. The household structure of Phase 1 society in which almost all adults marry, have children, and live in three-generation households, is most characteristic of the rural past of developed societies. For example, the first census of the United States, in 1790, shows that somewhere between 80 and 90 per cent of the households in the young country had children in them. A hundred years later, in the 1890 census, this had declined only by about 10 per cent, to somewhere between 70 and 80 per cent. The household structure of Phase 2, in which all adults marry, set up a new household, and have children, is most characteristic of the recent industrial past of developed societies.

Finally, the household structure of Phase 3, in which a quarter of adults choose not to have a family, with the result that children are concentrated in a smaller fraction of the households, and in which all adults except for some mothers earn incomes, is most characteristic of the emerging structure of developed societies. For example, at the 1980 census of the United States, only 39 per cent of the households had children in them. As Table 1 shows, if income were identical among all wage-earners or providers in each of the three hypothetical societies just described, the relative welfare position of children declines from Phase 1 to Phase 2 to Phase 3. Certainly the increase in absolute levels of living from Phase 1 to Phase 2, and the fact that children in Phase 1 were often required to be providers means that the relative position of children is overstated in Phase 1 compared to Phase 2. But the disadvantage in Phase 3 compared to Phase 2 is a real one.

The implications of this for the family's capacity to continue as the society's major means of redistributing income are serious. It means that a large portion of personal income in society comes to households in which no redistribution is possible: only adults are in the household, and each adult is a wage earner. The redistribution that households traditionally provide is left to a minority of households with a small fraction of the total personal income in society. Thus not only is it true in Phase 3 that adults have less interest in children; the relative ability of those adults with children to attend to their children's interests declines. Those adults without children in the household can afford an affluent style of life, while many needs of children go unmet. Modern welfare states have attempted to address this problem through child allowances and other policies but these never come close to compensating the cost of a child in the household.

2. CHANGES IN WORK INSTITUTIONS

Before describing the recent changes in work institutions which have a particular impact on youth, it is useful to look at some statistics on the relation of young people to work institutions. The statistic to which we would like to draw attention is the ratio of male teenage (15-19) unemployment to that of all adults (25 and over) in those OECD countries for which such data have been centrally tabulated by OECD (OECD, 1980). Table 2 shows the ratios.

Table 2. **Teenage male unemployment rates to adult unemployment rates, 1965 and 1979 in ten OECD countries**

	1965	1979
Australia	1.4	4.1
Canada	3.2	3.1
Finland	2.5	4.3
France	3.4	3.3
Germany	0.5	0.9
Italy	4.8	7.1
Japan	2.1	2.8
Sweden	2.2	4.5
United Kingdom	1.3	5.2
United States	4.0	4.4

Teenage unemployment has characteristically been much higher than that for adults in most developed countries, often two or more times that of adults. The column for 1965 in Table 2 shows that this ratio varies considerably between countries, variations that undoubtedly in part reflect merely differences in definition and measurement of unemployment (for example, Germany with its apprenticeship system shows *less* unemployment for teenage males than for adults). Thus comparisons between countries in this ratio connot be fruitfully made from these official statistics. The comparisons that are meaningful, however, are those *within* the same country, between the two points in time.

When those comparisons are made, it is evident that in a number of countries teenage unemployment has grown much more rapidly than adult unemployment. In Australia, Sweden, and the United Kingdom, the ratio has more than doubled. In Finland and Italy, the ratio has increased sharply. In the United States and Canada, on the North American continent, where it was already quite high in 1965, there has been little change in the ratio, and there has been little change in France. With these exceptions, the change has been extensive[5].

These changes suggest that there have been extensive and pervasive structural changes in the economies of these countries, affecting young persons and adults differentially. If this is so, and if these changes continue, the problems of youth unemployment might well be in their infancy, with greater difficulties in the future. In this section, we will outline economic changes that are occurring which have implications for change in the relation of youth to the labour force. The relative importance of these various changes is not known, but they do indicate changes in the economies and societies of developed countries that must be taken into account if youth are not to be even more disadvantaged as they attempt to join the employed work force.

Internationalisation of the Economy

The growth in international trade and in temporary labour migration from less- to more-developed countries has created a situation in which workers of developed countries must compete not only with others in their own countries, but also with workers in less developed countries. This competition is most severe for low-skilled, entry-level jobs. Because the exports of goods and services from developed countries to less developed ones are based on labour with a high technical or educational component, and their imports of goods from less developed countries is based on labour with a lower technical and educational component, an increase in the volume of international trade means that the demand for domestic low-skilled entry-level labour continually declines in developed countries. This internationalisation of the economy has taken place largely through multinational corporations, and more generally through the export by developed countries of skilled management to less developed countries. Thus cars may be manufactured in Spain for the European market, with Spanish labour, but with German engineering and German management. Or shoes for the United States market may be manufactured in Brazil by Brazilian labour but United States design and United States management[6].

Youth, who are in competition for jobs with adult workers in their own society paid at about the same level, now find themselves in competition with workers in another country, paid at a much lower level.

A second, though less important, way in which internationalisation of the economy has an impact on youth employment in developed countries is through the import of temporary workers from less developed countries. If one pattern is exemplified by the manufacture of cars in Spain for the European market with Spanish labour and German engineering and management, another is the manufacture of cars for the European market with Spanish labour and German engineering and management, but in Germany rather than Spain. The pattern of using guestworkers will probably never expand again to the level it reached in the late 60s and early 70s, but immigration from less developed to more developed countries will probably continue to increase, not least for the menial jobs in the service sector.

A comparison of youth in developed countries with immigrants to the same countries can show especially well the position of youth in developed countries vis-à-vis workers in less developed countries. [By immigrants we include not only permanent immigrants, such as Hispanics in the United States, and those from the Indian sub-continent in England, but also guestworkers in Europe and illegal immigrants (mostly Hispanic) in the United States.] A central observation is that immigrants are more successful than native youth; they will take jobs that native youth will not; they will work for lower pay; they are less likely to quit a job after a short time; and they work harder.

All these differences can be summed up as one: their expectations about what they should give relative to what they should get in employment are sharply higher than those of native youth. There is one principal source of this, and several subsidiary sources: the principal source is that the society from which they have come is one in which they had to give much more and received much less in wages or other benefits. A subsidiary source is their lower living costs – even living side-by-side with natives in developed countries – partly because initially their expected living standard is low, and partly because they often belong to strong families in which pooling of income and joint consumption increase efficiency. Another subsidiary source is the fact that in some cases immigrants are eligible for a smaller set of welfare benefits than are natives, and thus the alternative to labour force success is more grim for them.

All these elements give immigrants a competitive edge over native youth within a developed country; but internationalisation of the economy puts youth equally into competition with the immigrants' countrymen who have remained at home.

One way of viewing the policies of developed countries toward internationalisation of the economy is that these policies have benefited the consumer role of the average citizen at the expense of the worker role. This trade-off may have a source in political democracy, for the import of lower priced goods benefits the majority as consumers at the expense of a minority of workers whose jobs are lost. The trade-off certainly has in part an origin in economic analysis which can show that all countries are better off with free trade. That economic analysis, however, contains two assumptions: that each country maintains an overall balance of trade, allowing the international value of its goods to fluctuate until that balance is reached; and that each country is able to solve without cost the distributional problems that arise from a high degree of specialisation. It is clear that neither of these assumptions is met in modern national economies.

Internationalisation of the economy differentially benefits different segments of the population; and it is likely that because of distributional problems, it constitutes an absolute harm to certain segments. Ever since economists showed, via Kaldor's compensation principle, that *everyone* could be made better off by free trade, the arguments for increasingly free trade have been irresistible. However, since compensation is not carried out in practice, free trade also leads to economic dependency of increasingly large segments of the population. But the same processes operate to a lesser extent in developed countries. What is necessary, if economic efficiency is not to be bought at a price of increased welfare dependency, is to devise policies which will have an economically compensatory effect, without stopping the engine of economic development that results from free trade.

Altogether, the internationalisation of the economy – which will increase rather than subside – has serious and negative consequences for youth in developed countries. It has the implication that youth should do increasingly less well relative to adults in the labour force in developed countries. Similarly, lower levels of the labour force should do increasingly less well relative to higher levels. Internationalisation of the economy should have an effect in increasing the income inequality in developed countries. In less developed countries the reverse should be true: the relative position of lower levels of the labour force should be improving, and income inequality should be declining.

Another important implication of this increasing internationalisation of the economy is that many traditional paths of entry into the labour force for youth are being closed off in developed countries. Those youth whose counterparts in the 1950s and 60s entered the labour force with a willingness to work and a healthy body as their principal credentials find that a number of the sectors which they would have entered are gone from the country. They find educational attainment as an increasingly important credential today in a society which increasingly is becoming meritocratic. This means that the difference between the labour force opportunities for those youth who do well in school and to those who do poorly tend to be greater today. Educational inequalities are translated more directly into income inequalities, so that policies which affect inequality of educational opportunity have an increasing effect on inequality of opportunities in work.

Perhaps the most important implication of the internationalisation of the economy for youth is, however, the overall one: these changes mean that the work opportunities for youth relative to that for adults is continually declining, so that any policies designed to aid youth opportunities are addressing a problem that can be expected to grow progressively worse – from a cause no more complex than the fact that international trade is increasing[7].

The internationalisation of the economy is only slightly in advance of another change which makes the labour of youth less competitive: the robotisation of manual tasks. As more and more repetitive manual tasks are successfully mechanised, and in many cases computer-designed and computer-steered, the level of abilities needed by persons to remain competitive in the labour force increases. These developments will, of course, increase the overall affluence and economic activity, so that as in the past, they may create as many jobs as they displace. But what is essential for youth is the kind of jobs created and the kind displaced. The jobs displaced are those at which relatively unskilled and inexperienced persons could compete; the jobs created tend most often to be technical jobs requiring skills which are ordinarily obtained only through extensive training and prolonged formal schooling.

The fact that robotisation of manufacturing will follow rather soon after internationalisation of manufacturing means that a stopgap solution of trade barriers will not be effective as an aid to employment of youth and others disadvantaged in the labour force. What is clearly necessary in our view is much more fundamental solutions to the problem of how to achieve income distribution without dependency.

The Organisational Context of Work

A. The normalisation of work

There is a continuing change in economic organisation in modern society away from informal work arrangements to more formal ones. The informal work arrangements characteristic of the past were less often "full-time jobs"; they had room for persons on the fringes of the labour force, and those in the process of entering it or leaving it. The formal work arrangements characteristic of the present and probable future have little opportunity for such marginal persons and marginal activities. The effect of this change on unemployment of black youth is shown well by Cogan (1981), who records that the timing of the rise in unemployment rates of black youth coincided with their migrations from the rural South, with its informal work arrangements, to the urban North, with work characterised by full time jobs in formal organisations.

Informal work settings are often family enterprises. Young persons, growing up within the family or as part of family-and-neighbourhood networks, easily move into these enterprises, with the principal credential being the family connection. The more formalised work settings that replace these have little or no connection to the family, require formal educational credentials in lieu of family credentials, and entail more extensive job search. In addition, they sometimes have a direct exclusionary effect on youth. For example, insurance regulations in the United States exclude workers under age 18 from certain jobs and certain workplaces. Yet far more dangerous jobs have been regularly carried out by youth under 18 in the more informal work arrangements of the past. The insurance regulations are probably beneficial to the safety of youth under 18, but they are harmful to their employability.

One implication for youth of this change in the organisation of work away from the family and from informal work arrangements is that youth are disadvantaged at the point of entry into the labour force. Their principal social connection is to the family, and as the family is increasingly separate from the workplace, youth are increasingly deprived of natural connections to employment. This increased separation of work from the family has probably also reduced the exploitation of youth and has probably helped free a youth's life chances from the confines of social origin. The result is likely a replacement of some aspects of social class inequality by inequality between ages, with youth taking on some of the attributes of an underclass.

B. The increase in employment-related benefits

The fringe benefits associated with employment have been increasing at a faster rate than wages themselves in developed countries. These benefits can be regarded as the fixed costs associated with adding a new worker to the payroll, in contrast to wages as variable costs. As the fixed costs have risen, the attractiveness to an employer of hiring youth of uncertain work habits, no matter how low the wage, is less. It becomes more efficient to have a smaller set of highly-trained and well-paid employees than a larger set which is less experienced and lower paid. This is especially so when there are wide swings in prosperity, for a smaller and more efficient workforce allowed less reduction in the workforce with economic depression. Thus a side effect of the increase in non-wage benefits for workers is its exclusionary effect on youth.

In this respect, many experts are also arguing that:

- Legislation on minimum wages for youth and the policy on the part of the trade unions to equalise wages have jointly contributed to unwillingness on the part of employers to employ young people without experience and with low productivity;
- Legislation on job security and social security whereby young workers are affected already at the age of sixteen has also, according to many experts, contributed to the reluctance on the part of employers to employ young people.

C. The "hidden economy" and its implications for youth

In much of the developed world, a second economy has developed in attempt to escape the heavy hand of the State. In Western countries, the second economy, apart from the sector involving illegal activities (such as theft, prostitution, illegal gambling) involves principally exchanges of services and purchase of services in such a way as to escape the increasing tax burdens of the welfare state. Two questions may be asked about this second economy in relation to youth employment. One is its current impact on youth employment, and the second is how various government policies toward it might affect youth.

Because this is a "hidden economy" there are few statistics that could help in assessing its impact on youth. However, because it has more of the characteristics of the informal community economies of the past (no full-time jobs, no formal credentials, no fringe benefits, few official regulatory constraints), it is likely that youth participate in this economy disproportionately.

Assuming this to be the correct answer to the first question, the answer to the second question follows: government policies which inhibit the hidden economy differentially restrict youth employment. Conversely, government policies which facilitate the hidden economy – especially policies which facilitate this only for youth – aid youth employment. It is probably useful here to establish a principle of the following sort: Tax receipts lost through operation of a second economy are thereby unavailable for redistribution to aid a dependent segment of the population. Certain hidden economic activities, however, shift some persons from dependency on state welfare to economic independence. A policy which encouraged those legitimate activities which reduced dependency more than enough to offset the welfare aid lost via foregone tax receipts would have a net effect in improving economic welfare.

Thus modified taxation policies for those currently unreported economic activities engaged in by certain segments of the population could both increase economic efficiency and reduce dependency in the population. Such policies, of course, could be instituted only after careful investigation of current patterns of unreported economic activity, and of potential side effects of differential treatment.

The idea of introducing such policies may appear bizarre; yet in a more piecemeal and clumsy way, current policies such as a differential minimum wage for youth or subsidies to employers for youth use a similar strategy. The principal added element here is the notion of making use of existing and potential hidden economic activities to serve this end.

D. *Sectors with increased opportunities for youth*

As women have increasingly moved into the paid labour force, some new opportunities have opened up for youth, taking over activities vacated by women. This is particularly evident in food preparation. Some home preparation of food is being replaced by that of fast food restaurants. This is most pronounced in North America, but is apparent in Europe as well. These restaurants are staffed almost entirely by teenage youth, whose work in effect replaces a portion of women's household labour.

This area of employment is expanding rapidly, providing jobs for youth, often part-time work by youth who are still in full-time school. In some respects, this partly offsets other changes in the economy which reduce opportunities for youth employment. This sector of the economy can be expected to continue its expansion, as the family's internal functioning declines. Thus the prospects are for continuing expansion of jobs for youth in this sector.

However, these jobs have certain characteristics which reduce their value as entry-level jobs for youth. They are primarily part-time, and they are in an industry in which there is almost no natural ladder for advancement. Thus for most youth who hold such a job, the job is not the first rung on a career ladder, but only a means of earning spending money and gaining some work experience before any serious entry into the labour force.

Summary of Changes in Work Institutions

Most of the changes in work institutions in recent years imply a more difficult role for youth entering the full-time labour market. This difficulty is especially pronounced for youth who could enter the labour market relatively young, at unskilled or semi-skilled jobs. It is these jobs which, for a variety of reasons, are declining in developed countries, and these jobs which are unlikely to be revived. There are some changes which benefit youth employment, such as the movement of some food preparation from the home into fast food restaurants. These jobs, however, appear to serve primarily as temporary income supplements for an increasingly consumption-oriented youth, and are seldom the first rung on a career ladder.

3. CHANGES IN SCHOOLING

Some of the most dramatic recent changes in social institutions in OECD countries have been changes in education. Within education, the most dramatic change has been its democratisation. This has taken two forms: reduced differentiation among different streams of students, and prolonged formal schooling for a large fraction of the cohort. Educational systems have moved away from stratified tiers toward comprehensive secondary schools and greater mobility between those sectors that remain. A single dominant education ideal, academically-oriented in character, has replaced previous multiple ideals and multiple paths in education.

A second part of the democratisation has been the extension of education, with a broad segment of youth now continuing in education through a comprehensive secondary school, and a large segment into university as well. The structure of post-compulsory education that was once the province of a privileged few has become the structure through which large numbers of youth make the transition to adulthood.

The nature of these institutions has not, however, changed as dramatically as have the numbers of students passing through them. There has been some relaxation of standards, but in general, it can be said that in most OECD countries the character of what had been the academic tier of the system has become the standard for the system as a whole, or most of it.

This democratisation, this change from a structure with multiple, and often parallel paths to a single path, does not mean that differences in educational credentials have vanished. It means rather, that the difference in credentials of different paths is replaced by a difference in credentials according to *how far* one has gone on the (single) educational path before leaving it. This does undoubtedly constitute a gain in equality of educational opportunity, for it defers the time of selection to the point at which the student leaves school, in place of earlier selection (e.g. at age 9 or 11), which is more dependent on family background. It may not, however, yield educational outcomes that are any nearer to equality.

Sources of the Growth and Democratisation

The extension of education to the cohort of youth as a whole, and the extensive broadening of its academic sector, have resulted from several changes which followed World War II. Perhaps chief among these was economic growth, which freed large numbers of families from the necessity of sending their teenage children out to work. Further education for their children was seen by many families to be the most desirable investment they could make as their economic situation improved. Human capital theory in economics would see this also as a social investment that would constitute an engine for further economic growth (see Part I). However, what seems to have occurred instead in some countries is an oversupply of persons with high levels of academic education (Freeman, 1976). It seems likely that the high demand for further education was in some part an aspiration for the status conferred by academic diplomas, and that neither the families exercising that demand nor educational policy-makers tailored the educational choices and opportunities toward expanding occupations[8]. It is quite possible, of course, that no such tailoring would have been effective, and that an oversupply of youth in the labour force merely resulted from other constraints on economic growth. This is the contention of those economists with a "demand-limited" model.

Consequences of Educational Changes

There is a major consequence of the economic affluence and the educational democratisation which has risen since World War II, one which has implications both for the transition of youth from school to work and for the economy itself. This is that the expectations and aspirations of both youth and their families are incompatible with many industrial jobs. There is a disjunction between the character of the economy and the character of the educational system designed to prepare for it. The educational system has shifted increasingly toward a single academically-oriented mode, while the economy remains highly differentiated, with many jobs toward which an earlier educational structure was directed. For many industrial blue-collar jobs in non-expanding industries, there is no visible payoff to the educational investment represented by extended schooling, yet the principal measure of qualification that the student carries with him from school is how far he was able to go. Thus the closer the educational system is to a single egalitarian path, the more this means that there is no education which points to certain sectors of jobs. The single educational path points only to jobs near the top of the income or status ladder, and anything less is missing the mark. This is not an educational system that leads to pride in work, except at higher levels, nor one that engenders high self-esteem, except among those who end up near the top.

Many blue-collar jobs have career trajectories that are incompatible with the expectations generated by a single-path academically-oriented educational system. A worker at age forty-five has no higher status, and not much higher wages, than he had at age twenty-five. It is true that for many wives, whose primary aspirations are associated with raising a family, and who see their jobs principally as sources of additional family income, lack of income and status growth with age creates few problems. For men who remain in rural settings, maintaining some part-time connection with the land or with other informal work, the industrial job may represent only a source of income for a life whose quality is defined in other terms. But for the large numbers of young persons who fit neither of these categories, the prospect of an industrial job with no expectation of advancement is not attractive. Their education has given them expectations of an upward occupational career, not to be content with continuing in the same job. The disjunction is magnified by the relaxation of standards in educational systems of some countries, so that the actual level of preparation of youth in secondary schools and colleges in no way corresponds to the credentials they have received.

Relation Between School and Work

Education is ordinarily seen as a "preparation" for work, providing training, skills and credentials which will aid in one's occupation. This is the traditional conception of the role of education, and certainly captures a large part of the relation between education and occupation. However, education takes place in institutions – in schools and universities – and this fact means that there are at least two other relations between education and occupation. First, full-time education displaces full-time work. Extended schooling is a principal means by which surplus youth are absorbed and held off the labour market. Second, schooling *reduces* the qualifications of a person for certain work. Extended schooling brings about other changes in addition to increased skills. These changes result from the nature of educational institutions as currently constituted.

One change is in work habits. School, particularly post-secondary education, is less "demanding" in terms of work discipline than are most jobs. For example, absences and tardiness have less severe consequences, especially in post-secondary schooling. Schooling imposes less continuous work requirements than do most occupations. The schooling day is broken into numerous segments, and in universities there are no "regular hours" comparable to regular work hours. It is often possible to arrange course times to fit with a rather casual style of life.

A second change is in expectations. Schooling broadens horizons, and thus makes certain traditional jobs, especially those with low potential for growth, less attractive. School can, in fact, increase expectations to a greater extent than it increases skills. Even if it does not, it may increase expectations beyond the job opportunities in a given economy.

Third is a change in attitudes. Schooling generates negative attitudes toward certain occupations and certain activities, particularly those involving manual labour and physical exertion. This comes about purposively, as part of the process of creating an incentive for children and youth to achieve well in school, so that they can avoid certain kinds of jobs.

A fourth change concerns responsibility. The role of student is one with few responsibilities for others or even responsibilities that concern others. Thus extended schooling as presently organised means extended deferral of responsibility.

The implications of the first three points are that extended schooling makes youth both less well equipped for, and less attracted to certain jobs (manual labour, jobs with a high degree of supervision, and jobs with low income growth potential) than would be true if schooling were less extended. The implication of the fourth point is that extended schooling as school is presently organised makes youth in some respects less well equipped for any job. Recognising these effects of extended schooling implies recognising that the balance between the educational system and the economic system can be off in either of two directions: an insufficient number of youth provided with the extended schooling necessary for those jobs that require the cognitive skills and knowledge which school can impart; or too many youth provided with the extended schooling which creates work habits, expectations and attitudes incompatible with those jobs in the economy that do not require the cognitive skills and knowledge imparted by extended schooling.

An imbalance of the first kind means that the economy grows less rapidly than it otherwise would. The consequence for youth is that they are employed at jobs for which they are underqualified, and in the best circumstances, will learn through on-the-job training or return to part- or full-time schooling. An imbalance of the second kind means that lower-level jobs in manufacturing are exported or immigrant labour is imported. It also implies that insofar as unemployment is structural, the society will have unemployment at the same time that low-level jobs are being exported (through net import of manufacturing goods) or unskilled labour is being imported for low-level jobs. The consequence for youth is that their unemployment rate is high despite the existence of low-level jobs. It is ordinarily those who do least well in school (although their schooling may have been extended), and have learned bad work habits through school that are the most frequently unemployed.

Relation Between School and Family

Public schools have in most countries had an ambiguous stance toward the family. They have attempted to gain the family's support in enforcing their demands upon children: not missing school nor being late, doing homework, working hard to get good grades. At the same time, they ordinarily keep parents at arms length, discouraging parental observation of the classroom, limiting parents' knowledge of everyday school activities, and seldom making use of parents for voluntary work at school. These distancing actions of the school increase substantially as the child proceeds from early grades into secondary school.

Private schools, especially those associated with a religious body, show much less ambiguity in their orientation to parents. They make stronger attempts to gain the family's aid in enforcing their demands, and they are more receptive to parental involvement in the school itself. They make more extensive use of volunteers, and they are more likely to encourage parents' involvement in their children's education[9].

A portion of the tendency in the schools to maintain distance from parents arises from the professionalism to be found in any school staff. However, the difference between public and private schools indicates that something more is involved. Schools associated with a religious body tend to grow out of a functional religious community of which the family is the basic unit. Involvement of the parents in the school is a natural part of such a setting. In other private schools, this is not so, but their dependence on parents for support of the school very likely induces a more open stance toward parents.

The maintenance of distance from parents by public schools has part of its roots in the antagonism between the family's goals for the child and the school's goals characteristic of Phase 1 societies and Phase 1 families. The families' interests lay, at least in part, in keeping their children's horizons narrow, while schools attempted to broaden them. This has been a source of tension between school and family since schools came into being. It was especially important for the school to maintain that tension, when most families had narrow horizons for their children. The school was the principal avenue by which a child could be freed from the constraints imposed by the family and the immediate neighbourhood. But a stance on the part of schools that was appropriate for Phase 1 is not necessarily so as developed societies move toward Phase 3. Here the major problem is not that of countering the narrow interests of the family in limiting their child's opportunities. It is, rather, the problem of strengthening, supporting, reinforcing and when necessary, substituting for, the family's interest in the child's development.

4. CHANGES IN RIGHTS AND RESPONSIBILITIES

Some of the changes in the transition of youth to adulthood concern rights and responsibilities. Emerging from childhood, youth also emerge from a status of being dependent, of being the responsibility of others and under the authority of others. There are two kinds of recent changes in this set of rights and responsibilities. One is in the timing of the shift from others' authority over and responsibility for a youth's activities, toward independence, with authority over and responsibility for the person located in that person himself. The second is in the distribution of rights and responsibilities concerning the dependent child or youth between parents and the State.

From Others to Self

In general, changes in timing of the shift have gone in one direction for responsibility and in the other for authority. Authority over the youth is more quickly relinquished than before, with the youth assuming full civil rights earlier than before. Table 3 shows this. In most OECD countries, the age of first voting has been lowered in recent years to 18. In some countries, the age at which alcoholic beverages can be first purchased legally has been lowered (though the change has been partially revoked in some countries or localities). In some countries, the age at which parental consent is required for marriage has been reduced. Quite generally, the age at which full adult civil rights are acquired has been lowered from 21 to 18.

In universities in some countries, the university administration has, during the past two decades, relinquished the role of *in loco parentis* over students, as parents have lowered the age at which they claim authority over their children's behaviour from the end of university to the end of secondary school. In the United States, this change is reflected in recent legal changes which allow college students to claim residence for voting purposes at their college rather than at their parents' residence.

There are signs in some countries, such as the United States, that parents' claim of authority during high school is eroding, for children in United States secondary schools now have closer to a full set of adult civil rights than earlier (e.g. the right to due process before being expelled), and some youths move out of their parents' household while still in secondary school. It is also probably the case that in the increasing number of single-parent households, parental authority is weaker than in the standard two-parent household.

At the same time that the transfer of rights of authority from custodian to the youth himself is occurring earlier, responsibility for the youth as economically dependent is being maintained longer. This is most evident in the increase in the school-leaving age in some countries over the past two decades, such as the United Kingdom. It is also evident in the continued financial support of youth by parents for a longer period, as the period of schooling prior to entry into the full-time labour force has lengthened.

These changes in opposing directions create some disparity in the locus of responsibility for and authority over youth. The earlier acquisition of authority over one's actions than of responsibility introduces into the young person's life a period of invitation to irresponsibility.

A Period of "Socially Induced Irresponsibility"

There is a period during which youth have authority over their own actions, but are protected from full responsibility for those actions. This period, which largely coincides with the early period of post-secondary education, is an invitation to irresponsibility. Unlike elementary and secondary school, and unlike later work life, there are no penalties for being absent from daily activities, nor penalties for being late. There are no continuing demands for performance, but only occasional assessments through examinations. Misbehaviour which

Table 3. **Changes in timing of the shifts in authority over youth and in the assumption of responsibilities by the young themselves**

	United Kingdom	Canada	Finland	Sweden	Germany	Spain	Norway	Italy	Japan	France	Australia	United States of America
1. Complete full-time education	16↑	15-16	16	16	16	14-16	16	14	15	16	15	13-18
2. *a)* Full-time employment	16	15-16	15	18	15	16	16	15	15	16	15↑	16
b) Part-time employment	13		14	16	14	16	13		6	16		14-16
3. Unemployment/social security for self	16↑	18	16	16	18	16		15	15	18	16	Usually 18 or when worked to establish benefits
4. Marriage restrictions	16	12-16				12-14				18M 15F	16M 14F	14-20
Unconditional as from	18↓	18-19	17-18	18	18	18	18	18	18M 16F	18	18M 16F	16-21
5. Vote	18↓	18↓	18↓	18↓	18↓	18↓	18↓	18↓ (1975)	20	18↓	18↓	18↓ (1971)
6. *a)* Own property/land	18-19↓		18	18	18	18	Any age	18	20	18	18	18
b) Obtain mortgage	18-19↓		18	18	18	18	18	18	20	18	18	18
c) Become a tenant	18-19↓		18	18	18	18	18	18	20	18	18	18
d) Hire purchase	18-19		18	18	18	18	18	18	20	18	18	18
e) Credit card	18-19		18	18	18	18	18	18	20	18	18	18
f) Cheque book	18		18	18	18	18	18	18	20	18	18	18
g) Place bets	18		18	18		18	18	18	20	18	18	– (Often banned)
h) Purchase alcohol	18	16-19↓↑	18-20	20	14-18	16-18	18-20	18	20	16-18	18	18-21↑↓
i) Purchase cigarettes	15	18	16			16-18	16	18	20	–	16-18	14-18 (Where statutes exist)
j) Attend cinema without restrictions	18	18	18	15	18	18	18	18	18	18	18	17
7. Driving licence (car)	18	16-17	18	18	18	18	18	18	18	18	16-18	15-18

under other conditions could be regarded as serious is dismissed as merely "student pranks". If the youth are still regarded as children, this treatment indicates that they are experiencing neglect from adults, a neglect which masquerades as freedom. If they are regarded as adults, they are deprived of any useful function, merely held in storage for a few years.

All this provides a period of respite for those youth who continue past secondary school, before they take on full responsibilities of adult life. But the question is, what are the effects of this period for the youth who experience it? What is learned about work habits, about standards of performance? For some youth who bring with them a strongly internalised set of standards, there will be little impact, but for others, the impact may be great. And since most of these youth will enter occupations in which strict and regular work habits are required and in which performance to some standard is demanded, this suggests that the period of irresponsibility which society has imposed on them is harmful to their futures.

From Parents to State

The second change concerning rights of and responsibilities for youth is a change in the agent which holds that authority and responsibility. The State, presuming to act as the child's agent, has made continued attempts to wrest authority over the child from parents.

The change is a long-term one: under the Napoleonic code, the father and household head had total authority over those in his household, while under recent Swedish legislation the State has in principle the right to prevent the parent from administering physical punishment to the child. And by law the child may be taken from natural parents for "state care" based on the judgement of an agent of the State, a social welfare officer.

One source of the State's increased authority over children and youth is the increasing number of child and youth care professionals. These persons have a financial interest in expanding the scope of the State's authority, for it is they who exercise that authority. Altogether, there is a growing imbalance in the struggle between parents and the State for authority over children and youth. The State and its agents are both highly interested and well-organised to take legal actions against parents who fail to provide good care for their children. There is nobody, however, that is both interested and organised to take legal action against the State when it fails to provide good care for the children under its authority. Yet just as there are blatant cases of neglect or abuse of children by parents, there are blatant cases of neglect or abuse of children who are under the care of the State.

There will very likely be a continuing trend away from parental authority, toward increased authority by the State, as some parents want to relinquish this authority early, and as more families become unstable and incapable of caring for children. However, the increasing imbalance referred to above suggests that the forces in government itself accelerate the trend more rapidly than optimal for youth.

In general, the principle of *subsidiarity* should be observed with respect to responsibility for children and youth. This principle states that responsibility for personal care should rest with that level closest to the person being cared for. The principle is based on the assumption that on balance, the benefits of a close personal bond between the responsible party and the dependent child outweigh those provided by the professional expertise of an employee working for pay. This assumption may not always be valid, and it would be valuable to examine those circumstances in which it is not valid (for example, possibly a summer camp setting, or the close community of a boarding school) as a guide to establishing, for those youth under State care, institutions that are most beneficial to them.

A Client-Oriented Society

It is useful to conclude this part of the report with an examination of a broad set of changes in society that have important consequences for youth. The welfare society of the mid-20th century, as depicted, for instance, by Marquis Childs (1936) has developed into what could be labelled a "client society" where a rapidly growing class of publicly employed people are expected to take care of the rest of the citizens from cradle to grave. Old institutions, such as the school, have been given largely widened scope with regard to the objectives they are supposed to achieve, and new institutions, particularly those taking care of children of pre-school age and elderly after retirement, have emerged. There has been an enormous expansion of health and medical care institutions. As pointed out earlier, the services provided by professional experts have become increasingly specialised with nobody taking care of the "whole client". The ensuing fragmentation of contacts between caretakers and clients both in terms of kind and continuity can be observed in various types of institutions.

Many institutions have a high turnover of staff which affects particularly children in day-care centres and other pre-school institutions. Studies have shown that children at that age grow afraid of emotional investment in particular individuals because they repeatedly have the experience that their contacts with caretakers are of a short-range character. (See, for instance, a study conducted at day-care centres in Gothenburg.)

Even at the stage where traditionally the teaching has been conducted within the framework of a self-contained classroom, the contacts with the class teacher have tended to become increasingly fragmented. Leaving aside the increased turnover of class teachers during the school year, the change from almost complete containment by the class teacher has occurred because of specialised teaching in, for instance, music, art and gymnastics.

Over the last few decades new categories of staff have emerged and proliferated in many urban school systems, particularly in schools with a large enrolment: school nurses, school psychologists, social workers, school lunch personnel, librarians, technicians and – not least – administrative personnel assisting the principal, such as secretaries, etc.

New public institutions have been established to take care of young people during out-of-school hours, such as youth centres with specialised staff trained for leisure activities. Attempts to serve the whole child run up against what in the Ditchley Park Conference in 1976 (Cerych, ed., 1977) was referred to as "the iron web of bureaucratisation". Youth problems, such as unemployment, drug abuse, and delinquency, are dealt with by specialised agencies enviously protecting their own turfs or empires. A "thoroughly co-ordinated overall therapy" administered to individuals or groups is difficult, not to say impossible to achieve under such circumstances. Actions are taken along "vertical" instead of "horizontal" lines. Willard Wirtz (1977) reporting from the Ditchley symposium of youth education and employment comments:

> But the easiness of the rhetoric was not permitted to camouflage the difficulty of what is involved here. For what this means is that the society's institutionalisation – along ... vertical lines – in terms of education, for example, and unemployment and health and law enforcement, has exacerbated the problem under consideration and must yield to the development of an institutional structure and approach here along a horizontal line, taking youth – the passage to adulthood – as the area of responsibility.

The bureaucratic rigidities constitute an "iron web" when it comes to attempts to deal with the "youth problem". The public sector today has a series of such built-in rigidities that prevent the implementation of concerted policies designed to cope with the unemployment problem as well as with various social pathologies that beset the youth, such as apathy, drug abuse, delinquency, and vandalism. It is, however, in the interest of the growing class of public functionaries to keep young people not only from the streets but from productive life as well, because this would imply a diminishing role for themselves and put their jobs in jeopardy.

The Ditchley group in 1976 reached the conclusion in Wirtz's words that "there is in contemporary industrial society a youth problem of extraordinary, unprecedented, and worsening proportions – lying beyond the reach of macro-economic counter-cyclical measures and defying established institutional approaches".

One of the consequences of a fragmented caretaking with compartmentalised services by professional experts is a bureaucratised and impersonal climate. In such a setting, for instance in a school with large enrolment, social control is difficult to exercise. Fragmentation of contacts with adults, for instance compartmentalisation between school and home, leads to inconsistencies of norms and the possibilities to inculcate norms. In a study of primary school discipline teachers' and parental attitudes with regard to upbringing were measured on a scale assessing the degree of strictness vs. permissiveness. It was found that the discrepancy between teacher and parents on that dimension correlated substantially with the degree of disciplinary problems with the child in school[10].

The client-oriented society easily becomes a control-oriented society where the step between what is motivated by the caretaking duties and self-serving convenience for the caretaker can be rather short. The tasks performed by the caretakers easily develop into a network of elaborate control actions.

Part III

INSTITUTIONAL INNOVATIONS TO AID YOUTH

1. REFORMING SCHOOLS

Because new generations are growing considerably smaller than their predecessors, the productivity of each member is especially important to all members of society. Thus investment in young human capital is especially important, which appears to imply greater investment in each young person's education. This, however, is too narrow a view. The crucial question is *what* kind of investments in young capital will have the greatest payoff. When that question is asked, there are various alternatives, some within education, others outside.

School is the principal institution through which society beyond the family facilitates the process of transition to adulthood. Because of this, the natural first move in designing policies of youth is some modification of the school. For example, when in the 1960s the swollen cohorts of the baby boom came close to the point of entry into the labour force, and jobs were too few, the number of upper secondary school and university places was expanded greatly – not merely to acccommodate the same fraction of the larger cohort, but even to accommodate a *larger* fraction. "Educational" policy was in reality economic policy in disguise, making use of the schools to solve a problem of mismatch between number of jobs and number of people.

Although the move to use the school as a universal policy instrument for problems of youth is natural, it arbitrarily limits policies and may cause certain potential remedies to be overlooked.

For this reason, some of the innovations we will examine involve changes in school policies, and some involve other institutional changes. The explicit recognition of possibilities for institutional innovations other than those involving the school is important if the problems of youth are to be fully addressed. In the sections below, we will focus first on changes within education, and then on changes in other areas. As will become apparent, the boundary becomes a rather blurred one.

First, we may ask, how would one go about "reforming" the school? This is what several recent reports on the American high school have asked, advancing lists of improvements all the way from administration and teacher training to curricula and methods of instruction. Our hesitation is inspired by two circumstances. In the first place, and perhaps most obviously, the distance between blueprints for sweeping reforms and the rigid, hard reality is usually enormous. Quick cures can easily be recommended but they tend to defy any reasonable implementation. Secondly, and more important, reforms cannot be confined to the school as an institution. As we have suggested above, problems besetting secondary school are problems besetting modern urbanised and industrialised society at large. Many of these problems are unforeseen side-effects of recent social and economic changes, especially the increased standard of living that has taken place over the past few decades. Other problems have their roots in changes that pertain mainly to the school as an institution.

Thus, "reform" proposals by necessity affect the entire social fabric. Tinkering with certain marginal changes more or less exclusively in the classroom or within the individual school is often not very helpful. In spite of this, changes and institutional arrangements proposed would mainly have to be confined to the school and the close community. Some of the proposals advanced here were discussed more in detail in the late 1970s at a series of international seminars sponsored by the Aspen Institute for Humanistic Studies (Husén, 1979, Chapter 9).

In discussing possible changes of the institutional arrangements in school there are certain considerations which constitute our point of departure. Genuine care with responsibility for the single individual has historically been performed by the family and the close community. The smaller the unit in charge the more close and humane the attention. Yet the trend in developed societies for a long time has been to allocate care to larger and larger units, particularly when it comes to health care and schooling. The responsibility for the individual child has been split up between various agencies with a corresponding reduction of the responsibility left with the family. This implies conflict between on the one hand the family and the close community, and on the other the welfare society represented by the State. It implies also a conflict between on the one hand personal service and attention and on the other hand equality of service.

These are some of the issues that must be addressed in considering how schools and schooling might be changed to benefit youth. In the sections below, we examine several specific changes, some because they are commonly advanced as remedies for problems involving youth, some because they appear to be particularly promising.

Lengthening School as a Universal Remedy

The most common pattern of response to youth labour market difficulties has been to extend the period of education. Legislation passed in recent years providing some kind of "youth guarantee" has in its implementation heavily relied on the formal school system. This response has been effective in keeping youth out of the labour force, reducing the supply of labour, and thus reducing youth unemployment. Whether it has also been effective in creating a more able labour force is, however, not clear. Some of the problems created by extended schooling that were discussed in Part II – over-expanded expectations, poor work discipline, poor attitudes toward work – may nullify the benefits that arise from increased skills.

In addition, the skill benefits of extended schooling are concentrated among those youth who do well in schoolwork, while the negative effects of extended schooling are greater for those who do poorly in school. The end result is that insofar as there are benefits from extended schooling, these are differentially distributed in favour of advantaged youth[11].

Thus, although extended schooling keeps youth out of the labour force longer, this policy implemented alone may do little to make youth more effective in the labour force after completing education.

If, as has been suggested in Part II, the extension and democratisation of schooling have created over-expanded expectations and poor attitudes toward work among those who do not "succeed" at school, the question arises as to what policies may compensate for these negative effects. There are several alternative policies which could remedy such a situation. One is to restructure education in a way that the distribution of expectations is consistent with the distribution of jobs relative to the educational levels of youth competing for these jobs. A portion of the problem of disjunction results from expectations based upon the correspondence between educational levels and type of job that existed in the past rather than which exists in the present. The rapid expansion of education changed that correspondence, without changing equally the perception of it.

A second policy alternative is to restructure the occupational system so that there is a greater consistency between opportunities and expectations. Although this might seem to be impossible because of economic constraints themselves, the contrast between Japan and the West in the career patterns of manual workers in the upper tier of industry gives some indication of the latitude available. The blue-collar vs. white-collar distinction in the industrial structure of the West may well be outmoded, with the extensive growth of "technician" jobs which do not fit neatly in either category. It is quite conceivable that different structuring of work and careers could go some distance toward reducing the gap between expectations and occupational reality.

A third alternative is suggested by the fact that for some persons the principal expectations on which their perceived quality of life depend are not related to work. (Wives in the labour force and men in a rural setting who work in industry are the examples used earlier.) This alternative lies in somewhat modifying styles of life so that the status and future opportunities of a job mattered less to a worker than is true for most persons now.

None of these policy alternatives is simple or straightforward. The problem which they address is one created by the lengthening of schooling for all, and the creation of a single ideal for educational success. The extent of this problem suggests that the lengthening of schooling is not by itself a wise policy for the future.

Intensifying Education

A second possible response to the problems of youth employment is intensification of education, by raising scholastic standards. The historical trend in education has, of course, been the opposite: with the democratisation of extended education to cover increasingly large fractions of the total cohort of youth, the standards as measured by average performance have tended to be lowered, and the content watered down. There are, however, periods of revolt against this trend, as is currently true in North America and in some countries in Europe. In principle, "intensification" of education will bring about higher levels of achievement, and research results indicate that it does so in practice. The problem, however, lies in the side effects. If educational programmes are "intensified" solely through increasing demands on students, without changes that affect either motivation or environmental constraints that inhibit achievement, it also brings with it increased failures. Since the most evident problems of youth are found among those at lower achievement levels, the group that we have above referred to as the "new educational underclass", such a policy would exacerbate these problems rather than alleviate them.

Policies designed to increase achievement by raising standards and increasing demands on students provide a useful illustration of a major dilemma in educational policy. The dilemma is this: any policy which increases the threat of failure as a way of bringing about higher achievement must, by definition, increase the failure rate in school. It has thus increased the problem of transition to work and to adulthood generally for those youth for whom the problem was already greatest. This may be offset in either of two ways:

i) the capabilities of those who are just above those who fail may be sufficiently increased to greatly facilitate their transition to work; and

ii) the increased achievement of those at the high end of the achievement continuum may be sufficiently great that they will spur economic growth through new enterprises and new developments, thus through a very *indirect* process expanding the economic opportunities.

But if neither of these effects is sufficient to offset the direct effect on rate of failure in school, the policy of raising standards has negative, rather than positive effects. It appears likely that for a given organisation of education, there is an optimum level of standards from the point of view of the preparation for entry into the labour force, in which the direct effect of the failure rate is balanced off against the two indirect effects in the other direction. How to determine where that optimum is, however, is not altogether clear. It is certainly true that different OECD countries have chosen different levels of standards in their educational policies (see Passow *et al.*, 1976).

Changing the Content of Education

A third response to youth's labour market difficulties is to change the content of education. Vocational education is at the centre of controversy in many countries. It appears especially difficult, within the context of a single country's educational system, to determine the relative effectiveness of vocational programmes in secondary schools, because of differential selection of students into different programmes. For that reason, and because vocational education plays a quite variable role in different OECD countries' educational systems, comparative research on the effectiveness of vocational and technical education would be particularly valuable. In the last section of this part, we shall discuss such research.

The rapidity of technological change raises questions about how new work-related content can be best introduced into education. This is presently exemplified by the so-called "high tech" industries. The question of just what skills are necessary for those industries, as well as what profile of educational experiences and activities provides the best training for employment remains open.

It is, however, true that the increase in various types of technician occupations, and in the fraction of new jobs in these occupations, is blurring the distinction between manual or blue collar labour, and non-manual or white collar labour. This blurring may be beneficial for secondary and post-secondary education, for it makes more feasible the intermixing of traditional academic education with technical training. The situation corresponds loosely to an earlier period in which traditional secondary and post-secondary education was largely humanistic studies, involving very little science, and only classical mathematics. The intermixing of science and broadened forms of mathematics with the humanistic curriculum has been largely accomplished, with results that have been on the whole beneficial. The opportunity now exists to carry out a comparable broadening of the curriculum in technical directions, with the most obvious candidates for inclusion at the secondary level being statistical data analysis and computer programming.

Towards a new Transition System

Until recently, particular institutional arrangements for the 16-19 year olds – the stage now often referred to as the upper-secondary – were made only for a socially and/or intellectually selected minority. A few decades ago at most, 5-10 per cent of those who had completed mandatory schooling proceeded with their education in full-time institutions, most of which were preparing for university and thereby for the professions.

What we now can see emerging in all OECD countries is a transition system of post-compulsory education and training for *all* – or almost all – young in the age bracket that concerns us here, those 16 to 19 years of age. Thus, there is, as has been argued above, ample reason to talk about a "new" age of pre-adult status which follows after what earlier was referred to as adolescence.

The emerging provisions on the part of the society for young people at this stage tend to be comprehensive in two major respects. In the first place, provisions are made for all of them, and for instance, the idea of a youth guarantee implies that everybody has the right either to further education or to a job. Secondly, institutional arrangements, such as those in the schools, are made for every category of young people, be they "academic", "vocational" or "general" in their orientation.

In other words: we can in the OECD countries note how a deliberate policy of creating a *transition system* from youth to adult status or from schooling to working life is emerging. In order to be successful such a system has to be flexible and innovative so as to avoid tight institutional straitjackets. There is a growing realisation that we cannot cope with the "youth problem" just by providing more of the same. Further institutionalisation, that is to say, simply to keep the young in school, is not the panacea. We attempt in this part of our report to exemplify new approaches which would facilitate the transition to adulthood under the new conditions that will prevail in the foreseeable future.

A new flexible transition system must also pay attention to the sizeable *individual differences* among young people who at about the age of 16 have completed their mandatory school attendance. The spread in the competencies that they have acquired over a period of 9 to 10 years of schooling increases over time and covers at the end of the period the equivalent of several school years. The disadvantaged among the students, those whom we have referred to as the "educational underclass", tend to leave school after having "stuck out" the mandatory minimum of years. Within this group is the hard core group of unemployed youth, whereas the majority stay on in school for some more years and have brighter job prospects.

Recent concerns with the quality of the outcomes of basic schooling which have been aired in several countries, in particular concerns about the quality of the "basics", have tended to focus either on the academic students, those who are candidates for further formal education or on the marginal students who are the lowest achievers. But raising standards of one group does not automatically lead to enhanced standards in the other. The cross-national studies conducted by the International Association for the Evaluation of Educational Achievement show wide variations between countries, not only in terms of mean achievements but in spread of achievements between students and schools as well (Husén, 1979).

Therefore, a major challenge faces educational policy-making: namely to reconcile on the one hand the concern for quality in basic schooling, and on the other to develop a new and flexible transition system for all 16-19-year-olds in such a way that those who have had problems in basic schooling will be given meaningful options in a new transition system that has to cater for more or less all 16-19-year-olds.

2. THE SCHOOL AND THE FAMILY

With respect to the family, the matter may be put this way: Schools had their origins at a time when most families were at a subsistence level, described in Part II of this report as "Phase 1". The appropriate – and even necessary – stance of the school in relation to the family under these conditions was one of distance, providing for the child a path toward greater opportunity that was not too greatly impeded by the family's narrow interest in keeping the children "down on the farm".

The school has characteristically maintained that distance, in part because this task of "rescuing" a child from narrow interests of the family has always remained for at least a few children in school.

However, with families in Phase 2, where their interests lie in investing in their children, such a stance is no longer necessary. And in society where many families move toward Phase 3, and where interest in children lessens altogether, the opposite of such a stance is necessary. The school can help children and youth by helping to strengthen, support, and reinforce the family's interest in the child's development.

The implication for educational policy is a reversal of the school's orientation to the family. This means encouraging parents' involvement in the school, even at the cost of taking their views into account. It means, above all, reinforcing or recreating a community among the parents of children in the school. With the reduced importance of neighbourhoods as communities, due largely to ease of transportation, the school itself may be the most likely candidate as the focal point around which to create a community. Throughout their history, schools have been one of the principal means by which the larger society can invade a constraining community and provide the children with paths out of it. But now, when the larger society has invaded and disrupted the local community in so many ways, the school is one of the few institutions through which the benefits of a community for children and youth can be regained.

As models for how this is and can be done, schools created around a religious body probably are most useful. Their constituency is already a community, making their task an easier one. Because the community is a resource, religious schools make extensive use of it, both to strengthen the school (for example, by parent-sponsored activities) and to strengthen the families within it – for example, by encouraging joint activities involving parents of children in the same class or grade in school.

Whatever the specific steps, there is a single central point: a new and important role for the school, ultimately to aid it in accomplishing its goals for youth, is to strengthen and reinforce the family and community whose children the school serves.

More particularly, a major problem in the future for social policy is the creation of institutional arrangements for taking care of the whole child. For example, since the family has in certain respects begun to wither away, should there be created a system of boarding schools with full responsibility for the young persons? Or is it possible to bring about institutional changes which would reduce the role of the State and restore more of the responsibility of the close community and the family? Most of what is briefly suggested in the following could be subsumed under the general heading "de-institutionalisation".

First, genuine education pertaining to the whole child has to occur in small, close communities, such as the family, the neighbourhood and the small workplace. In a society with large-scale institutions and enterprises a person's life tends from the outset to be segmented into many roles and to be influenced by many, often contradictory, forces. Under such circumstances more attention tends to be drawn to superficial qualities, such as manners or dress, and less to deeper character traits. In the large setting genuine educative values which affect the deeper levels of the personality have difficulties in being realised.

Any attempt to come to grips with the troubled school would in our view have to consider ways and means of establishing more self-directed and self-responsible units with closer ties to some form of community. By doing so one contributes to de-institutionalisation.

The two overriding goals in trying to achieve a better education for young people and thereby facilitating their transition to adult responsibility would be to make school less "action-poor" and to provide an educative milieu for the whole young person and not only segments. Responsibility among young people can be developed by their being accepted – and participating – as members of a community that has responsibility for its own destiny. Such an arrangement is increasingly in jeopardy in a "client-oriented" society. What has to be seriously considered are steps conducive to what we would like to call "functional responsibility", that is to say, young people being given the opportunity to be in charge of tasks that are within their capacity and by taking reponsibility for the outcome of their efforts. As youth have come to be in school longer and longer, the pedagogy which leaves them in a passive and non-participatory role is increasingly inappropriate, leaving them unable to take responsibility. They cannot be expected to acquire more genuine independence and responsibility than they are allowed.

3. RESTRUCTURING YOUTH'S POSITION

A major problem which youth increasingly suffers in modern society is the problem of irrelevance. As physical strength – which is acquired early in life – is replaced by various learned skills, cognitive and others relevant for self-sufficiency, as less labour is necessary altogether in society, and as work is increasingly carried out as full-time jobs in large formal organisations, youth have lost their functions in society (except in wartime) until they are approved for entry into one of these organisations.

The question arises, how can some function be regained by youth? We see three kinds of opportunities, each following a different principle. The three principles can be illustrated by three extreme kinds of policy approaches.

A. Work organisations as extended family: age-balanced organisations

One means of reintegrating youth in society is through the transformation of work organisations into age-balanced organisations, with age structures approximating to that of the society as a whole (Coleman, *et al.*, 1974). Thus the responsibilities for aiding and strengthening the parents' care of the child, as well as responsibilities for its schooling, would lie within the work organisation (which in a market economy would necessarily be subsidised by public funds). The direct responsibility for children and youth by the organisation, and the necessity to insure an effective transition to adulthood (as well as its own interests in replenishing its labour force in the next generation) would lead the organisation to be particularly attentive to its youth and their preparation.

The idea of a society composed of age-balanced organisations as modern counterparts to the extended family may appear extreme, and would certainly constitute an extensive change in the organisational fabric of Western societies. It should be noted, however, that the evolution of the modern corporation in Japan, an evolution which has a strong base in the patriarchal family, has created a social structure which is intermediate between the European and American corporation and the age-balanced corporation.

B. School as a self-directed community

In some schools, and in some activities in many schools, something of a self-directed and self-responsible community exists. Beginnings of this can be found in some extra-curricular activities of many schools, where the school newspaper staff, the drama club, athletic teams, and the yearbook staff all constitute in part small communities, partly self-directed and self-responsible. Some private boarding schools, such as the Putney School in Vermont and the *"École d'Humanité"* in Switzerland, were established to implement this principle of self-direction and self-responsibility extensively. All well-functioning boarding schools, even the most regimented military academies, implement it to a degree.

Yet this principle of creating a largely self-directed and self-responsible community of children and youth as a means of developing independence, responsibility and positive qualities of character has never come to play a large part in educational philosophy and theory, even as the school moves ever farther from the adult world[11].

One possible element in the construction of schools as self-directed communities is structured interdependence between children and youth, exemplified by youth in the role of teachers or tutors. This could function through youth shifting to the role of tutor upon meeting certain achievement criteria. After meeting these criteria in a given year, the student might "graduate" to employment as a tutor or teacher for younger children[13]. It would be quite possible, with such a system, to structure matters so that youth would earn entitlements toward the cost of higher education (or under the voucher system discussed elsewhere in this report, toward support for a broader range of activities).

C. Youth enterprises

In China, an institution has been developed to cope with the problem of transition for youth who finish secondary school with few prospects of employment. The institution is the youth co-operative, consisting of 4-10 young people under the guidance of an adult retired worker. The co-operative begins a small enterprise, called, for instance, a "street business" which may be a small restaurant or retail store. The adult aids in getting the co-operative launched, and it then becomes autonomous if it can be self-sustaining or fails if it cannot[14].

In some American high schools, there is a club called "Junior Achievement", in which young persons are encouraged to begin profit-making enterprises, often building upon a hobby or special interest. Most such enterprises do not last long, and very few extend beyond the high school period. But within this period, some young people develop successful small enterprises, and gain valuable experience toward forming or running businesses after they finish school.

In the evenings on *Kurfürstendamm* in Berlin, numerous young entrepreneurs sell small items or give small performances. For some, it is a way-station between regular jobs in organisation; for others, it is a starting point for a larger enterprise.

In Silicon Valley (California), in Bellevue (Washington), and in numerous cities and towns throughout the United States and Canada, youth with some knowledge of computer programming or electronics are beginning small computer software or hardware firms. The industry, which is growing explosively, was begun by young people, some high school graduates, others college graduates, working in basements and garages, and was only later taken up by adults and by large organisations.

All these examples illustrate youth enterprise, and suggest its potential for aiding the problems of transition. There are certain unique virtues that this mode of human capital development has, including these:

- For successful enterprises, the problem of employment is automatically solved for the enterprising youth; whereas, for investments in education, employment remains problematic.
- Successful investments create jobs for the entrepreneurs, thus directly expanding the number of jobs in the system.
- Successful enterprises create jobs for non-entrepreneurs, as they employ others (and they will reverse the usual bias of not employing youth). Thus the labour demand for youth is expanded.
- The enterprises will in many cases be in expanding sectors of the economy, providing a spur to economic growth.
- The disadvantage that youth suffer in high-wage societies is reduced, because such enterprise at its outset does not confront stringent wages-and-hours legislation; yet there is not direct "unfair" competition with adult workers, because new jobs have been created.
- Reward is directly connected to effort, which is a spur to the intense effort that benefits both individual and the economy.
- The continuing supply of such small enterprises provides an antidote to an economy in which concentration of industry proceeds, as it were, inexorably.

Facilitating the Connection of Youth to Small Business

As a kind of addendum to the policy strategy of facilitating youth enterprise is a strategy of facilitating employment of youth by small business. Most economic growth in recent years in Western society has not been among large corporations but among small businesses. For example, there were 1.5 million fewer jobs in 1982 in the largest 1 000 firms in the United States than there were in 1977, while there was a growth of 8.6 million jobs in smaller firms. These statistics indicate that ways of facilitating contact and employment of youth in small and medium-sized firms can be especially productive.

Transition Vouchers

A policy alternative which has been discussed in some countries but never fully implemented is the use of "transition vouchers". The idea of a transition voucher is not greatly different from the various educational and other benefits provided in many countries to veterans of military service. In its broadest form, such a voucher could be used for education, for subsidising wages provided by first employers, or for financial aid in setting up a small business.

Current public financing of education beyond the minimal school-leaving age in most developed countries is inappropriate on several counts. First, it is regressive, benefiting those from more affluent backgrounds at the expense of those from less affluent backgrounds. (This is true so long as family income is positively related to length of education, as it is quite generally.)

Second, current financing provides money to institutions which supply education, rather than to the youth consuming it. The institutions are consequently not disciplined by choice of the consumer, as they would be in a fully competitive system. The supplier of education is too free in his decisions about what to supply[15]. This defect of institutional financing is, however, reduced in a system with an excess of educational places, in which scholarships are placed in the hands of prospective students who can choose a university.

Third, the limitation of public subsidy to educational activities distorts the choice of youth, biasing it in the direction of more schooling than would be chosen if non-educational activities that aided the youth's path to adulthood (e.g. employment, or starting a small business) were comparably subsidised. The end result is that public funds are spent in keeping youth out of productive activity, in a rather passive role (that of student) at the cost of personal initiative and productive activity. At the same time, many young people are in an activity they would not prefer if it were not free or artificially cheap. Youth in extended education in modern society is in somewhat the same position as the heir who can claim his inheritance only by meeting some criterion deemed desirable by the father, such as entering the family business, or not marrying until after age 21.

Programmes of National Service

National service is an activity that began as an extension of military service, but has come to gain attention on its own merits as an aid to the transition of youth.

National service may be seen either as an independent activity, or as interleaved with a university or other post-secondary educational programme of a traditional type. Ideas about programmes of national service have arisen primarily as variants of obligatory military services. Yet this origin should not lead the approach to be neglected.

A number of OECD countries continue to have obligatory military service for young men. In some countries, this is moving toward a more generalised national service, civilian or military. For example, in the Federal Republic of Germany, since 1st January 1984 young men may choose either 20 months of civilian service or 15 months of military service.

The broadening of compulsory national service to include civilian service also makes more acceptable the inclusion of young women. The possible benefits of such service range from those of specific skill training to those of providing a transition period between the classroom and the workplace during which youth are in positions of responsibility with some authority, yet are not subject to the whims of economic fluctuations easily leading to unemployment.

For countries without compulsory military service, a programme of voluntary national service paralleling the existing voluntary military service might be instituted. In such programmes, a portion of earnings might be deferred, in the form of entitlements toward the cost of higher education or toward subsidising post-service employment. Alternatively, a combined programme, alternating post-secondary education and national service for periods of something like six months each, could be initiated, with educational costs covered as an entitlement based on the service.

It should be added that in some countries the prospects for such service are not particularly bright. For example, in the United States, the Congress recently voted against establishing a commission to examine various forms of voluntary national service. However, the costs of a university education have grown to the extent that there might well be support for a policy which interleaved university attendance with national service, the latter giving entitlements for the former.

4. RESEARCH AS AN AID TO POLICY

Throughout this report, there have been points at which the need for research to inform possible policy has been apparent. In this section, two of these research questions are further elaborated to indicate ways in which broadened information from researchers may provide the basis for policy.

A. *Economic accounts with jobs as the unit*

A simple policy to address the loss of youth-accessible jobs that occurs through international trade with less developed countries is the erection of trade barriers. Such a policy of course not only reduces the distributional problems resulting from international trade, thus aiding youth, but also acts as a brake on economic growth.

More sophisticated policies which would not have this undesirable side effect are difficult to implement in the absence of information about how many jobs of various kinds are eliminated by increase in a given import, and how many of various kinds are created by increase in a given export. Thus one important research task that could facilitate such policies would be to carry out a set of international economic accounts, in which the unit is not money, but jobs with different characteristics. Through such a set of accounts it would be possible to see just what kinds of jobs are lost and what kinds are gained by each component of a nation's international trade, and through each change in international trade.

For example: assume a hypothetical developed economy with two types of jobs, type A requiring a high level of training and ability, and type B without these requirements. Assume also an equilibrium between types of workers and types of jobs, with all jobs filled and no workers unemployed. Then there is a change in international trade, giving an increase in import of goods based on labour of type B, and an increase in export of goods based on labour of type A. This eliminates some jobs of type B, and leaves a surplus of workers of type B. It creates additional demand for workers of type A. Unless workers of type B can without great difficulty change to type A, the new job potential is either filled by overtime on the part of type A workers, or remains unmet, with the economic potential lost. From the perspective of youth unemployment, the central point is this: in developed countries, the nature of goods exported and imported is such that the type B jobs, those lost through imports from less developed countries, are those for which a large fraction of youth are qualified, while the type A jobs are those for which they are not qualified, and for which extended education or training is necessary.

If a set of economic accounts for jobs of different types were carried out, accompanied by a system of money accounts, it would be possible also to determine the money gains or losses corresponding to changes in international trade. This would allow raising the question, just how much of the increase in per capita real income arose from increase in number of jobs of particular types, how much arose from increase of wages in the sectors where there was increased demand, and what was the cost in losses of jobs of particular types.

Such systems of national accounts, covering jobs as well as money, would constitute a basis for four kinds of policies:

i) First are policies involving education or training of workers who have characteristics suitable for lost jobs to equip them with the additional characteristics necessary for jobs which are gained;

ii) To carry out Kaldor's compensation: taxation of the economic activities which have gained at the expense of others, economic activities that replace the jobs lost by the changes in international trade[16].

iii) Third are economic policies which are the obverse of the educational policies of *i):* to carry out work restructuring which will modify jobs to fit the characteristics of young people whose prospective jobs are lost (or workers whose current jobs are lost) through the increase in trade;

iv) To create, through public works, a set of jobs that have characteristics that make them accessible to young people whose prospective jobs are eliminated through increase in trade. These may be temporary jobs while retraining occurs. The difference in skills between the jobs gained by increase in trade and the jobs lost constitutes the skills gap to be overcome in training.

The essential aim of the development of a set of economic accounts with jobs as a unit would be to provide an information base for policies to cope with the distributional problems that arise from internationalisation (and automation) of the economy. These distributional problems fall disproportionately upon those among the young who enter the labour force early, with a weak educational background. Information to aid these distributional readjustments will be especially valuable for these youth.

B. Vocational vs. general education

In all OECD countries there has been a move, justified by educational philosophy but principally generated by consumer demand and by expanding egalitarian values, toward a single comprehensive stream of secondary education, away from the multiple tiers and multiple paths in post-elementary education of the past.

These policies have constituted a move in the direction of equal educational opportunity, but a move that is not without side effects. One negative side effect is that the single path substitutes a single definition of success, accompanied by many levels of failure, for multiple definitions of success. Such a substitution may have serious negative consequences for self-esteem of those who do not succeed in the dominant academic mode.

A second negative side effect is that because the school system does not aim toward certain manual and technical occupations, but aims uniformly away from them, it leads to disdain of these occupations, and unwillingness to work in them. The simultaneous existence of youth unemployment and importation of guestworkers in some OECD countries is at least in part due to this side effect.

Because the move toward a single mode of secondary and post-secondary education has had both beneficial and negative effects, and because it is unclear what are the next steps in such policy (in most OECD countries, the move toward comprehensive secondary schools has slowed or stopped), it is important to carry out research which focuses on these questions. Because different OECD countries have rather different degrees and kinds of differentiation remaining in their systems of secondary education, cross-national research on these questions can be readily carried out. For example, Sweden made early and extensive moves toward a single-track comprehensive system, while the Federal Republic of Germany (which also has extensive differences among different *Länder)* has showed much less change away from a differentiated system.

Such cross-national research is especially valuable in this instance relative to research carried out within a country, because of the self-selection of youth into vocational and technical programmes. When research is done within a country on a problem of the sort posed here, the effects of such self-selection are especially difficult to disentangle from effects of the educational programme itself. This difficulty is much less pronounced when studying youth comparatively across countries with different systems.

NOTES AND REFERENCES

1. The most prominent studies were those of Margaret Mead (1928, 1953), which showed a far calmer and more orderly process of transition among Samoan youth. Freeman (1976) has recently challenged the correctness of their picture. Although Mead seems clearly to have overstated the case, the differences between the transition period for youth in modern industrial society and youth in many primitive societies are extensive.
2. Mention should also be made of Japanese comparative studies on socialisation (Teichler, 1976).
3. In a 1980 survey of achievement among a national sample of high school students, it was found that the students' achievement showed zero relation to parents' income, when parents' education and other subjective indicators of parental interest were statistically controlled (Coleman, Hoffer, and Kilgore, 1982, Table A.6-A.9).
4. The question of whether this decline in strength of the family will be reversed by future social developments cannot of course be answered. Phase 3 could well be followed by another phase in which the social conditions for strong families were once again present. We are not proposing that the decline is irreversible, but only that there is now a decline, that there is no evidence of an imminent reversal, and that educational policies should recognise the decline.
5. German statistics in Table 2 are an anomaly; we will not attempt to account for their deviation from the others.
6. Recently, one has seen micro-computers made in the United States but with silicon chips in it from seven countries: Malaysia, Taiwan, Indonesia, Korea, Japan, Philippines and El Salvador.
7. The lack of perception of a connection between difficulties of youth in the labour market and internationalisation of the economy is a very general one. It is well exemplified by two adjacent items in the Ford Foundation *Letter* for 1st February 1984. One is titled "Barriers to Trade", and reports Foundation activities designed to help reduce these barriers. The other is titled "Youth's Rocky Road", and reports the Foundation's activities designed to help youth make the transition to work. There is no recognition that the first of these activities works at cross purposes with the second.
8. There were earlier a number of attempts in OECD countries to do just this, by differentially expanding the number of places in post-compulsory education, to fit projected economic demand. However, the lack of correspondence between courses of study and later occupation that is observed in developed societies made these efforts appear fruitless. The educational systems shifted from such attempts to make them supply-directed, toward simply responding to demand.
9. In a recent national study of high schools in the United States, it was found that the ratio of volunteers to students in the public schools was less than half that in the Catholic schools, and less than one eighth in other private schools (see Coleman, Hoffer, and Kilgore, 1982, p. 79).
10. This research, covering 15 primary schools in the city of Stockholm in the mid-1950s also found extensive differences in disciplinary problems between grades 2 and 6 (L. Husén *et al.*, 1959). Comparisons were made between grades 2, 4, and 6. It was found that both subjective complaints and objectively recorded incidents of undisciplinary behaviour were strikingly few in grade 2 as compared to grade 6.
11. When there is explicit policy to extend schooling among disadvantaged youth, as there has been for blacks in the United States, for example, then that may lead to differential benefits for the targeted group. For example, in the United States, college enrolment among blacks, which until recently was considerably below that of whites, is now about the same level as that of whites.
12. There are some elements of this principle in John Dewey's writings, though Dewey emphasized continuity between the outside community and the community within the school. That continuity, feasible in many places in Dewey's day, is much less so today. Except where it is feasible, it must be foregone for the benefits and costs of a more detached, more autonomous community of children and youth.
13. "Graduation" could also be to other activities, principally in the personal service area (such as care of the sick), for which the youth would receive the same form of compensation as for teaching younger children.

14. We are indebted to Professor Qui Yuan of the Institute for the Study of Comparative Education at East China Normal University in Shanghai for information about these co-operatives.

15. This judgement that the educational supplier should be constrained by choices of the educational consumer in decisions about what content to supply is based on the assumption that the consumer's choice is likely to be better. While university students would hardly be good at designing curricula, our assumption here is that they are good at selecting among curricula.

16. Note that this is in direct opposition to economic policies which subsidise export industries. Those policies are designed to increase economic growth at the cost of increases in distributional inequities and (at least in the short run) welfare dependency. Many countries, developed and less developed, have such an industrial policy, perhaps the most widely-known being that of Japan, with identification of "sunrise" and "sunset" industries. Japan's two-tiered economy, however, with a lower tier of firms with poorly-paid jobs, allows a fluctuation of wages in the lower tier which can spread the distributional disadvantage among workers in the lower tier.

BIBLIOGRAPHY

Andersson, Bengt-Erik (1969) *Studies in Adolescent Behaviour.* Stockholm: Almqvist & Wiksell International.

Adamski, Wladyslaw W. (1980) *Social Structure versus Educational Policy: Polish and European Perspectives.* Warsaw: Polish Academy of Sciences.

Arrow, K. (1973) "Higher Education as a Filter", *Journal of Public Economics,* Vol. 2, No. 3, July 1973, pp. 193-216.

Barnes, S. H. and Kaase, M., Beverly Hills, CA: Sage, (1979) *Political Action: Massachusetts Participation in Five Western Democracies.* .

Becker, G.S. (1964) *Human Capital.* New York: Columbia University Press for the National Bureau of Economic Research.

Berg, Ivar (1971) *Education and Jobs: The Great Training Robbery.* Boston: Beacon Press.

Bergedorfer Gesprächskreis zu Fragen der freien industriellen Gesellschaft (1979) *Jugend und Gesellschaft. Chronischer Konflikt – neue Verbindlichkeiten?* Protokoll Nr. 63, 1979.

Boyer, E., (1983) *High School: A Report on Secondary Education in America.* New York: Harper & Row.

Brown, Frank B. *et al.,* (1973) *The Reform of Secondary Education: A Report to the Public and the Profession* (The National Commission on the Reform of Secondary Education, established by the Kettering Foundation). New York: McGraw-Hill.

Bühler, Ch. (1921) *Das Seelenleben des Jugendlichen.* Jena: Gustav Fischer.

Carnegie Council on Policy Studies in Higher Education. (1979) *Giving Youth a Better Chance.* San Francisco, Ca: Jossey-Bass.

Cerych, Ladislav (1977) "Youth Employment: the Potential and Limitations of the Education System". Proceedings of an International Symposium held at Fère-en-Tardenois (France) 27th-30th April 1977. *Youth-Education-Employment,* 25-34. Amsterdam: European Cultural Foundation.

Childs, Marquis W. (1936) *Sweden – The Middle Way.* New Haven: Yale University Press.

Cogan, John (1981) "The Decline in Black Teenage Employment, 1950-1970", *American Economic Review.*

Coleman, James S. (1961) *The Adolescent Society: The Social Life of the Teenager and Its Impact on Education.* Glencoe: The Free Press.

Coleman, James *et al.* (1974) Panel on Youth. *Youth: Transition to Adulthood.* Report of the Panel on Youth of the President's Science Advisory Committee. Chicago and London: The University of Chicago Press.

Coleman, J., Hoffer, T., and Kilgore, S. (1982) *High School Achievement.* New York: Basic Books.

Cremin, Lawrence A. (1980) "Changes in the Ecology of Education: The School and Other Educators". In: T. Husén (ed.), *The Future of Formal Education.* Stockholm: Almqvist & Wiksell International.

Deutsche Shell (1975) *Jugend zwischen 13 und 24 – Vergleich über 20 Jahre – Sechste Untersuchung zur Situation der deutschen Jugend im Bundesgebiet.* Jugendwerk der Deutschen Shell.

Deutsche Shell (1977) *Jugend in Europa. Ihre Eingliederung in die Welt der Erwachsenen.* Eine vergleichende Analyse zwischen der Bundesrepublik Deutschland, Frankreich und Grossbritannien. Siebente Untersuchung zur Situation der Jugend, anlässlich "75 Jahre Shell in Deutschland".

Dore, Ronald (1976) *The Diploma Disease: Education, Qualification, and Development.* London: Allen and Unwin.

EMNID-Institut (1966) *Jugend, Bildung und Freizeit.* Bielefeld: EMNID-Institut für Sozialforschung.

Erikson, E. (1968) *Identity, Youth and Crisis.* New York: Norton.

Freeman, Derek. (1983) *The Making and Unmaking of an Anthropological Myth.* Cambridge, Ma: Harvard University Press.

Freeman, Richard B. (1976) *The Over-educated American.* New York: Academic Press.

Goodlad, John I. (1984) *A Place Called School.* New York: McGraw-Hill.

Gordon, C. Wayne. (1957) *The Social System of the High School.* New York: The Free Press.

Gottleib, D. and Reeves, J. (1963) *Adolescent Behavior in Urban Areas.* New York: The Free Press.

Gow, McPherson, (1980) *Tell Them from Me.* Scottish School Leavers write about School and Life Afterwards, Aberdeen University Press.

Gray, J., McPherson, A.F. and Raffe, D. (1983) *Reconstructions of Secondary Education.* London: Routledge and Kegan Paul.

Habermas, Jürgen (1970) Thesen zur Theorie der Sozialisation. In: Habermas, J. *Arbeit, Erkenntnis, Fortschritt.* Amsterdam.

Hall, G. Stanley (1904) *Adolescence and its Relations to Physiology, Anthropology, Sociology, Sex, Crime, Religion and Education.* I-II. New York & London: Appleton & Co.

Hall, G.S. (1923) *Life and Confessions of a Psychologist.* New York: Appleton.

Hanushek, E. (Fall, 1981) "Throwing Money at Schools." *Journal of Policy Analysis and Management".*

Havighurst, Robert J. and P.H. Dreyer (eds.) (1975) *Youth.* The Seventy-fourth Yearbook of the National Society for the Study of Education. Chicago: University of Chicago Press.

Hill, John P., and Franz J. Mönks (eds.) (1977) *Adolescence and Youth in Prospect.* Guildford, Surrey: IPC Science and Technology Press.

Hollingshead, A.B. (1949) *Elmtown's youth.* New York: Wiley.

Horkheimer, M. and Adorno, Th. W. (1969) *Dialektik der Aufklärung.* Frankfurt am Main: Fischer.

Hurrelmann, K. (1976) "Gesellschaft, Sozialisation und Lebenslauf". In: Hurrelmann, K. (ed.), *Sozialization und Lebenslauf.* Reinbek: Rowohlt, 1976, 15-33.

Hurrelmann, K. (1978) "Programmatische Ueberlegungen zur Entwicklung der Bildungsforschung". In: Bolte, K.M. (ed.), *Verhandlungen des 18. Deutschen Soziologentages,* 531-564. München.

Hurrelmann, K. (1983) "Kinder der Bildungsexpansion", *Zeitschrift für Sozialisationforschung und Erziehungsoziologie,* Vol. 3, N° 2, pp. 263-283.

Hurrelmann, K. and D. Ulich (eds.) (1980) *Handbuch der Sozialisationsforschung.* Weinheim: Beltz.

Husén, Lennart *et al.* (1959) *Elever – Lärare – Föräldrar.* Stockholm: Almqvist & Wiksell.

Husén, Torsten *et al.* (1973) *Naturorienterande ämnen.* Stockholm: Almqvist & Wiksell.

Husén, Torsten (1974) *Talent, Equality and Meritocracy.* The Hague: Martinus Nijhoff.

Husén, Torsten (1979) *The School in Question. A Comparative Study of the School and its Future in Western Societies.* London & New York: Oxford University Press.

Husén, Torsten (1983) "Are Standards in U.S. Schools Really Lagging Behind Those in Other Countries?" *Phi Delta Kappa* March 1983, 455-461.

Inglehart, R. (1977) *The Silent Revolution: Changing Values and Political Styles among Western Publics.* Princeton: Princeton University Press.

Johansson, Bror A. (1965) *Criteria of School Readiness. Factor structure, predictive value and environmental influences.* Stockholm: Almqvist & Wiksell.

Johansson, Sten (1971) *Laginkomstutredningen.* – Omlevnadsnivaunder-sökningen. Kapitel 1 och 2 i betänkande om svenska folkets levnadsförhallanden att avgivas av laginkomstutredningen. Stockholm: Statens offentliga utredningar 1971.

Kandel, D.B. and Lesser, G.S. (1972) *Youth in Two Worlds.* San Francisco: Jossey-Bass.

Keniston, Kenneth (1968) *Young Radicals.* New York: Harcourt, Brace and World.

Keniston, Kenneth (1968) "Youth, Change and Violence". *The American Scholar,* Vol. 37:2, 227-245.

Keniston, Kenneth (1970) "Youth – A "new" Stage of Life". *The American Scholar,* Vol. 39:4, 631-654.

Keniston, Kenneth (1971) *Youth and Dissent.* New York: Harcourt, Brace and Jovanovich.

Kerr, Clark (1977) "Education and the World of Work: An Analytical Sketch", in: James A. Perkins and Barbara Burn (eds.), *International Perspective on Problems in Higher Education.* New York: International Council for Educational Development, 133-42.

Kerr, Donna (1983) "Teaching Competence and Teacher Education in the United States". *Teachers College Record,* Vol. 84, No. 3, Spring 1983, 525-552.

Key, A. (1885) *Läroverkskommittens underdaniga utlatande och föslag angaende organisationen av rikets allmänna läroverk och därmed sammanhängande fragor.* Bilaga E. (The School Committee's Report and Proposals for the Organisation of Secondary Schools and thereto Related Questions. Supplement E.) Stockholm: Kongl. Boktryckeriet P.A. Norstedt & Söner.

Kitwood, T. (1980) *Disclosures to a Stranger.* London: Routledge and Kegan Paul.

Krappmann, Lothar (1975) *Soziologische Dimensionen der Identität. Strukturelle Bedingungen für die Teilnahme an Interaktionsprozessen.* Stuttgart: Ernst Klett Verlag.

Kreutz, H., (1979) *Die Dreigeteilte Welt des Studenten,* Hanover.

Kreutz, H. and Grete Fürnschuss (1971) *Chancen der Weiterbildung Vienna.* Österreichischer Bundesverlag für Unterricht, Wissenschaft und Kunst.

Kreutz, Henrik and Ulf Wuggenig (1980) "Hochschule, Gesellschaft und die Rolle des Studenten". *Angewandte Sozialforschung.* Vol. 8, N° 1/2: 7-32.

Lynd, Robert S. and Helen Merrel (1929) *Middletown. A Study in Contemporary American Culture.* New York: Harcourt, Brace and Company.

Lynd, Robert S. and Helen M. (1937) *Middletown in Transition: A Study in Cultural Conflicts.* New York: Harcourt.

Mannheim, Karl (1928) "Das Problem der Generationen", *Kölner Vierteljahresheft für Soziologie,* Band 7, 1928

Mead, Margaret (1928) *Coming of Age in Samoa.* New York:

Mead, Margaret (1953) "National Character". Pages 642-667 in International Symposium on Anthropology, New York, 1952. *Anthropology Today: An Encyclopedic Inventory,* edited by Alfred L. Kroeber, University of Chicago Press.

Mincer, Jacob (1974) *Schooling, Experience, and Earnings.* New York: Columbia University Press.

The National Commission on Excellence in Education (1983) *A Nation at Risk.* Washington, D.C.: U.S. Government Printing Office.

OECD (1975) *Education and Working Life in Modern Society.* A Report by the Secretary-General's *ad hoc* Group on the Relations Between Education and Employment. Paris: OECD.

OECD (1980) *Youth Unemployment: The Causes and Consequences.* Paris: OECD.

OECD (1981a) *Youth Without Work: Three Countries Approach the Problem.* Paris: OECD.

OECD (1981b) *The Welfare State in Crisis: An Account of the Conference on Social Policies in the 1980s.* Paris: OECD.

OECD (1983) *Education and Work: The Views of the Young.* Paris: OECD/CERI.

Oevermann, U. (1976) "Programmatische Ueberlegungen zu einer Theorie der Bildungsprozesse und zur Strategie der Sozialisationsforschung. In: Hurrelmann, K. (ed.), *Sozialisation und Lebenslauf.* Reinbek: Rowohlt, 34-52.

Psacharopoulos, G. (1981) "Returns to Education: An updated International Comparison". *Comparative Education* (October 1981).

Passow, A. Harry (1975) "Once Again: Reforming Secondary Education: *Teachers College Record,* 77:2, 161-87.

Passow, A. Harry, Harold J. Noah, Max A. Eckstein & John R. Mallea (1976) *The National Case Study: An Empirical Comparative Study of Twenty-One Educational Systems.* International Studies in Evaluation, Vol. VII. Stockholm: Almqvist & Wiksell, and New York: Wiley & Sons.

Roberts, Kenneth (1983) *Youth and Leisure.* London: George Allen & Unwin.

Rosenmayr, Leopold (1976) *Jugend.* 2. völlig neubearbeitete Auflage. Band 6, Handbuch der empirischen Sozialforschung, herausgegeben von René König. Stuttgart: Ferdinand Enke Verlag.

Rosenmayr, Leopold (1979) Referat in *Jugend und Gesellschaft, Chronischer Konflikt – neue Verbindlichkeiten?,* 5-13. Bergedorfer Gesprächskreis zu Fragen der freien industriellen Gesellschaft, Protokoll Nr. 63.

Schultz, T. (1961) "Investment in Human Capital". *American Economic Review,* 51 (March 1961).

Silberman, Charles E. (1970) *Crisis in the Classroom. The Remaking of American Education.* New York: Random House.

Sizer, Theodore R. (1984) *Horace's Compromise. The Dilemma of the American High School.* Boston: Houghton Mifflin Company.

Spence, M. (1974) *Market Signalling.* Cambridge, Mass: Harvard University Press.

Spranger, Eduard (1924) *Psychologie des Jugendalters.* Leipzig: Quelle & Meyer.

Stern, William (1929) *Reifende Jugend. 1. Anfänge der Reifezeit.* 2 Ed. Leipzig: Quelle & Meyer pp. 14.

Super, D.E. (1957) *The Psychology of Careers.* New York: Harper and Row.

Tanner, J.M. (1962) *Growth at Adolescence.* Oxford: Blackwell.

Taubman, P. and Wales, T. (1973) "Higher Education, Mental Ability, and Screening", *Journal of Political Economy,* Vol. 81, Nº 1, January-February, 1973, pp. 28-56.

Teichler, Ulrich (1976) *Das Dilemma der modernen Bildungsgesellschaft.* Stuttgart: Klett.

Teichler, Ulrich, Dirk Hartung, and Reinhard Nuthmann (1976) *Hochschulexpansion und Bedarf der Gesellschaft.* Stuttgart: Klett.

Thurow, L.C. (1972) "Education and Economic Equality". *The Public Interest,* 2866-81 (Summer 1972).

Timpane, Michael *et al.* (1976) *Youth Policy in Transition.* Santa Monica, Ca.: Rand Corporation.

Trow, Martin (1979) "Reflections on Youth Problems and Policies in the United States", in Margaret Gordon (ed.) *Youth Education and Unemployment Problems.* Washington D.C., Carnegie Foundation for the Advancement of Teaching.

Walker, David A. (1976) *The IEA Six Subject Survey: An Empirical Study of Education in Twenty-One Countries.* International Studies in Evaluation, Vol. IX. Stockholm: Almqvist & Wiksell, and New York: Wiley & Sons.

West, J. (pseud.) (1945) *Plainville, United States of America.* New York, Columbia University Press.

Wirtz, Willard (1977) "The Ditchley Park Conference: A Report". Proceedings of an International Symposium held at Fère-en-Tardenois (France), 27th-30th April 1977. *Youth-Education-Employment,* 7-21, Amsterdam: European Cultural Foundation.

Wright, A.F. and F. Headlam (1976) *Youth Needs and Public Policies.* Melbourne: Department of Youth, Sport, and Recreation, Victoria.

Yankelovich, Daniel (1973) *Changing Values on Campus: Political and Personal Attitudes on Campus.* New York: Washington Square Press.

Yankelovich, Daniel (1974) *The New Morality: A Profile of American Youth in the 70s.* New York: McGraw-Hill.

Yankelovich, Daniel (1974) *Changing Youth Values in the 70's.* New York: The John D. Rockefeller III Fund, Inc.